CW01511691

Outposts on the Frontier

Outward Odyssey

A People's History of Spaceflight

Series editor

Colin Burgess

Outposts on the Frontier

A Fifty-Year History of Space Stations

Jay Chladek

Foreword by Clayton C. Anderson

UNIVERSITY OF NEBRASKA PRESS • LINCOLN & LONDON

To Alice Anderson
and
James M. Busby
One helped to teach future generations,
while the other helped them to learn
the history of the future

Contents

Illustrations

Foreword

I have known Jay Chladek for about a decade. Essentially acquainted through the Strategic Air Command and Aerospace Museum, located just off Interstate 80 in my home town of Ashland, Nebraska, Jay and my mother Alice were frequent attendees at various museum events. Recognizing in each other in a simultaneous zeal for space, along with an overwhelming desire to occupy front-row seats (Mom didn't want to look at the back of Jay's head), they continued to strike up conversations at book signings and lectures, most notably those by NASA astronauts. It was through their developing friendship—further enhanced by my standing as Nebraska's first and only astronaut—and the completion of my first mission to the space station in 2007 that we would meet and ultimately become friends.

As an astronaut, I had the wonderful privilege of flying in space . . . twice. While the hours spent orbiting Earth were memorable, far more hours were spent between the countries of the United States and Russia—key training locations for any astronaut destined for the International Space Station (ISS). How interesting that during the time of my training and spaceflights, the United States and Russia would work together as space-faring partners rather than competitors. These were two prideful space programs, the envy of the entire world. Driven almost solely by the enormous number of dedicated people toiling behind the scenes, they were all working together, half a world apart. They had a single—and remarkably easy-to-state—goal: the safe delivery and return of humans to outer space, including protecting and sustaining these beings known as astronauts and cosmonauts as they lived and worked side by side in space for months at a time. A goal not so easy to execute, these prideful nations were faced with political and financial woes on the ground, catalyzed by the oft-antagonistic relationship of the upper levels of their respective governments.

I was given a remarkable gift: the chance to see things from both sides. Working at NASA from the very early days of the space shuttle program, before one day joining the astronaut corps and flying in space, I was a ticket-holding spectator of sorts, watching two rival superpowers come together through a program we called the ISS.

Jay's book, *Outposts on the Frontier*, chronicles a large part of this historical transformation. Much as he does while building intricate scale models of flying and space vehicles, Chladek provides the reader with intricate, precise details, covering nearly everything happening in space over the past fifty years. Analogous to the tight-fitting pieces and in-depth paint jobs of one of his model projects, Jay gives the astute reader a deep, interlocking look behind the scenes, clearly illustrating the processes that allowed the program to fall into place. In Jay's own words, "It is kind of like a big tapestry of international developments . . . the relationship between Russia and America seeming like an episode of *The Odd Couple*, with the European Space Agency, Canada, and Japan coming along for the ride. Everyone seems to be working together because they don't have the ability to go it alone anymore."

You will put down this book a much smarter reader than the day you started!

Live long . . . and prosper!

Clayton C. Anderson

Acknowledgments

When I was first approached to write this book in late 2007, the task seemed like a relatively easy one that would only take maybe three years at most. I jumped at the challenge, to write my first space-history book. Little did I know what I was in for, but honestly I wouldn't change anything for the world. It has been quite an adventure.

As I write this, it has been nearly a decade since it all began, and finally the fruits of my labor now sit in your hands. The work took longer than expected, but the story was able to benefit from recent events, specifically the decision by the National Reconnaissance Office to declassify portions of the Manned Orbiting Laboratory program. So I would say the wait was worth it.

I have many people to thank for their assistance, the first one being Colin Burgess, who sent me that first email all those years ago asking me to take on this project. Another shout-out goes to Robert Pearlman, the creator and moderator of collectsPACE.com, who I am pretty sure provided some input about me to help influence Colin's decision. Sincere thanks also go to the University of Nebraska Press not only for approving me as Colin's choice but also for continuing to expand the Outward Odyssey book series. Even if I hadn't been asked to personally take part in it, I still would have every other book sitting on my shelf at home, as the work done by the other authors in the series has been first-rate.

I also extend thanks to the tireless work of the media-relations people at both the Johnson Space Center and the Kennedy Space Center, who helped me to obtain the proper credentials and access to the facilities. One person in particular I am eternally indebted to is Gayle Frere. She helped to schedule and coordinate my interviews at the Johnson Space Center on multiple occasions and did what she could to accommodate my wishes, even though I was a relatively unknown author at the time.

Another shout-out goes to the library and records people who are the keepers of the archives for NASA's Johnson Space Center, at the University of Houston–Clear Lake. They were most helpful in locating research on my behalf and helping to find materials that perhaps have not seen the light of day in over three decades. Others I have to thank include the engineers, space workers, and astronauts I interviewed in person, since this story is ultimately theirs.

As for others whom I need to thank by name, they include Bert Vis, who was most kind to allow me access to his astronaut and cosmonaut interview transcripts through Colin Burgess. Special thanks also go to Francis French, whose numerous contributions include helping grant me access to the archives section at the San Diego Air and Space Museum. Still others include Clay and Susan Anderson, T. J. Creamer, Emily Carney of the Space Hipsters Facebook group, Kim and Sally Poor of Spacefest, Geoffrey Bowman, and Manny Gutsche. A special shout-out goes to Bill Grush, proprietor of the gone, yet not forgotten, Star Realm bookshop. There are plenty of others I want to thank as well, but I don't have enough space to list them all.

Finally, I would like to save the last bit of thanks for my family. My parents, Gary and Mary Ann, have been most patient and supportive, even during the hard days. My sister, Joanie, and my brother, Jeff, have also found ways (both direct and indirect) to keep me on the right path. Thanks to you all.

Outposts on the Frontier

Introduction

The surface of the earth is the shore of the cosmic ocean. On this shore, we've learned most of what we know. Recently we've waded a little way out, maybe ankle deep . . . and the water seems inviting.

—Carl Sagan (1934–96)

When the Soviet Union and the United States began flying humans into space in 1961, it seemed as though the path was clear for the exploration of our new frontier. To most Americans who grew up reading *Collier's* magazine and watching Walt Disney's *Man into Space* television series hosted by Dr. Wernher von Braun, the logical path laid out after the first proving flights would include the construction of manned outposts in Earth orbit. These outposts would serve to look down on our planet for study, for reconnaissance of threat countries, and to help build and fuel spacecraft for longer expeditions. Only after this logical step was taken would humankind begin its journey to the moon and beyond.

Of course, it didn't quite happen that way. On 25 May 1961, during a special joint session of Congress, U.S. president John F. Kennedy firmly put America's sights on "landing a man on the moon and returning him safely to the earth" within the decade. At that point, the goal was not so much to just explore space but to try to achieve a space first over the Soviet Union, who at the time seemed to be firmly in the lead with the accomplishments of Sputnik's launch in 1957 and Yuri Gagarin's single orbit of Earth a few short weeks before Alan Shepard's suborbital spaceflight. The goal would have serious repercussions in both the United States and the Soviet Union, many of which are felt to this day. Both countries would have to sidestep the seemingly logical approach to spaceflight and expend their efforts into reaching the moon as soon as possible. As history has shown, the United States succeeded in the lunar-landing goal, while the Soviets fell short of that achievement.

By the start of the 1970s, the focus had turned to Earth orbit and long-term manned missions with space laboratories. For the Soviets, it would initially be another way to score a space first and divert attention away from their lack of lunar-mission success. For the United States, it would be a way to utilize hardware drawn from the Apollo program in a cost-effective manner, keeping NASA's astronauts in orbit and the workforce intact while development work began on the space shuttle. The hope of NASA at the time was that the shuttle would serve as a transportation and assembly vehicle, ferrying equipment to and from the next generation of space stations. After many years, that finally happened, but not before the shuttle became a premiere space laboratory of its own.

Along the way, some interesting things happened. The path initially envisioned by the Americans and the Soviets didn't quite take place as each side expected. At one point, both countries became partners in an initial joint space mission before political ideologies forced them to again pursue different directions for another two decades. It wouldn't be until after the Soviet Union fell that both NASA and the Russian space agency would again look to performing missions together. Additional countries would join in with the superpowers to make space exploration a truly international endeavor, with each one contributing something unique to the venture.

There were also parallel space station programs at work with more basic goals in mind—offshoots of the Cold War built with the intention of gathering intelligence at altitudes where one couldn't easily shoot down the vehicle collecting it. For the United States the vehicle was known as the Manned Orbiting Laboratory, or MOL. For the Soviets, the vehicle was called Almaz. Ultimately, only one of these observation platforms would orbit the planet, but both programs eventually provided valuable contributions to the stations and laboratories that would come along later, in ways likely not dreamed of by their creators.

At a glance, many don't consider the space station programs to be all that worthwhile compared to what had been done before. Public perception tends to regard station and shuttle missions as little more than astronauts floating around Earth for several weeks spending taxpayers' money with little to show for it. Our cultures have been brought up to regard progress or success as being something tangible, with a prize at the end, as opposed to something open-ended for the purpose of knowledge itself. A destination on some new world is something we can understand, while Earth orbit seems more like a place already visited, complete with a giant billboard that says, "Yuri Gagarin and John Glenn were here."

So the construction and operation of space stations and laboratories tend to generate about as much interest among the general public as a new office building going up in a city. After they are built, not many people really understand what goes on inside or care too much, unless they happen to work in the building. To put it into perspective, think about some of the petroleum and pharmaceutical companies that sponsor public-television programs in the United States, such as the Public Broadcasting System's *Nova*. You see the advertisements that showcase the flashy graphics and imagery while a scientist talks about something he or she might be working on that could improve how we live. But other than that, can people really describe what these companies do in their labs? Day in and day out, all anyone sees are the people going to work in the morning and coming home in the evening. It is no different with space-program coverage, as only the launches and landings tend to attract live coverage, while the day-to-day activities might only yield thirty seconds on the evening news or the back page of the newspaper if there is anything "important" or unusual to report.

The problem, at least in the United States, isn't just with the public perception of space stations; it has also been with the policy makers who decide on the space agency's budget. While the public likes to follow high-profile missions, politicians need to secure a quick return on their investment, preferably before the next election year. New space hardware and missions can take a long time to develop, and there is reluctance to provide funding for the time required. Tangible success has to be shown early, or a program risks cancellation before hardware is ready to fly. And, of course, there are always the budget fights between those in favor of more space funding and those who push for the money to be distributed elsewhere.

The station and laboratory programs of the first half century of space exploration embrace a vast number of stories—episodes filled with excitement, danger, humor, sadness, success, and failure to rival any of those experienced on other manned missions, even ones that flew to the moon. At the heart of these tales are people, both of greatness and of modesty.

Some of these people are already well known to many, while others are not quite as well known, except to a select few.

Naturally, the people who have flown in space tend to draw most of the attention. But for every astronaut and cosmonaut who has reached space, there are thousands of behind-the-scenes support people doing work in every imaginable capacity to get them there, to assist them on orbit, and to bring them home safe. Yet while all have a story to tell, only a few can be told within these pages. So come along for what promises to be a most informative, enlightening, and hopefully enjoyable ride as we push the veil back on the dynamic history of mankind's first outposts on the frontier of space.

1. Humble Beginnings

To fully appreciate the story of manned space outposts, one has to look back through history to a time before people even left the ground. Since the dawn of time, mankind has looked up into the heavens and wondered just what was up there. As our knowledge of the physical world increased over the centuries, dreamers and writers began to speculate about trips to the heavens, writing stories about fictional journeys to other planets and stars.

By the mid-1800s, astronomers had a pretty good understanding of the movement of planets. At the same time, Isaac Newton's laws concerning gravity, objects in motion, and objects at rest had some wondering if an object might be propelled at a velocity so fast that it could fall around Earth and not be pulled back to its surface by gravity. Down low, air friction causes an object to slow down. From barometers first invented in the 1600s by various individuals, including Galileo Galilei, there came a scientific understanding that air gets less dense the higher up one gets. With this knowledge, others began to wonder if it were possible to get something to orbit Earth in a manner similar to the way in which planets orbit the sun, at a height where there is virtually no air resistance.

One of the greatest creators of science fiction, Jules Verne, in the writing of his 1865 story *From the Earth to the Moon*, managed to inspire a lot of people. His fanciful story featured a group of intrepid explorers being shot out of a cannon in a projectile to visit Earth's nearest celestial neighbor. Toward the end of the nineteenth century, many theoreticians were studying what it might take to achieve such a feat, and several concluded that accomplishing a trip to the moon would not be possible in a single, direct shot. But if some sort of nearer orbiting base camp could be set up, an outpost as it were, then a craft heading to the moon would only have to visit it first to refuel and then continue its journey.

Russian space pioneer Konstantin Tsiolkovsky was one who derived inspiration from Verne's works. Many credit Tsiolkovsky as being the father of modern spaceflight, as he laid the foundations for what problems and challenges needed to be overcome and potentially how to do so. One concept included a space station with a rotating section to generate artificial gravity. Such a station could be used to investigate Earth and refuel ships proceeding to other destinations, such as the moon.

In Germany, a rocketry pioneer named Hermann Oberth, who was born just before the turn of the century, formed similar conclusions to Tsiolkovsky. Oberth wrote a dissertation for his physics doctorate at the University of Heidelberg, where he tackled the challenges of spaceflight, starting with getting a rocket off the ground, thrusting it into space, and having it stay there. He also wrote about the need to have a space station in orbit to use as a refueling base for a craft to go elsewhere. His dissertation was rejected, as it was considered to be too speculative. But rather than give up on the years of work, Oberth raised some money on his own and had the dissertation published as a ninety-page book in 1923 titled *The Rocket into Planetary Space*. Eventually, this book gained him a professorship at a different university. It wasn't the same as a doctorate, but it would do.

The book sold slowly, but eventually it gained acceptance in science circles. The work of American rocketry pioneer Robert Goddard helped to prove many of Oberth's theories about rocketry in practical applications. Oberth expanded on the concepts in later pressings of his book and also introduced concepts from other early rocketry pioneers such as Austrian-born Max Valier and German American space historian and advocate Willy Ley in his own writings. In his later publications, Oberth expanded on the role of space stations as he theorized that stations placed in higher orbit could be used for observation of the stars. Others could be used as space-based platforms for spying on distant countries or perhaps be fitted with weapons to use against an aggressor nation.

Oberth's book would have quite an impact on impressionable, young people, as many rocketry clubs were founded all over Germany thanks to his work. Max Valier along with nine other men helped to found a rocketry club known as the Verein für Raumschiffarht (German for "Society for Space Travel"), or the vfr. Oberth was asked to join the vfr, as his presence would give their group legitimacy. For his part, Oberth saw it as an opportunity to gain publicity for his ideas and to locate funding to help develop them. He joined up and served as the group's first president. Another person who joined the group early on was an eager eighteen-year-old named Wernher von Braun from Wirsitz in the Province of Posen, then part of the German Empire.

Anyone who has a passing interest in the history of rocketry knows that von Braun would eventually help to develop rocketry in the VfR to a point where it was directly funded by Adolf Hitler's Nazi-led German government. Germany's leaders understood the untapped weapons potential of rockets. Eventually, von Braun and a team of German engineers and scientists would develop their A-4 rocket into a practical weapon known as the V-2 (Vergeltungswaffe 2), which made its presence known to those fighting against Germany in the Second World War. Any possibility for space exploration and space stations would take a backseat for a number of years as the rocket simply became something that could deliver a bomb beyond the range of the most powerful artillery cannons of the day.

Toward the end of the war, von Braun and his team knew that Hitler's days as Germany's leader were numbered, and they made efforts to surrender to advancing U.S. military forces fighting Germany in the west, figuring it was preferable to surrendering to the Soviet Union forces fighting in the east. Von Braun succeeded as his team and their families were later brought to the United States to continue their work after the war. The Germans started their work for the U.S. Army at Fort Bliss, Texas, and test fired captured V-2s at White Sands, New Mexico, before settling down in Huntsville, Alabama, at the U.S. Army's Redstone Arsenal. After a few years, they became naturalized American citizens. Oberth and Willy Ley immigrated to the United States themselves and, in the following years, lent their expertise to various engineering projects. Oberth lived a long and fruitful life before he passed away in December 1989, at age ninety-five; yet he only acted as a consultant to one postwar rocketry project, the U.S. Air Force's Atlas missile in the 1950s.

While von Braun's job for the U.S. Army was to design rockets as weapons delivery systems, his true desire was in the more peaceful endeavor of space exploration. Like Oberth, von Braun also saw the need to increase public awareness. In 1951 he wrote a visionary article for the *National Defense Transportation Journal* titled "Next Stop Mars," detailing a project to send men to the Red Planet using available technology. A principle component of this mission was ferrying parts of an interplanetary spacecraft into Earth orbit and assembling it at an orbiting space station. When assembly was finished, the crew would launch from Earth to the space station and transfer over to the craft before heading to Mars.

Excerpts from von Braun's writings also appeared in article form in *Collier's* magazine. With the help of illustrations from artist Chesley Bonestell, von Braun's ideas struck a chord with the American public. The public perception was reinforced further in 1954 when entertainment icon Walt Disney showed a three-part series about spaceflight on his new Disneyland television show, with von Braun introducing the segments.

As shown in von Braun's book, the *Collier's* articles, and the Disney programs, space exploration would proceed along a logical progression. First, an unmanned satellite would be launched into orbit atop a massive rocket. This satellite, and others like it, would collect data on the space environment. Next, astronauts would launch into orbit aboard a spacecraft, preferably one with wings so that on its return the vehicle could glide down to land back at its base of origin. After each mission, the shuttle spacecraft could be refurbished and loaded on another rocket to fly again.

After these first steps into space, with the duration of each visit lasting progressively longer in order to determine its affect on health and equipment, the next step would be to construct and occupy an orbiting space station. To achieve this, cargo rockets and winged shuttles would ferry men and material into orbit to build a rotating space station capable of generating artificial gravity for its occupants via centrifugal force. At the station, preparations would be made to launch spacecraft (manned and unmanned) to the moon and to Mars. This effort would take years to accomplish, with each step being a small one. If a setback were to occur at one stage, it wouldn't derail the entire program. Even if a trip to the moon or Mars proved unsuccessful, there would still be a station in Earth orbit to help prepare the next attempt.

While von Braun publicly proposed ideas for a space station to be used in peaceful exploration and support of planetary missions, he also knew that the U.S. military would likely be the primary backer for such an expensive endeavor. In the 1950s the ideological differences between former World War II allies the United States and the Soviet Union resulted in a rift as

1. Wernher von Braun's wheeled space station concept from 1952 as illustrated by artist Chesley Bonestell. Courtesy NASA.

each country used its influence to show the rest of the world that their system of government and way of life was superior to the other.

At first the United States had the upper hand, as it was the only country to possess the atomic bomb. But the stakes changed dramatically in 1949 when the Soviet Union detonated its own atomic bomb. This incident, coupled with the Berlin blockade of 1948, and the Korean War led to the Soviet Union and its allies being considered serious adversaries by the United States. A cold war had begun, and one of the mandates of the U.S. government and the military was to keep things from flaring up, becoming a shooting match. A hot war with nuclear weapons was guaranteed to be a very brutal conflict with massive casualties on both sides. As a consequence, the war of wills would take place in other ways.

In late 1952, not long after his first *Collier's* article was published, von Braun was a guest at a public speaking engagement in Washington DC before an audience that included leaders of various corporations and government agencies. Von Braun's presentation was called "Space Superiority as a Means for Achieving World Peace." The presentation was similar to the views he expressed in his *Collier's* articles but slanted a bit more toward military applications. In his talk, von Braun proposed that his space station concept could be used as a platform to perform the reconnaissance of threat countries. It would also be equipped with missiles for either attack or defense. The station would operate in a polar orbit at about one thousand miles above the earth and be capable of photographing every square mile on the planet. As von Braun stated in his speech, "An orbital reconnaissance station can pull up any Iron Curtain, no matter where they lower it!"

If an enemy nation should perhaps try to launch an attack against the United States or its allies, the station or a pair of stations, with one acting as the bomb aimer, could launch a missile at a target on the ground. Rocket thrust would be used to counteract the missile's orbital velocity, and it would theoretically drop on its target with an accuracy greater than that which could be achieved by a ballistic missile fired from halfway around the world. Von Braun conceded that the greatest threat to such a station might come from a missile interceptor on the ground releasing a cloud of shrapnel in front of the station to puncture it with high velocity debris. But he felt that the challenges of developing such a system would make it too easy to detect from space before it could be launched and that a station could be armor plated against such a threat.

Von Braun urged those with the capability to fund what he was proposing, as he knew the Soviets would likely be pursuing similar projects themselves. But while the speech did help to generate interest from private companies to look into space technology applications, not much else came of it.

For the next few years, rocketry development in the United States continued at a steady pace. Von Braun's team in Huntsville, Alabama, developed the Redstone tactical missile and began work on the Jupiter ballistic missile—both based on lessons learned from the captured v-2 rockets. Elsewhere, work was progressing on more-powerful ballistic missiles, such as the Thor and the Atlas ICBM (intercontinental ballistic missile). Sounding rockets, including converted v-2 missiles, had also been gathering scientific data at the boundaries of space, and some of them carried biological payloads in the form of insects and mice. But the game changed considerably on 4 October 1957 when a Soviet R-7 ICBM placed the world's first artificial satellite, Sputnik, into orbit.

With the launch of Sputnik, the world realized that the Soviets indeed had their sights set on space. The public believed a "missile gap" existed. If the Soviets had a rocket capable of reaching orbit with a satellite, they could also reach the United States with a nuclear warhead. The administration of President Dwight D. Eisenhower knew that the missile gap perception was false, but plans were accelerated to equal Sputnik's achievement.

After a major launch failure of the U.S. Navy's Project Vanguard on its first attempt, von Braun and his team were given permission to use their Redstone booster in the form of the Jupiter-C rocket to launch a satellite. They succeeded in January 1958 by placing *Explorer 1*, a far more sophisticated satellite than Sputnik, into orbit. Ultimately, von Braun's group left the U.S. Army to join the newly created National Aeronautics and Space Administration (NASA) in the early 1960s to develop rockets for the peaceful exploration of space. But it wouldn't be von Braun or NASA who would be responsible for the creation of America's first space station program. Instead, that task would go to another government agency, one that likely listened to von Braun's September 1952 speech with great interest.

The Road to the MOL

On 29 July 1958, six months after the United States orbited its first satellite, President Dwight Eisenhower signed the National Aeronautics and Space Act, creating NASA, which was charged with exploring space publicly and peacefully. As a result, several civilian and military research facilities around the country fell under the administration of the newly formed agency. NASA was also given priority to embark on a man in space program, which eventually became known as Project Mercury.

In military circles, the creation of NASA and Project Mercury struck many as being a waste of resources and a politically driven decision. Many felt the U.S. Air Force already had the capability to send people into space. A logical choice for spaceflight was the x-15 rocket plane, created in partnership with NASA's predecessor, the National Advisory Council on Aeronautics. The x-15 was a research aircraft designed to examine speeds above Mach 3 (three times the speed of sound), and it could potentially reach altitudes above fifty miles, which the U.S. Air Force considered space. However, the Fédération Aéronautique Internationale set their own boundary of space at 100 kilometers (62.1 miles altitude), and the x-15 did not carry sufficient fuel to reach orbit.

The U.S. Air Force also proposed a program known as Man in Space Soonest (MISS), which they drew up one month before NASA was created. The proposal was to launch a single astronaut into orbit in a capsule atop a Thor or Atlas missile, hopefully beating the Soviets to that goal. Ultimately, the Man in Space Soonest proposal was canceled, as the task was given to NASA. The x-15 would spend the next decade providing invaluable research data, and several military test pilots were awarded air force astronaut wings, flying the winged craft to the edge of space and back. Project Mercury ultimately achieved its goal of flying Americans into space and then into orbit, although not before the Soviets did so on 12 April 1961 with Yuri Gagarin's single-orbit flight aboard the first Vostok spacecraft.

The next logical step after man had taken his first steps into space would be to find ways to spend more time in orbit. The construction of a space station was the most likely way to achieve that. The Mercury spacecraft design was limited in its capabilities since it did not have the ability to change orbit. A follow-up spacecraft was needed. But there were still limits to what even a larger capsule could do. Having in orbit a station that a capsule could dock with and that men could occupy for longer periods seemed like a worthy goal NASA should pursue, but it didn't happen quite that way.

Shortly after astronaut Alan Shepard became the first American to fly in space, U.S. president John F. Kennedy put NASA and the country on a path to land on the moon before the Soviets at the end of the 1960s. His advisors felt that this was the best goal to exploit. It would force the Soviets into a sustained test of capabilities where the limitations of their design philosophy and political ideology would potentially be exposed. At the same time, it would showcase the technical expertise of the United States to the rest of the world.

A prime reason for a space station's existence in fictional works was to service spacecraft headed out to explore elsewhere. A logical idea was that if NASA were to place a station in orbit, it could act as a floating fuel stop for a craft headed to the moon. But again, a station was not in the cards. Instead, two approaches were considered in Project Apollo's lunar-landing goal: direct ascent and Earth-orbit rendezvous. In direct ascent, a massive rocket launches one spacecraft to the moon with all the fuel it needs. The spacecraft on top lands, and the astronauts perform their tasks and then return home. Earth-orbit rendezvous, on the other hand, required the launching of two or more smaller rockets with a part of the lunar spacecraft. The ships would rendezvous and dock in Earth orbit to transfer the fuel between them, and the completed spacecraft would fly its mission to the moon and back. Neither method required a space station, although an Earth-orbit rendezvous might leave some hardware in orbit for later use.

While NASA received a sizeable amount of funding for Apollo, it wasn't unlimited, and the agency still had to get creative in terms of what mission profile to use. Ultimately, a new mission profile known as Lunar Orbit Rendezvous was created to help save fuel by landing a purpose-built craft on the moon while the heavier launch and recovery spacecraft remained in orbit. This approach meant that nothing would be left in Earth orbit once the spacecraft was on its way to the moon.

NASA's plate was going to be full for almost a decade with the lunar-mission goal. The follow-up program to Mercury, known as Project Gemini, tested many of the techniques needed for a flight to the moon while also testing the endurance of men and equipment for periods up to fourteen days in orbit. Mercury's short-duration spaceflights and Gemini's goals provided experience on living and working in space. But beyond two weeks for a flight to the moon and back, there were still plenty of questions about long-term stays in space that needed to be answered.

The air force still had a major interest in space. They regarded it as the next possible battleground, especially if the Soviets had similar thoughts. Concerns about the effects of long-duration spaceflight weren't necessarily limited to just human beings, as no one quite knew what the effects would be on equipment. While American satellites had been flying since 1958, their limited battery life and endurance, plus the fact that they couldn't be recovered, meant that there was no way to see exactly how the space environment was affecting their systems.

Hostile to man-made objects, space is an environment of extremes. The temperature change alone between direct sunlight and shadow provides challenges of thermal control to ensure that the internal contents of a spacecraft, be they manned or unmanned, don't overheat or freeze solid. There are also the very harmful effects of solar radiation and cosmic rays to consider. The microgravity environment also means that objects designed for use in a 1-g environment might not behave quite the same. Loose balls of solder undetected on a circuit board might float free to touch electric terminals, potentially causing an electrical short. Other particles floating free could get into places where they weren't supposed to go. Heat convection doesn't act the same either in zero gravity, since heated air can't rise above cool air. Even moisture expended by astronauts from sweat, urine, and exhalation can cause condensation on critical equipment. Unforeseen problems can creep up when least expected as missions grow longer.

There were also the questions of what people do when they reach orbit and how such tasks could be exploited for military gain. To the air force, space was an extension of their territory. The air has always been seen as a strategic high ground, going back centuries to when an army might place lookouts on hills to see what an opposing force was doing. During the American Civil War, observation balloons were first used to look far in the distance. When World War I broke out, the airplane made the tethered observation balloon all but obsolete.

Reconnaissance airplanes became a primary source for gathering visual intelligence of other countries both during and after the Second World War. As jet engines were created and refined, these aircraft became more specialized and could fly higher than defending fighters. Yet as the Soviets developed their own high-performance jet aircraft, recon planes became more vulnerable.

It was the Central Intelligence Agency (CIA) who would fund and develop one specialized aircraft that could potentially fly out of range of any antiaircraft defense at the time, and that was the U-2 spy plane. However, when the Soviets successfully managed to shoot down a U-2 flown by Francis Gary Powers deep over their territory on 1 May 1960 with an S-75 Dvina surface-to-air missile (SAM), the days of reconnaissance aircraft directly flying over the Soviet Union were numbered.

Development of much faster aircraft in the form of the A-12 and SR-71 Blackbirds would make photographic intelligence gathering by airplane viable for many more years. But the mission of direct overflight was largely abandoned in favor of flying along a country's border so as not to risk being shot down. Beyond what could be seen from that distance at extremely high altitudes, there was no direct means of observation. A manned spacecraft orbiting overhead would have little fear of being intercepted.

Weapons delivery from space, as von Braun predicted, was another possible use. But international politics had made the deployment of offensive space weapons something that the West either did not want to consider or at least make public. In October 1961 the Soviet Union air-dropped and detonated a fifty-seven-megaton hydrogen bomb (since nicknamed the Tsar Bomba). It was a prototype for a one-hundred-megaton nuclear device that could be placed in Earth orbit to remain dormant until a war broke out. It has recently been revealed that Soviet premier Nikita Khrushchev considered the one-hundred-megaton orbiting bomb to be more of a political bluff, rather than something seriously considered for operational deployment. In 1961, however, the idea that such a weapon might be flown operationally was very real.

With the role of an orbital offensive weapons platform effectively out of the picture, the air force looked to other military applications that it could perform in orbit. Several internal air force studies were conducted in a proposal to fly a vehicle known simply as the Manned Orbiting Laboratory, or MOL. With NASA's manned efforts focused on the lunar-mission mandate, having the air force develop a manned space station seemed like a good idea, although it had to receive Department of Defense (DOD) approval and funding first.

In years past, when the air force had received funding to develop dedicated experimental aircraft, its research goals were pretty clear-cut to fly higher, farther, or faster. Long before the Department of Defense was created in 1947 along with the U.S. Air Force as a separate branch of the U.S. military, dedicated test airplanes had been built to push the frontiers of their capabilities. Following the creation of the DOD, dedicated military programs now had to justify their existence. So a proposal to investigate the long-term effects of spaceflight and to test applications of military technology with no specific goal or mission in mind seemed like a potential waste of military dollars.

The lack of a clear-cut goal had already affected another air force program. This was the Dyna-Soar space plane, later known as the X-20. The name of the vehicle was an abbreviation of the term "Dynamic Soaring." It was designed to launch into space atop a Titan booster in a high suborbital trajectory but not quite at orbital velocity. In order to keep itself in space, the craft would skip off Earth's atmosphere in a similar fashion to a thrown stone skipping off the surface of the water, until its speed dropped low enough to reenter the atmosphere and land on a conventional runway.

When conceived in 1957, the Dyna-Soar had two planned missions: nuclear strike and strategic reconnaissance. Eventually, the nuclear mission became obsolete as ICBMs matured to become effective weapons. The reconnaissance capability was also given a lower priority due to the development of the first-generation Corona satellites, initially launched as part of a cover project known as Discoverer. Even with the very high speeds and altitudes at which the X-20 was designed to fly, its original missions were still considered high risk, as a threat nation might employ a shoot-down capability. As a result, the air force was left trying to find a new mission to justify the X-20. They were able to keep the project going a little longer by proposing its use as a rescue vehicle in the event of Gemini or Apollo astronauts becoming stranded in orbit due to a malfunction. Giving the vehicle an experimental "X" designation in 1962 also meant that it was being pitched to the DOD as a research craft rather than as its original concept as an operational vehicle with a specific military mission.

In December 1963, U.S. secretary of defense Robert S. McNamara cancelled the X-20 project. As a replacement, the Air Force decided to focus its efforts on the MOL program. The Department of Defense complied, altering its policy a little to allow the program to be funded, even though its goals were considered to be more open-ended, at least in public. In August 1965, President Lyndon B. Johnson directed the Air Force to commence work on the project. While the decision was regarded as a good one in some circles, elsewhere it left more questions than answers as the Air Force was somewhat evasive in discussing exactly what the MOL would do once it achieved orbit. Since this program wasn't being flown by NASA, the Air Force did not have to publicly disclose its objectives, and various elements of the program remain cloaked in secrecy to this day.

The Douglas Aircraft Company became the main contractor for the MOL design. But rather than assigning a prime contractor to build the MOL and having it coordinate subcontractors, as with an aircraft contract, the air force assumed the coordination role. The air force would define the specifications, the contractors would build the hardware, and the air force would take delivery and perform the final assembly work themselves. This was done to help maintain a higher level of secrecy.

Anatomy of the MOL

In some aspects, the MOL borrowed heavily from hardware that was already built or in development to help save both time and money. Stacked on top of the MOL was a modified version of the two-man Gemini capsule, known as the Gemini-B, which was built by McDonnell Aircraft for NASA (McDonnell and the Douglas Aircraft Company would merge into one company in 1967 to form McDonnell Douglas Corporation). Behind the capsule and its mission adaptor section was the MOL itself, housed in a cylinder measuring 10 feet in diameter by 56.5 feet long. Including the Gemini capsule, the entire complex was just over 72 feet in length.

Just behind the modified Gemini spacecraft with its adaptor section was an eight-foot-long equipment section containing fuel and consumables tanks, plus reaction control thrusters for the laboratory. The pressurized habitat section was located just behind the equipment section. It measured about twelve feet in length, with the interior layout set up with the front of the spacecraft being the top and the rear being the bottom, as opposed to a horizontal layout seen on many subsequent space stations. Behind that was a thirty-seven-foot-long mission module section.

The entire vehicle, complete with astronauts in the Gemini capsule, would launch into orbit aboard a modified Titan III rocket. The Titan III was a stretched version of the Titan II ICBM, which itself was being used for NASA's Gemini program. Improvements were made to the newer Titan III's engines, and additional power for the rocket was provided in the form of a pair of five-segment solid-rocket motors (SRMs) strapped to the booster, one on each side. The Titan III system, both with and without the SRMs, would become a workhorse space launcher for DOD payloads until the early 1980s, with versions of the rocket also being used to fly unmanned NASA probe missions to Mars and the outer planets.

For the MOL, the Titan III-M booster would use a set of slightly longer seven-segment SRMs for added power. In this configuration, the Titan III-M could loft a payload weighing sixteen tons into a polar orbit or nineteen tons into an equatorial orbit. While the Titan III-M never flew operationally, the longer SRMs were paired with a modified version of the Titan III

2. Manned Orbiting Laboratory concept, with Gemini capsule in front.
Courtesy U.S. Air Force.

core to create the Titan IV launcher, which flew from 1989 until 2005. The Titan series rockets were eventually retired from service, in part because they used hypergolic propellants. Hypergolic fuels combust on contact with one another, meaning they don't need a spark to ignite. The fuels are highly efficient and storable. But they are also very toxic, and it doesn't take much exposure to kill an unprotected human.

While the Gemini capsule was designed for docking operations, it was only intended to dock with a test target vehicle in orbit and not a pressurized craft. Internal crew transfers between docked spacecraft wouldn't be carried out until the Apollo spacecraft began flying. That left the problem of how to get the crew from the Gemini capsule into the MOL and back again before undocking and reentry. Various proposals were considered. One proposal had astronauts spacewalking to and from the lab. Another had them entering the lab via an inflatable, external transfer tunnel. Still another had the capsule pivoting around on a giant hinge until it lined up with a transfer hatch on the laboratory.

The method ultimately selected was fitting a hatch in the capsule's heat shield and having the crew transfer between the Gemini and the habitat section of the MOL via a pressurized transfer tunnel that went through the Gemini's heat shield and adapter section. While this method was considered the best, it had a few problems since it meant the internal configuration of the Gemini spacecraft had to be altered somewhat to accommodate both items. Another worrisome aspect was that no one knew if a hatch could be designed to withstand the heat of reentry without compromising the integrity of the spacecraft's heat shield. It would require testing to prove that the concept was a sound one.

Once the crew transferred into the MOL's habitat section, the Gemini would remain shut down in cold storage. It would not be used again until the return to Earth. The astronauts would spend the entire mission inside the MOL. The MOL's habitat section could also act as an airlock module, should the mission call for a space walk, also known as EVA (extravehicular activity). Inside the module, the crew would have all the comforts of home in the form of sleeping berths, a food-preparation area, an exercise bicycle, and a zero-g vacuum toilet. A control panel would give the astronauts full control of the orbital attitude of the MOL and the mission module's equipment. Unfortunately, there was no window present in the MOL's habitat module, so astronauts would have no view of what was going on outside except for what they could see through their mission module's periscopes.

Even before the *Apollo 1* fire occurred in late January 1967, the air force had misgivings about using pure oxygen in a spacecraft due to the potential fire danger, and there were medical concerns about the possible side effects of breathing pure oxygen for long periods. A two-gas oxygen-nitrogen system was the most ideal solution, but it would have added weight and complexity. Eventually, the air force settled on a two-gas system featuring 31 percent helium and 69 percent oxygen at 5 psi (pounds per square inch). Helium is an inert gas that will not contribute to combustion, and it is lighter than normal air. One potential drawback was the fact that helium affects the pitch of people's voices, causing them to sound like Donald Duck when talking. This side effect may have been considered a positive one as it would be almost impossible to identify who was flying an MOL mission from their voice transmissions.

Power for the MOL would come from either solar panels or fuel cells.

Both systems were considered for their advantages, and provisions were made to equip the MOL with either one, depending on the length of the mission. For missions lasting less than thirty days, fuel cells would be carried because of their simplicity as they utilized oxygen and hydrogen to produce electricity. As a by-product, they produced water, which could be used for drinking and cooling electronics. But fuel cells can only produce power for as long as they have reactants. Solar panels, on the other hand, are better suited for longer missions since they generate electricity as long as they have sufficient sunlight. However, storage batteries are needed to maintain electrical power when a spacecraft is not in direct sunlight, and relatively large solar arrays are needed to generate an equivalent amount of power to what fuel cells can generate. Solar arrays would only be used for missions intended to last longer than thirty days.

At the end of a mission, the MOL would be shut down; once the crew had transferred into the Gemini spacecraft, they would prepare for a return to Earth with the results of their experiments. Once the Gemini had undocked, the MOL would be commanded to deorbit and burn up since there were no provisions for other spacecraft to dock with it (at least not initially, although the idea of docking two MOL's together back-to-back was proposed). This approach seemed somewhat wasteful, but it had the added benefit of newer MOLs being steadily improved in quality over the earlier versions as the U.S. Air Force became more proficient at building and operating them.

Astronaut Selection

To fly the MOL, the air force needed its own corps of astronauts. While the astronauts selected by NASA were almost all active-duty military or civilians with previous military experience, the air force could not draw on NASA's ranks to perform the MOL missions. But they could at least recruit from the same pool of talent.

Every astronaut selected for the MOL had experience as a test pilot, with all of them having been trained at the Aerospace Research Pilot School (ARPS) at Edwards Air Force Base (AFB). Today, the ARPS is known as the U.S. Air Force Test Pilot School. But from 1961 to 1972, as reflected in its name, its curriculum included much more than that. There were two phases of study at the school, with the first phase focusing on the original test pilot curriculum and the second phase focusing on studies in orbital mechanics, astronautics, thermodynamics, and the various engineering disciplines required for aerospace projects. To complete both phases would take one year, and the knowledge provided was so thorough that it was considered equivalent to two years of engineering study at a major university. There were very high requirements to get into the school; once accepted, the students were subjected to even higher scrutiny as they progressed. Only the best students were selected to take part in the phase-two program.

Most of the first group of candidates selected to become MOL astronauts came from the 1963 and 1964 classes. Not all members of the group were U.S. Air Force, either, as the U.S. Navy also sent its eligible candidates to ARPS. In November 1965 eight MOL astronaut candidates were publicly announced. From the canceled Dyna-Soar program came Maj. Albert Crews, an early ARPS graduate. Next was test pilot Maj. Michael J. Adams. Prior to his MOL selection, Adams graduated from ARPS with honors and had been one of four aerospace pilots from Edwards AFB to have taken part in early moon landing studies at the Martin Aircraft Company. The third person was U.S. Navy lieutenant John Finley, who was invited to stay on at Edwards as an ARPS instructor after he graduated in 1963. Capt. Richard Lawyer, another member of the 1963 class, was fourth. Capt. Lachlan "Mac" Macleay was the fifth selection, while Francis G. Neubeck was the sixth. Both men attended the test pilot school in 1960 but returned to attend ARPS when the aerospace curriculum was added. Capt. James Taylor, a 1963 ARPS graduate, was the final air force selection for the first MOL group. The final candidate was Lt. Richard Truly of the U.S. Navy. Truly was the youngest of the first group; like Lieutenant Finley, he was serving as an instructor at Edwards after graduating from the ARPS program in 1963.

Over the next couple of years, two more groups of astronaut candidates were named to the MOL program, and many of their names are familiar today. The second MOL astronaut group included the following: Capt. Karol "Bo" Bobko; Lt. Robert (Bob) Crippen, from the navy; Capt. Gordon Fullerton; Capt. Henry Hartsfield; and Capt. Robert Overmyer, from the U.S. Marine Corps. The third MOL astronaut group, which was selected in 1967, consisted of Maj. James Abrahamson, Lt. Col. Robert Herres, Maj. Robert H. Lawrence Jr., and Maj. Donald Peterson. Prior to attending ARPS, James Abrahamson had been the U.S. Air Force project officer on the Vela nuclear detection satellite program. Robert Herres attended the Air Com-

3. Fourteen of the seventeen MOL astronauts selected. Pictured (*from left to right*) are (*back row*) Herres, Hartsfield, Overmyer, Fullerton, Crippen, Peterson, Bobko, and Abrahamson and (*front row*) Finley, Lawyer, Taylor, Crews, Neubeck, and Truly. Courtesy U.S. Air Force.

mand and Staff College prior to attending ARPS, making him unique among the MOL astronauts as he was being groomed for a command assignment. Donald Peterson was a nuclear systems analyst and a flight instructor prior to becoming a test pilot.

Maj. Robert Lawrence was among the youngest candidates selected for MOL, as he had only just graduated from the U.S. Air Force Test Pilot School in June 1967. He was the first person of African American descent to be selected as an astronaut and was already a respected scientist with a PhD in physical chemistry from Ohio State University. Unfortunately, his time as an MOL astronaut candidate was very brief.

All the crewmembers selected for the MOL did another six-month tour of duty at ARPS to help gain as much experience as possible. Major Lawrence was taking part in this training when he was killed in a plane crash in December of 1967. He was in the backseat of a TF-104, a two-seat trainer version of the F-104 Starfighter, while chief of operations at ARPS, Maj.

Harvey J. Royer, was flying a high angle of attack, high-energy, power-off approach to the runway at Edwards. This flight profile is very similar to that used in the x-15 program, and it was under consideration for the space shuttle proposals. But this method of approach has practically no margin for error, since things happen fast. The aircraft flared too high on the approach, ballooned up into a slight climb, and then came back down hard. On the second touchdown, the nose gear collapsed. Since the F-104 is a very slender airplane with stubby wings, mission rules called for ejection in the event of a gear failure, as the aircraft would likely end up tumbling after that. Major Royer ejected safely, but the aircraft rolled on its side before Major Lawrence's seat fired, killing him instantly.

Interviewed for the "Astrospies" episode of the American PBS Television program *Nova* in 2007, fellow class-three MOL candidate Robert Herres talked about his recollection of the accident: "That morning, Bob wanted to change his flight and wanted me to fly in his place on that particular flight. That's about all there is to say. I should have been in the backseat of that airplane instead of him."

Barbara Lawrence, Robert's wife, who was also interviewed for the same episode, continues, "I was standing at home changing buttons on a dress, and I looked out the window and saw Bob Herres coming up the walk. I thought, 'I don't have to ask.' That's what they call a life-changing experience, you know. Suddenly, you know, in the morning everything seems okay, and then a few minutes later it's all over."

Robert Lawrence wasn't the first fatality of an MOL astronaut candidate, as Michael Adams had been killed less than a month earlier. Major Adams was dissatisfied with the slow pace of MOL's hardware development and asked to be removed from the program so he could participate in other assignments. His request was granted, and he was selected to become part of the x-15 program, a joint effort of the U.S. Air Force and NASA.

Flying an x-15 mission on 15 November 1967, Major Adams was descending from a flight path that peaked at an altitude of 266,000 feet when the aircraft broke apart. The accident investigation determined that Adams was likely distracted by a previous electrical failure in the aircraft's control system and might also have been suffering from vertigo. He failed to make corrections when the aircraft began to deviate from its intended flight path. The aircraft yawed sideways in practically zero atmosphere. When the plane continued its descent sideways into the thicker atmosphere, it entered a violent spin at Mach 5 and finally broke apart at sixty-five thousand feet. Adams—likely unconscious—failed to eject. The air force posthumously awarded Major Adams astronaut wings, since he had exceeded the fifty-mile altitude mark on what became his final flight.

Crashes while flying such cutting-edge aircraft were commonplace among test pilots the world over, and they served as sobering reminders of the fate that can be dealt on a high-risk assignment. Indeed, Mike Adams himself had managed to cheat death a few years before. He was riding backseat in a TF-104 with a fellow ARPS student up front when the engine failed and the aircraft ended up dropping at a fast rate. The aircraft impacted with the runway at a higher-than-normal speed. The main gear snapped off, and the jet went careening down the runway before sliding off and coming to rest in the desert sand and scrub brush. Mike Adams ejected, but the frontseat pilot did not. Yet both pilots survived to fly again, as both had made the correct decision. At impact, the jet engine broke loose and punched through the back wall of the rear cockpit. It would have killed Adams if he had stayed with the plane. As for the pilot in the front seat, the ejection seat rails had buckled on impact; if the seat had fired, it would likely have blown up in the cockpit, killing the occupant instantly. The pilot in the front seat was Dave Scott, who would be selected as part of the third class of NASA astronauts. Dave Scott would eventually fly with Neil Armstrong on *Gemini 8* and help test docking procedures in Earth orbit on the *Apollo 9* mission before finally commanding the *Apollo 15* mission to the moon.

First Flight Test of the MOL

In 1966 at the Cape Canaveral Air Force Station in Florida, a new rocket was unveiled at Launch Complex 40. Unlike the five previous Titan III-Cs that had flown, this one had a long cylindrical extension and a Gemini capsule on top. The rocket was assigned to fly a mock-up of the MOL. This boilerplate module, made from a Titan I rocket stage, contained several experiments and pieces of MOL hardware to test in orbit. The Gemini capsule was a used spacecraft, as it had originally flown unmanned as part of the *Gemini 2* mission in January 1965. For this MOL test flight, the Gemini spacecraft had a crew transfer hatch inserted into its heat shield to test how well the system would work.

4. The first MOL test launch took place in November of 1966 with a modified Gemini capsule and a mock-up station on a Titan III-C rocket. Courtesy U.S. Air Force.

On 3 November 1966 the test rocket ascended from the launchpad at 13:50 GMT. After separation of the solids and both the first and second stages, the Titan's Transtage pitched the spacecraft stack downward and fired its engines to accelerate the capsule to orbital reentry velocity before it was jettisoned. Once the capsule was released, the Transtage changed its flight path again and continued to carry the boilerplate MOL into orbit. Along the way, it also launched three small research satellites. The Gemini capsule was recovered near Ascension Island in the Atlantic Ocean after its thirty-three-minute suborbital flight. The heat shield with the hatch performed flawlessly. The hatch had fused solid with the rest of the heat shield, just as it was designed to do. The experiments aboard the boilerplate MOL were meant to transmit data back to Earth for seventy-five days, but the onboard telemetry system failed after just thirty. Eventually the payload, given the designation OPS 0855, continued to orbit until it reentered Earth's atmosphere and broke up on 9 January 1967. Even with the telemetry failure cutting the orbiting mission short, the launch of the boilerplate MOL and recovery of the modified Gemini capsule were important successes.

MOL Training

Work was progressing toward a manned launch sometime in 1968, and the MOL astronauts were actively involved in each phase of the program, performing similar duties to their NASA counterparts as they trained for spaceflight, monitored development of the hardware, and provided valuable input to the engineers. They also flew with MOL equipment in cargo aircraft to gain zero-g experience.

The cargo plane would fly parabolas during the training flights. On the dive phase, the astronaut candidates would experience about thirty seconds of weightlessness as they tried to perform their tasks. Many of these tasks were quite complex, involving an evaluation of the interior of the redesigned Gemini craft with its heat shield hatch and crew transfer techniques. During their brief periods of weightlessness, the trainees would have to move out of their Gemini seats, open the hatch, shimmy through the transfer tunnel, stow the hatch, and continue back before the end of the weightless cycle, when the aircraft would begin its two-g climb for the next parabolic simulation run.

Weightless simulations in an airplane could only achieve so much, and something better was needed, especially for EVA training. During the end of the Gemini program, NASA determined that by performing some activities underwater, an astronaut could be weighed down so he would be neutrally buoyant, meaning he would neither float nor sink. A similar system was used by the MOL program.

Rather than using a pool, the air force decided instead to build a training facility just off an inlet known as Buck Island near St. Thomas in the U.S. Virgin Islands. This particular area was selected in part because it was federally owned land and the U.S. Navy already had a UDT (underwater demolition team) training base on the island. It took approval from six federal agencies to get permission to build the MOL program's underwater training facilities there. Former air force test pilot Norvin C. "Bud" Evans, who had been working for General Electric at the time, was placed in charge of the facility.

The clear waters and favorable weather of the Caribbean were almost perfect for neutral-buoyancy training, and security could be maintained in this remote part of the world. The crewmembers used modified Gemini-style pressure suits for these training missions. The tight-fitting helmets used for spaceflight were replaced by two-piece Plexiglas bubble helmets that were joined by a diagonal flange and held together by C-clamps. Rather than pressurizing the suit with air, water was used instead to maintain a pressure differential with the outside seawater. This practice simulated the pressure differential between the interior of a space suit and the vacuum of space, as a pressurized suit doesn't bend or flex as easily. Crewmembers would wear scuba masks inside the helmet, receiving air from the surface via a lifeline and breathing it through a regulator mouthpiece. It was a clever system, although the closed helmet meant that an astronaut could not easily clear water from his mask or mouthpiece should either develop a leak during a five-to-six-hour training session. Nor could an astronaut talk. Hand signals and written signs had to be used to convey information. The underwater facility was used not just for training but also for evaluating equipment and procedures. According to Bud Evans, in an interview conducted for "Astrospies," "We had to know how long these tasks were going to take. This was one way to get some real timeline studies."

The equipment briefings and the training had the MOL astronauts going to places all over the country. Sometimes they would travel under aliases, and usually they would wear civilian clothes instead of their military uniforms. Occasionally their assignments would take them to Houston, Texas, home of NASA's manned space program. As Richard Truly explained in his "Astrospies" interview, "You know, the NASA astronauts back in the sixties were all good friends of ours. We knew them all. We went to Houston, and when we went to Houston, they'd tell us all about what they were doing, and we wouldn't tell them anything about what we were doing."

Being out of the public eye meant that the MOL astronauts received none of the creature comforts that some of their NASA colleagues enjoyed, such as magazine deals or leases for Corvettes. This was the way the Department of Defense wanted things, and the men had no problem with that, as they felt their mission was worthwhile. As "Mac" Macleay described in his "Astrospies" interview, "I think everybody was tickled. I mean, it was something that we really thought would contribute. We weren't going to go check how the African fruit fly worked under zero gravity, you know, we were going to do something worthwhile—okay—that we thought was worthwhile." Richard Truly adds, "Although we did have a joke in the program that one day there was going to be a little article back on page 50 of the newspaper that said, 'An Unidentified Spacecraft Launched from an Unidentified Launchpad, with Unidentified Astronauts, to Do an Unidentified Mission.' That's the way it was."

The MOL's Actual Mission

While there are plenty of details available regarding the MOL's pressurized laboratory and the Gemini-B spacecraft, the mission module, over thirty feet long, is still clouded in secrecy to this day. While suspected for many years, it has only recently been confirmed that the MOL's primary mission involved reconnaissance and intelligence gathering of the Soviet Union from Earth orbit.

While surveillance satellite technology had matured in the 1960s after the first Corona satellites had flown, there was still no really effective way to control exactly what the satellite was shooting pictures of as it circled the globe. Many of the images taken by the first Coronas were obscured by cloud formations over the Soviet Union, which blocked details on the ground that they were trying to photograph. It also took time for a photograph interpreter to examine developed film of an area of interest in order

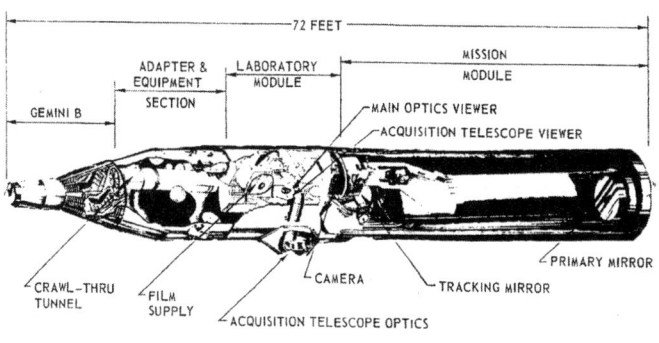

MOL BASELINE SYSTEM

FIGURE 1

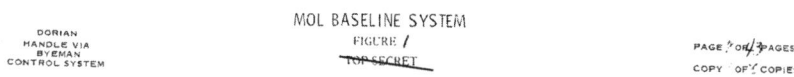

PAGE OF PAGES

COPY OF COPIES

5. This recently declassified illustration of the MOL shows its primary mission being one of photographic reconnaissance with a large telescope-based system. Courtesy National Reconnaissance Office.

to identify something. If images of a specific subject of interest were shot at the beginning of a roll and the film canister had not been returned to Earth until a month or so later, the subject itself might not be there by the time another satellite was sent up to shoot pictures of the same area. With the large volume of film taken by a satellite's camera, it might take weeks for interpreters to go through all the developed film to prioritize every target of interest. This was assuming the film canister had even been recovered safely in the first place.

It was felt that by putting a man in space with a camera system, he could more easily locate targets of opportunity. From orbit, he could observe a great many things on the ground and make decisions in real time about what to photograph. If an area were obscured by clouds, the astronaut could save precious film or use it to photograph a different location.

The primary camera system intended for the MOL was code-named

Dorian, also known in some publications as KH-10 or Keyhole 10. "Keyhole" was one designation used by the National Reconnaissance Office (NRO) to identify its intelligence-gathering satellites in memos and documents. The Corona series were designated KH-1 through KH-4. Camera resolution and mission potential increased with the KH-7 and KH-8 Gambit satellites, and advanced plans were underway to start flying the KH-9 Hexagon series satellite. But the later satellites had sacrificed high resolution in favor of being able to image a wider area, mainly because there was no capability at the time to direct a satellite to focus in on a tight area of study at a moment's notice.

The KH-10 camera was designed to have up to a three-inch resolution, meaning an object about the size of a miniature toy car would appear as one pixel. Something the size of a telephone would be about two pixels. At that resolution, it is possible to identify the model of a car, a plane, or a tank and to potentially see what visible weapons were carried aboard them. A photograph analyst could easily count the number of airplanes on the deck of an aircraft carrier or at a base, likely identify them all by type, and perhaps even read the serial numbers on the wings. Looking at tire tracks on a road, it would be possible to see if an area had experienced increased activity, perhaps exposing military maneuvers or a hidden base. Still, viewing such things from Earth orbit is easier said than done when the spacecraft is travelling at about 17,500 miles per hour. Decisions needed to be made quickly to focus on a target for pictures before it sailed out of view, so the MOL was equipped with a sophisticated optical tracking system.

The crewmember operating the system would look at a wide image of the area being photographed with a periscope viewfinder system. Spotting a target of opportunity, the camera operator could then focus in tight to determine if a photograph might be taken, and the auto-tracking systems aboard the MOL would focus on the target long enough to cancel out the motion blur of the earth to shoot a photograph or two. While additional capabilities aren't fully known due to the classified nature of the system, the MOL likely also had the capability to transmit imagery back to Earth, similar to the way unmanned space probes to the moon and Mars were already doing back in the day. Recently declassified drawings also suggest that the MOL could have been equipped with film return capsules, allowing crewmembers to send exposed film home midway through a mission.

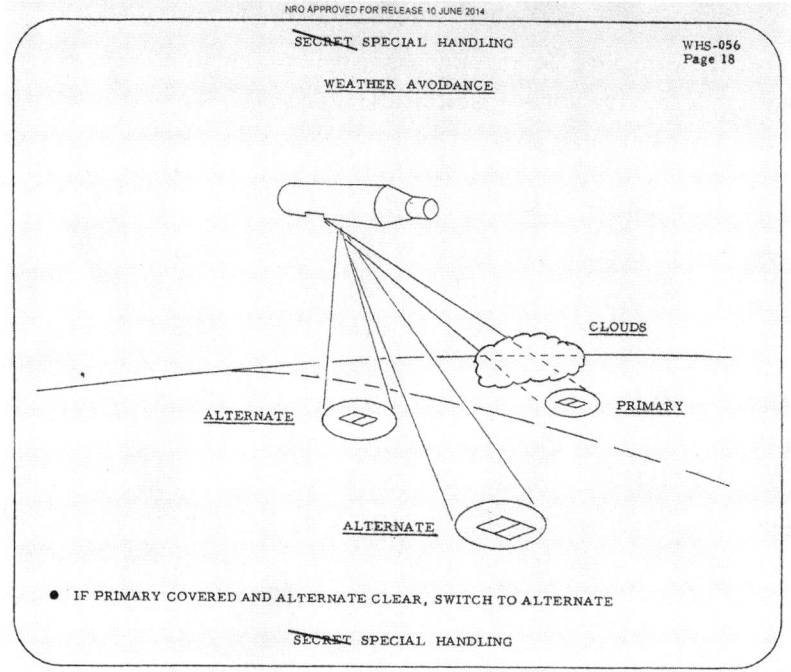

SECRET SPECIAL HANDLING

WHS-056
Page 18

WEATHER AVOIDANCE

CLOUDS

PRIMARY

ALTERNATE

ALTERNATE

● IF PRIMARY COVERED AND ALTERNATE CLEAR, SWITCH TO ALTERNATE

SECRET SPECIAL HANDLING

6. This recently declassified illustration shows how the MOL's optics could examine alternative areas on the ground if the primary target was obscured by weather. Courtesy National Reconnaissance Office.

Additional proposals were drawn up to utilize the MOL for other things. The air force had plans to conduct battlefield surveys, provided an orbit gave the MOL the opportunity to do so. Some of the more ambitious plans called for astronauts to erect a giant parabolic antenna in orbit that could be used to relay signals on Earth between military stations or perhaps to eavesdrop on electronic and communications signals from threat countries. There were also plans to equip the MOL with antennae for electronic intelligence (ELINT) and communications intelligence (COMINT) missions. The potential eavesdropping capabilities of the MOL weren't necessarily limited to observations of Earth itself, as a couple of mission studies also looked at using the MOL to get a close look at, and possibly even commandeer, a Soviet satellite in orbit. Such a mission likely would have required a different mission module than the one planned for the KH-10 camera, as additional fuel and more powerful engines would be needed for changing orbits.

Cancellation of the MOL

When the MOL program was announced by President Johnson in 1965, the budget for the program was projected to be $1.5 billion, a rather exorbitant amount of money to spend in the 1960s. The DOD funded the project, but ultimately its purse strings were controlled by Congress. American combat operations in the steadily escalating war between North and South Vietnam during the 1960s were causing strains to the DOD's annual budgets, forcing cuts to the MOL program's funding. The funding had been cut a little each year, while cost overruns were causing additional problems.

At the time the MOL was given approval, the plan was to have the first unmanned production MOL launch in April 1968, with the first manned flight coming a few months later. Due to budgetary and development issues, the schedule slipped to 1969 and then later to 1970. By the early part of 1969, many hardware elements of the MOL were well along in production and testing. McDonnell Douglas had delivered its first production version of the Gemini-B spacecraft, and everything else with the program seemed to be progressing steadily toward a mid-1970 launch. At Vandenberg AFB, the finishing touches were being put on Space Launch Complex 6 (SLC-6), a launchpad built specifically to handle the Titan III-M and the MOL. The MOL astronauts were also preparing to relocate from their main offices in Los Angeles to newly built crew quarters at Vandenberg.

On the NASA side, the Apollo missions had finally left the ground with the launch of *Apollo 7* in October 1968 and *Apollo 8* in December of that year. With the lunar-landing goal just months away, NASA began thinking more about flying a space station with leftover Apollo hardware. The MOL program's publicly stated goal of "space research" began to look like something NASA should be doing instead.

There were calls in Congress to merge the MOL with NASA's efforts, or at least utilize some of NASA's tracking-station assets rather than maintaining the air force's own network. If the MOL was indeed just an open-ended orbital research facility, there were potential advantages to that. But with its secret reconnaissance objective, using civilian tracking-network assets wasn't desirable. The air force was having a hard time justifying the MOL to certain members of Congress who didn't have a "need to know" the MOL program's actual mission.

Ultimately, the main factor working against the MOL was the NRO and its expanding surveillance satellite capabilities. The NRO was created in August 1960 by the Eisenhower administration due to the poor management and insufficient capabilities of the air force's first two intelligence satellite programs, SAMOS and MIDAS. The creation of the NRO was almost like a black-project version of NASA, as it incorporated some legacy organizations from the military side into its ranks, similar to how NASA gained some of its research centers. Unlike NASA, however, the NRO was part of the DOD and answered to the secretary of defense. The NRO was highly classified, and its name wasn't even revealed publicly until the 1990s.

While the KH-10 system for the MOL program was being developed, the NRO was hard at work on a parallel program. This system was known as KH-11, and it apparently had the same three-inch optical resolution as the MOL's KH-10 camera. Reports suggest that the KH-11 looked very much like an early version of the Hubble Telescope, except the optics would have been pointed at Earth instead of toward the stars.

Many of the MOL astronauts were reportedly allowed to see the KH-11, even though the NRO was apparently reluctant to let them do so until orders from very high up made it happen. There were two schools of thought for surveillance that were in direct competition with one another. The unmanned side said astronauts weren't needed, while the manned side said astronauts were absolutely necessary.

In January 1969 a new president, Richard Nixon, took office, replacing Lyndon Johnson, who had opted not to run for reelection, partly due to the state of affairs with the Vietnam War. Nixon likely had extensive knowledge of the state of orbital reconnaissance, given that he was vice president during the Eisenhower administration. With Nixon came a new secretary of defense, Melvin R. Laird. Nixon took a hard look at the MOL, as did his Budget Bureau director, Robert Mayo, and Secretary of State Henry Kissinger.

At first, the MOL did get increased funding from Nixon to help correct some of the underfunding problems it had incurred during the final two years of Johnson's presidency. But even with the funding increase, the damage was done. The launch schedule slipped further, with the first unmanned launch now scheduled for 1971 and the first manned launch scheduled for a year later. Another problem was a reduced number of missions. Ultimately, it looked like the DOD would only be able to fly five missions total, with just four of those being manned. A decision to fund additional missions would not to be made until later. With these factors in mind, plus the ramping up of NRO's KH-11 program, the MOL seemed to have run out of time and became obsolete before it even left the ground.

Finally, Nixon, Mayo, and Kissinger decided that it was best to cut their losses and made the decision to cancel the MOL program. Defense Secretary Laird reportedly was not in favor of the decision at all, but it was out of his hands. On 10 June 1969, a month before *Apollo 11* made its flight to the moon, the decision to cancel the MOL program was announced publicly by David Packard, the undersecretary of defense.

The astronauts involved with the program had no advanced warning that the decision was coming, and neither did the contractors. Richard Truly was at a contractor plant having an argument with an engineer about something, when "Mac" Macleay tapped him on the shoulder and whispered to him quietly that the program had been canceled. Hank Hartsfield listened to the news on the radio as he was driving in his car. Stories from the other MOL astronauts are about the same. All of them were shocked and dismayed when they heard the news.

Aftermath and MOL's Legacy

Even though the MOL program had been canceled, many of the program's astronauts stayed together to plan their future as a group. They continued to have regular meetings each week to exchange ideas about what to do next. Richard Truly and Bob Crippen made a trip to the Pentagon to see what the navy might have for them as far as programs that could utilize their expertise. It was "Bo" Bobko who apparently made the first suggestion to contact NASA to see if they had any openings. The support he got wasn't enthusiastic, since all the MOL astronauts knew that the Apollo program was also facing budget cuts with many of NASA's astronauts still not having flown in space.

The MOL group eventually traveled to Houston, where they were interviewed by NASA's director of Flight Crew Operations, former Mercury astronaut Donald "Deke" Slayton, who expressed similar concerns about potentially taking on so many new astronauts. Meanwhile, NASA officials in Washington DC felt that it would be a good idea to hire at least some of the MOL program astronauts, as NASA was starting to ramp up its space station program, which eventually became known as Skylab. Having worked on a similar program, the MOL astronauts had valuable experience they could contribute.

Ultimately, Slayton made the arbitrary decision that he would only accept astronauts from the MOL program aged thirty-five years or younger. This cut the number of eligible candidates from fourteen down to only seven. Slayton met with the seven and said that he obviously did not have any upcoming missions for them but that he would welcome their help on Skylab. There was also the space shuttle program in the works, even if it still hadn't been approved by Congress. Chances were that none of the seven MOL astronauts being given the opportunity to join NASA would even get a chance to fly for at least a decade, if at all. Nevertheless, all seven men said they were willing to join NASA and did so.

NASA's seven new astronauts were Richard Truly, Bo Bobko, Bob Crippen, Gordon Fullerton, Hank Hartsfield, Robert Overmyer, and Donald Peterson. All would provide invaluable support to the Skylab program, and all would fly missions as shuttle astronauts with six of the seven eventually becoming shuttle commanders. Al Crews, while being too old to join as an astronaut, still had an incredible wealth of knowledge to contribute, as he had been part of the Dyna-Soar program before he was selected for the MOL. Crews therefore joined NASA's Flight Crew Directorate in Houston and was involved in flight operations and systems testing for many years.

While the remaining MOL astronauts not selected to join NASA were obviously disappointed that they did not get a chance to fly in space, they enjoyed very fruitful careers both in the military and in the private sector. Naval officer John Finley had already left the MOL program in 1968 due to the project delays. He flew a tour of duty in Vietnam and served in many fighter squadrons throughout the 1970s in addition to other command-level postings before he retired from the navy as a captain in 1980. He died in 2006 after a long battle with diabetes and cancer. Richard Lawyer retired from the air force in 1982 as a lieutenant colonel after a long career as a test pilot. He continued flying as a civilian test pilot for many years until he suddenly passed away in 2005.

"Mac" Macleay would fly combat in Vietnam as commander of the Twenty-Third Tactical Air Support Squadron. After he retired from the air force in 1978 with the rank of colonel, he worked on missile systems for Hughes Aircraft Corporation. Greg Neubeck would fly combat in Vietnam as well. He then served as the vice commander of the Tactical Air Warfare Center at Eglin AFB in Florida before retiring from the air force in 1986 with the rank of colonel. He also attempted a run for the U.S. Congress but was not elected.

Lt. Col. James Taylor unfortunately had a very short life after the MOL program was canceled. He returned to Edwards AFB and became a test pilot instructor and deputy commandant of the test pilot school. On 4 September 1970 he and French Air Force exchange test pilot Capt. Pierre Dubucq were killed when their T-38 crashed at the Palmdale Regional Airport in Los Angeles, California, while performing touch-and-go approaches. The cause of the crash was attributed to wake turbulence from a cargo plane taking off. An award was given in Taylor's honor for the most outstanding student graduating from the ARPS phase-one course, but the award was eliminated after 1971 when the two-tier ARPS program was canceled.

James Abrahamson would eventually become the second-highest-ranking officer of the MOL astronaut veterans when he retired with the rank of lieutenant general (three stars) in 1989. He served on the staff of the National Aeronautics and Space Council in the early 1970s and was the air force's managing officer on the AGM-65 Maverick missile program. He became the associate administrator for the space shuttle program in 1981 and supervised the shuttle program's first ten successful missions. In Abrahamson's final assignment before retirement, he became the first director of the Strategic Defense Initiative Organization in 1984, a group tasked with using space- and missile-based technologies to develop a defensive shield against Soviet ICBMs.

Robert Herres would eventually become a four-star general and serve as the first vice chairman of the Joint Chiefs of Staff from 1987 until his retirement in 1990. Prior to that period, he served as commander of strategic bombing squadrons and also as commander of the Eighth Air Force. He later went on to become commander of the North American Aerospace Defense Command (NORAD) and became the first commander of U.S. Space Command when it was activated in 1985. After he retired from the air force, Herres served as chairman and CEO of United Services Automobile Association, a private insurance company for military veterans. He died in 2008 after a two-year battle with brain cancer at the age of seventy-five.

Most of the hardware designed and built for the MOL was never used. But NASA did take on some of the systems and hardware contracts to use in the Skylab program. The IBM command-and-control computer design was one such piece of hardware adapted for the station. The pressurized transfer tunnel from the Gemini-B to the MOL would form the basis of the crew transfer tunnel used for the European Space Agency's Spacelab used during the shuttle program.

The Titan III-M never flew from SLC-6 at Vandenberg, and the launch facility was mothballed after its completion. When the shuttle program began, SLC-6 was modified to become a shuttle launchpad for polar orbit missions. Following the loss of the space shuttle *Challenger* in 1986, the DOD decided to use unmanned rockets instead of the shuttle to launch its satellites from Vandenberg, so SLC-6 was once again mothballed. Coincidentally, former MOL astronaut Bob Crippen, by that point a veteran of multiple shuttle missions, was scheduled to be the commander of the first space shuttle mission from Vandenberg when it was canceled. Today, SLC-6 has finally become operational as a launchpad for Boeing's Delta IV rocket, with several successful launches taking place since 2006.

As for the reconnaissance mission goal intended for the MOL, indications are that the NRO programs that replaced the MOL have been doing their jobs efficiently. Today, many of the early Keyhole satellite programs have been declassified and revealed to the general public, with the latest one being the KH-9 Hexagon system. However, the MOL KH-10 is likely to remain classified for many years yet, since it apparently shares many of its design traits with the KH-11, which first began flying in 1976. Refined versions of the satellite, known in some publications as the KH-12, continue to be launched into orbit on a semiregular basis. Reconnaissance from Earth orbit is as important to protecting a country's interests today as it was when Wernher von Braun first proposed the idea over five decades ago, and it isn't likely to change anytime soon.

The MOL's legacy did not die with the program's cancellation in 1969. There were others watching the MOL's progress, and they were located halfway around the world in the Soviet Union. Despite the classified nature of the MOL program and its cover story, the Soviets knew exactly what the MOL's true purpose was. While not known publicly at the time, the Soviet response to the MOL would have a major influence on the following four decades of space station history in ways nobody could imagine, not even the Soviets themselves.

2. Chelomei and Almaz

The story of the Soviet Union's space station program is rather convoluted by Western standards. It is fully intertwined with enough politics, intrigue, confrontation, and behind-the-scenes maneuvering to make the story not seem out of place in a popular spy novel. Many engineers were involved, but ultimately it would be one man who got the project started. Soviet engineer and designer Vladimir Nikoleyevich Chelomei (pronounced *Chel-o-may*) isn't well known in the Western world, but without his efforts, Soviet, Russian, and even international efforts to build space stations would have evolved along a very different path.

Vladimir Chelomei was born on 30 June 1914 to a family of teachers in the Ukrainian town of Sedlets, located halfway between modern Warsaw, Poland, and Brest, Belarus. Today, the town is known as Siedlce, Poland. At the time of Chelomei's birth, the region was part of the Russian Empire.

When he was a few months old, Chelomei's family moved to Poltava in Central Ukraine to escape the fighting on the eastern front. Few accounts of Vladimir Chelomei's early life have been published outside Russia, so most details of his childhood are unknown. What is known is while growing up, he was encouraged to learn what he could from both a family of teachers and various neighbors who were also intellectuals. However, life could not have been easy for them when the October Revolution of 1917 removed Tsar Nicholas II from power and helped to create the Soviet Union. School closings were common in 1920 during the struggle for power among the various factions that saw the Communist Party eventually become the ruling body. When Chelomei was twelve, his family moved to Kiev, and it was here that he studied auto mechanics at a technical school. In 1932, now age eighteen, he attended Kiev Polytechnic Institute, which had an

7. Vladimir Chelomei was instrumental in the creation of the Soviet Union's first operational space station.

aviation branch. This was the same school that future chief designer Sergei Korolev had attended eight years earlier.

Specific details of Chelomei's time at Kiev Polytechnic are sketchy. He was considered quiet and analytical compared to Korolev's more hyperactive personality. He was a brilliant student and a prodigy with a keen engineering mind. In 1936, during his internship at the aviation plant in the industrial city of Zaporozhye, he proposed an unorthodox solution to deal with problems of mechanical failures in aircraft engines. He took quite an interest in vibration study. Vibrations can be harnessed, but they can also upset navigation systems and destroy equipment if not kept in check. That same year, he published his first book, *Vector Analysis*; by age twenty-four, he had published fourteen papers on his research. He graduated with honors in 1937 from Kiev Polytechnic and stayed on as a lecturer. By 1939 he had completed his candidate dissertation (equivalent to a master's degree) and received one of the fifty coveted Stalin postgraduate scholarships awarded annually to begin his doctoral program. He completed his doctoral dissertation and joined the Communist Party in 1941.

At the start of his country's involvement in what the Soviets refer to as the Great Patriotic War against Germany, Chelomei was named chief of the Jet Engine Group at the Baranov Central Institute of Aviation Motor Building in Moscow. With the support of the aviation minister, Dmitriy Volikov, Chelomei was able to form a research group of engineers to look into pulse-jet technology. His time as the head of this research group caused a minor scandal among other intellectuals, as the sign on his office door read, "Professor Chelomei." While he had completed his dissertation, he had not yet been awarded his doctorate. Eventually it came, though, making the title of professor official.

In 1942 Vladimir Chelomei and his research group created Russia's first pulse-jet engine. A pulse jet has few moving parts. Fuel is injected into a combustion chamber along with air, the air inlet closes, and the mixture is ignited, creating a combustion that is expelled out the back of the engine to propel the vehicle forward. This form of jet propulsion is characterized by a constant dull rumble or buzzing noise. Compared to other types of jet engine, pulse jets are somewhat crude, mainly because they can't be throttled. In 1943 Chelomei proposed using the engine to power a cruise missile, but his proposal was rejected by Soviet leadership. However, the Germans were also working on pulse jets at the same time, and the results of their research led to the engine for the Fi-103 cruise missile, more commonly known as the v-1, or buzz bombs. Chelomei created his pulse jet totally independent of the German design, which was a postwar point of annoyance for him when international aviation experts wrongly believed that early Soviet pulse-jet technology had emanated from German efforts. The truth was quite different.

In 1944, Britain wanted to enlist the aid of the Soviets in recovering materials related to the German v-2 rocket and the Fi-103 missile from retreating German forces. At the time, the Soviet leadership had little interest in the v-2, but they were quite interested in the Fi-103, since its shape and use was more familiar to them. The Soviets felt the design held great potential as a weapon.

When the British provided the Soviets with the remains of an Fi-103 missile and engine that had been recovered from a crash site in England, Josef Stalin immediately ordered a program to develop a similar cruise missile. On 14 June 1944 Chelomei met with Stalin's aviation minister, Georgy Malenkov, at the Kremlin to inspect the remains of the German missile. Chelomei was asked if such a weapon could be developed for the Soviet Union. He replied with an enthusiastic affirmation and reportedly gave an impressive speech advocating the potential of this technology. It is said that two days after that meeting, he had a hundred people at his disposal to begin work on a Soviet cruise missile. That was only the start; by the early fall of 1944, he had received all the resources he had requested. Chelomei was then appointed director and chief designer of Plant Number 51, a facility that was formerly part of the design bureau of deceased aircraft designer Nikolai Polikarpov.

By late 1944 Chelomei had succeeded in duplicating the Fi-103's engine, which was more powerful than his own engine design had been. Reportedly, ten different configurations for a cruise missile were considered before he settled on a design that resembled the Fi-103. Little remained of the missile that Chelomei had inspected months earlier. His team had to figure out on their own everything that made the missile work. By March 1945 a prototype of the missile was ready for testing, and its shape similarities with the Fi-103 did not go unnoticed.

That same month, Chelomei was summoned to a meeting with Stalin and the head of Soviet security, Lavrentiy Beria. Beria bluntly asked Chelomei if he had appropriated the design for his cruise missile from the Fi-103. It was likely an attempt by Beria to check if the designer was a loyal Communist, or even a German spy. Beria was feared among intellectuals, because his NKVD (Narodnyy Komissariat Vnutrennikh Del) secret police were personally responsible for the purges of the 1930s, which sent many aircraft designers, engineers (including Sergei Korolev), and other academics to prison and the gulags of Siberia. Those who didn't die from their time in custody had Germany's invasion of the Soviet Union to thank for early release as their talents were needed to counter the German war machine. Even then, these people continued their work from prison, and most weren't completely free to resume their peacetime lives until near the end of the war. To this line of questioning from Beria about his design's similarities to the German weapon, Chelomei bluntly replied, "I obviously could not have borrowed their ideas. Whether the Germans [stole] my ideas is a question for you, Lavrentiy Pavlovich." It was a bold statement from the young designer, and the matter was quietly dropped.

From March to August 1945, numerous test launches of the new missile (designated 10kh, or 10x) were conducted by airdrops from a specially equipped bomber. The results were only modestly successful, and there wasn't much hope that the missile could be employed in battle before World War II ended. Like Korolev, Chelomei reportedly made a trip to Germany at the end of the war to seek out any undamaged Fi-103 hardware and was apparently able to incorporate those findings into the 10kh design. However, successful results did not come as quickly after the war, especially as there were problems that took many years to solve. Nevertheless, Chelomei's team continued to refine the missile, working on variations for deployment from land, ships, and Soviet heavy bombers.

By 1953, political forces were conspiring to have Chelomei's design bureau taken away from him. Artem Mikoyan of the OKB (a Russian acronym meaning "Experimental Design Bureau") 155, more commonly known as the MiG Bureau, submitted his own proposal for a different cruise missile design based on the MiG-15 fighter jet. This design would become the KS-1 Kometa cruise missile. One of the individuals who worked for MiG at the time was a young engineer with the rank of colonel in the Soviet Army. His name was Sergei Beria, the son of Lavrentiy Beria. Having the son of the influential head of Soviet security working at the bureau meant that Mikoyan could make proposals higher up the political ladder than he was capable of doing otherwise. Artem Mikoyan also had other political allies, as he was also the younger brother of Anastas Mikoyan, for many years a high-ranking member of the Communist Party and Soviet Central Committee.

Mikoyan's bureau had the success of the MiG-15 on its side, as the jet was very much an equal to the U.S. Air Force's F-86 Sabre, which it met in the skies over Korea. One of the limitations with Chelomei's pulse-jet-powered missile was that it would be unlikely to break the sound barrier, while a pure jet-powered missile potentially could. Many considered the pulse jet a dead-end technology. The argument for the MiG Bureau to take over development of all cruise missile designs was very persuasive. In early 1953 Stalin signed a decree making Chelomei's Plant Number 51 part of the MiG design bureau and stripping Chelomei of his title of chief designer. Chelomei tried in vain to fight the decision. Following his dismissal, all he was left with was his teaching job at the Baumann Moscow Higher Technical School, where he had been a professor since 1952.

Very soon after, things would swing in Chelomei's favor once again. On 5 March 1953 Stalin died from complications related to an apparent stroke. Georgy Malenkov assumed leadership as premier of the Soviet Union with Lavrentiy Beria as deputy premier. But Beria only stayed in office for 113 days. In June of that year, a power struggle orchestrated by both Malenkov and party official Nikita Khrushchev caused Beria to be removed from office. Beria was arrested, interrogated by many of the same NKVD interrogators who had worked for him, and executed in December of that year.

Premier Georgy Malenkov remembered young Chelomei from his days as aviation minister and looked favorably on the engineer. Chelomei's theories also made an impression on Nikita Khrushchev, who had become First Secretary of the Central Committee after Beria was removed from power.

During Chelomei's brief period of unemployment as a designer, he fleshed out a design for a cruise missile with folding wings that was capable of being launched from a submarine. It wasn't just the missile itself that Chelomei came up with; his plan was to develop a fully encapsulated system that could be stored on the submarine in an inactive state until the order to launch was given. It was a radical concept, and this approach would be a hallmark of Chelomei's design philosophy.

Vladimir Chelomei was reported to be a charming and charismatic figure by those who knew him. This proved to be an advantage when he approached Nikita Khrushchev to discuss his proposals, and he received a favorable reception. Chelomei's line of thinking appealed to Khrushchev, who was supportive of cruise missiles and submarines to help keep a technological balance with the U.S. Navy, as opposed to building a large and expensive fleet of surface warships, which Soviet naval commanders desired at the time. In 1955 a decree was signed that created a new design bureau, the OKB-52, under the Ministry of Aviation Technology. The OKB-52 was tasked with developing submarine-based cruise missile technology, and Vladimir Chelomei was appointed as its chief designer. He subsequently staffed the OKB-52 with many members of his old team from the Plant 51 days and several promising young engineers straight out of the universities. Chelomei's new cruise missile designs were very advanced for their day and were successful when deployed operationally. His star was definitely on the rise. In 1958 Chelomei hired Nikita Khrushchev's son Sergei to join the OKB-52 as an engineer of guidance systems. Sergei Khrushchev was certainly a smart individual, but his hiring by Chelomei may have simply been a repeat of what Mikoyan did with Beria, in order to gain acceptance from Nikita Khrushchev, who had been appointed Soviet premier in 1958 in addition to retaining his role as First Secretary.

By this time, Chelomei had also begun to set his sights on ballistic missiles and spaceflight. As advanced as OKB-52's cruise missile designs were, Chelomei realized that ballistic missiles were the wave of the future, since there would likely be no defense against them for the foreseeable future. Sergei Korolev's OKB-1 already had a virtual monopoly on the development of the Soviet Union's first ICBM, but Chelomei felt he could design a better one. In the days of its early development, the R-7, or Semyorka (Digit-7), rocket had suffered numerous failures, and Chelomei voiced bold criticism that OKB-52 should be placed in charge of developing the new missile. But OKB-1 continued to refine the R-7 until it was finally ready to be deployed operationally. The rocket would also be used to launch the world's first artificial satellite, Sputnik, in October of 1957. Because of Sputnik, the world stopped and took immediate notice of the Soviet Union.

While Korolev completed his task of creating an ICBM, the limitations of the design as a weapon system immediately became apparent to the Soviet leadership. Given that the R-7 required fueling with LOX (liquid oxygen) and kerosene, the missile required hours of preparation to fly. In addition, all the operational R-7 ICBM sites were above ground and potentially easy to target for destruction. The need for LOX also meant that there were production, transportation, and storage challenges, especially at remote launching sites. So while Sputnik revealed to the world that the Soviets had an ICBM and space launch capability, it meant that at best they only had a handful of operational R-7 ICBMs capable of being launched against targets in the United States.

With the blessing of the Soviet government, Sergei Korolev's OKB-1 began to focus its efforts on manned spaceflight activities. Korolev's aspirations to continue with work in spaceflight should have meant a need for other design bureaus to take up the challenge of developing more capable ICBMs. Yet at the same time, Korolev lobbied Soviet leadership that OKB-1 should also continue to develop ICBMs as the only design bureau responsible for all ballistic missiles and rockets. It was at this time that Chelomei threw the resources of OKB-52 into the ring, since Khrushchev was not in favor of a single design bureau for all rockets. Khrushchev knew the Americans were focusing a maximum effort on rocketry development as well, with multiple American companies taking part. Chelomei was not alone in his aspirations, as Mikhail Yangel's OKB-586 was also working on a more capable ICBM. Yangel was a pioneer in the use of storable hypergolic fuels for the Soviets. His first operational success was the R-16 rocket, but something better was needed. As for OKB-1, they succeeded in developing one more ICBM, the R-9, before they narrowed their focus exclusively to spaceflight activities.

OKB-52 began work on two rockets, the UR-100 and the UR-200. The acronym UR stood for "Universal Rocket," as Chelomei came up with designs capable both of weapons delivery and of launching spacecraft. It was his attempt to try to cut a slice of the spaceflight pie from Korolev's bureau while also trying to secure a majority of the ICBM business. The UR-200 came first and was designed to deliver heavy nuclear weapons in addition to space payloads. The UR-100, by comparison, was a lightweight ICBM. While it couldn't loft the larger megaton-range nuclear weapons, the UR-100 made up for that with a very large number of missiles deployed operationally, making it all but impossible for a first strike against the Soviet Union to succeed in taking out all of them before they were launched.

One early obstacle with Soviet ICBM designs that didn't use LOX and kerosene was their dependence on acid-based hypergolic fuels. If the rockets were put on alert and fueled, they could not remain this way indefinitely, as the acid-based chemicals would damage the fuel tanks. This required the rockets to be overhauled after fueling if they weren't used. The need to fuel the rockets also meant that launch preparations after an order was given consumed precious time. Soviet solid-fuel technology was well behind that of the United States in the 1960s. While the Soviets found success in making small solid motors for use in air-to-air missiles and parachute-deployed retro-rocket landing systems, they did not have the capability of building a large-scale solid-fueled ICBM of the same capabilities as the U.S. Air Force's Minuteman I missile or the Titan III-C's SRMs. Chelomei's solution was to design the UR-100 to be fully fueled from the factory. Taking a page from his cruise missile designs, Chelomei's UR-100s were built with the fuel and oxidizer already loaded, and the missiles were transported to their silos in special encapsulated containers. The rockets could remain in this stored state for up to five, and later ten, years between overhauls. They could be launched within three minutes of receiving an order to fire. Many technological challenges had to be overcome to produce the UR-100 missile, and the process was long and troublesome. But Chelomei's methodical approach to design and testing meant that OKB-52 was able to solve the problems one at a time and that the operational UR-100 was an effective counter to the Minuteman I.

When OKB-52 started work on the UR-200, Yangel's group was already hard at work on their R-36 ICBM design to replace the R-16. The R-36 and the UR-200 were very similar to one another in their specifications. Khrushchev welcomed this competition and the different approaches, as he felt that the only way the Soviet Union would be able to negotiate on equal terms with the United States was to have a very capable force of nuclear-armed ICBMs. The better the ICBM design, the better the bargaining position and international posture. But Chelomei had grander plans for the UR-200. OKB-52 was also working on an antisatellite weapons system (ASAT) and a nuclear-powered communications-and-targeting satellite for the Soviet Navy for use with Chelomei's cruise missiles. Both were designed to use the UR-200 as a space launch vehicle.

As work began on the UR-200, Chelomei began to look at other plans to conquer space. Utilizing the technology for his encapsulated cruise missile, he proposed ideas for several *kosmoplans* (a Russian word that literally means "cosmic planes"). The idea was for a modular vehicle that could be powered by solar, chemical, or nuclear power systems depending on the mission requirements. The spacecraft could be loaded onto a missile and sent into space, where it would fly its mission. When this was completed, a small winged craft based on one of Chelomei's cruise missiles would be jettisoned for the return to Earth, protected from atmospheric friction by a heat shield. Low in the atmosphere, the return vehicle would sprout wings and return to land on a runway like a conventional aircraft. The most ambitious mission that Chelomei intended for the kosmoplan was a landing on the planet Mars. However, the radical concept never went further than the drawing board.

Chelomei also developed a weapons delivery variation of this idea and called it the *racketoplan* (Russian for "rocket plane"), using his UR-200 ICBM as the launch vehicle. During ballistic reentry, the vehicle would release a payload of multiple cruise missiles with warheads that had the capability of striking targets in multiple locations with either conventional or nuclear armament. It was a very ambitious idea. If it were deployed operationally, many experts believe there would have been no defense against such a system. As ambitious as the kosmoplan and racketoplan designs were, though, Soviet leaders were reluctant to provide additional funding after the early design studies.

It was during this time that Chelomei first encountered the individual who became his greatest political adversary. That man was Dmitry Feodorovich Ustinov. Ustinov was an engineer who in the 1930s mainly specialized in weapons development. During the outbreak of World War II, Stalin appointed him the "people's commissar of armaments." Ustinov's efforts were instrumental in the evacuation of the Soviet war factories from cities close to the front lines to east of the Ural Mountains, well out of range of German bombers. His management style and skills were considered very important to the war effort.

After the war, Ustinov's efforts were instrumental in helping the Soviets to acquire German rocket technology, set up the infrastructure required to build copies of the V-2, and come up with an equivalent ballistic missile capability. He became a Central Committee member in 1952. When Stalin

8. Minister Dmitry Ustinov would become Chelomei's biggest political adversary.

died the following year, the Ministries of Armaments and Aviation were combined into the Ministry of Defense Industry with Ustinov in charge. He was known as "Uncle Mitya" to the leaders of the design bureaus.

In those days, Soviet rocketry was considered to be a form of artillery. Since Ustinov's responsibility was more in line with ground armaments and artillery, he considered Chelomei's aviation-based design bureau to be outside its area of expertise. What annoyed him most of all was Chelomei's connection with Nikita Khrushchev, which allowed the designer to bypass the management chain of command if he wanted something, and Ustinov was powerless to do anything about it. Ustinov likely considered Chelomei to be a grandstander who had no respect for the minister's authority. He also realized that while Chelomei's designs were more advanced than the competition, the designs would also require more time and resources to develop, utilizing resources already stretched thin in the postwar Soviet economy.

Ustinov was also not in favor of having two design bureaus as prime hardware developers for the Soviet space program. He had already lost Korolev to the siren song of space travel and wasn't about to lose a second designer to an activity the minister apparently considered to be wasteful, except for propaganda purposes. Granted, the Soviet space efforts required more than one design bureau, but such arrangements were similar to the American approach with one prime contractor (OKB-1) in charge and several subcontractors developing hardware such as rocket engines for them. Ustinov knew that subcontractor efforts usually didn't consume as much manpower and resources as a primary contractor.

Proton

During the years that Nikita Khrushchev was in power, Chelomei had great success in having his designs approved and funded for study and development. This isn't to say that everything was approved, but Chelomei at least seemed to have a higher rate of success than some of his contemporaries. Over time, his relationship with Ustinov worsened, but OKB-52's leader had more than enough support to keep Ustinov in check.

With work far enough along on the UR-100 and the UR-200 ICBMs, Chelomei turned his attention toward designing a space station. In order to fly it, a new booster capable of lifting payloads more massive than what the R-7 could achieve was needed. Thus, work began on the UR-500. The UR-500, in keeping with Chelomei's universal-rocket strategy, was first proposed as an ICBM capable of lofting a thirty-megaton hydrogen bomb. Its use as an ICBM was a dead end, though, as Khrushchev decided it was a bit too big and expensive to develop for that purpose. However, Chelomei was able to gain approval to begin its development as a space launcher. The UR-500's early design approval occurred at a meeting of the Defense Council between Premier Khrushchev, representatives of the Soviet military, and the various design bureaus in February 1962. The design work proceeded quickly, and it was authorized for full hardware development by August 1964. Eventually it became a highly successful space launcher, known to the world by a name derived from its very first payload, Proton.

At the same 1962 meeting, Korolev submitted a proposal to develop his dream booster, the massive N1, with the goal of achieving a manned Soviet landing on the moon ahead of the United States. As initially pitched to Soviet leaders, the N1 was intended to launch a seventy-five-ton payload into Earth orbit. Korolev felt that the capability was sufficient for a spacecraft design intended to land one cosmonaut on the moon. The booster design was a monster in most every sense of the word. The first stage would be powered by twenty-six motors using liquid oxygen and kerosene. Soviet technology was not yet capable of creating rocket engines of the same size and power class as the F-1 engines used on America's Saturn V, hence the need for many smaller motors.

Vladimir Chelomei was allowed to present his plans for the UR-500 the day before Korolev, as he did not feel well after an airplane journey to the meeting's location. He felt he might be too ill to attend the next day, when the ballistic missile and space proposals would normally take place. Given this problem, Sergei Khrushchev sat in at the meeting as Chelomei's representative when Korolev pitched the N1. Khrushchev was mainly there to relay to his boss what other programs were being proposed. As he later wrote in his book, *Nikita Khrushchev and the Creation of a Superpower*, the younger Khrushchev enthusiastically told Chelomei about Korolev's N1 proposal that night. Chelomei thought for a moment and then responded, "I don't think the N1 will fly."

Surprised by this response, Khrushchev pressed his boss for more information as he questioned Chelomei's skepticism about the design. While

Chelomei didn't go into specifics, he felt that synchronizing the thrust of twenty-four engines and dealing with the resultant vibrations and oscillations from them would be an impossible task. He concluded his points by saying, "The devil himself couldn't bring it off."

Reports suggest Vladimir Chelomei and Sergei Korolev got along well. They were very respectful of each other, although highly competitive behind the scenes. This contrasted greatly with Korolev's attitude toward Valentin Glushko. Korolev believed it was comments from Glushko to Stalin and Beria that caused him to be arrested, tortured, and sent to a gulag in Siberia (later released to become a designer working for Glushko). Korolev usually never let his personal feelings get in the way of collaboration with Glushko's bureau, as he needed their engines to power his R-7 rocket. Behind the scenes, however, their relationship was considered tumultuous at times. By comparison, Chelomei's relationship with Glushko was a better one, as he also needed Glushko's engine designs to power his own rockets. For the N1, Korolev selected a different design bureau to build the engines for the first stage.

While Korolev preferred using liquid oxygen and kerosene to power the N1, Chelomei and Glushko both favored the use of the hypergolic propellants unsymmetrical dimethyl hydrazine (UDMH) and nitrogen tetroxide. These were essentially the same fuels used in the U.S. Air Force's Titan boosters. While the use of such fuels meant that the UR-500 would potentially not be as powerful as a booster fueled by LOX and kerosene, both Chelomei and Glushko felt that the advantages of this approach more than outweighed the disadvantages, thanks to the simplicity of the rocket engines used in the UR-500.

Where Korolev failed with the N1 and Chelomei succeeded with the UR-500 Proton was partly due to the approach each designer took toward engineering and testing. Korolev's approach was the one he had successfully used on his rockets back before the R-7. The design was finalized, built, and then tested. As problems cropped up in testing, the design was modified until success was achieved. While this approach worked well with less complex machinery, for something as massive as the N1, with an astronomical number of possible failure points, the problems became almost insurmountable. The N1 design had major issues from the outset. As its size and weight grew, the first stage gained six more engines for a total of thirty motors powering the first stage. This increase in thrust also increased the complexity of the design, which in turn increased the weight factor. It was a vicious cycle to overcome.

Chelomei, on the other hand, favored developmental testing. Once a rocket's specifications had been drawn up, hardware was built and tested to verify the soundness of the design, and any changes were made along the way. Many problems were resolved during testing, and the results were used in finalizing the hardware for the production vehicle. It was a design approach favored more in the United States, especially among aircraft corporations. The only drawback to this approach was that tangible results typically didn't come quickly, and often there was no production hardware available until years after the project had begun. In the case of the UR-500 Proton, though, work proceeded as quickly as possible, and the first versions were ready for testing by 1965.

The Proton booster would become a true workhorse of Soviet space efforts. To the casual observer, the first stage of the rocket resembled a single core stage with six strap-on boosters attached in an arrangement similar to Korolev's R-7, with its four boosters around a center core. In reality, those strap-on-like protrusions were integral to the core and contained the six first-stage motors. Chelomei's team designed the first stage in this manner to cut down aerodynamic drag and also to make it easier to transport the subassemblies over road networks. The second stage used elements of the UR-200 design. In its original two-stage form, the UR-500 lofted a series of cosmic ray astronomy satellites known as the Proton series, which in turn gave the rocket its name. In a four-stage configuration, Proton also opened up access to geosynchronous orbit for communications satellites. Many years later, when the booster was offered to international commercial firms for their payloads, it became one of the most successful commercial launch boosters ever built, and Proton rockets continue to be used to this day. A three-stage version would also be developed to loft Chelomei's next great idea.

Almaz

At the same February 1962 meeting in which the UR-500 and NI rockets were approved for study, Vladimir Chelomei submitted another project proposal. This was for the Almaz (Russian for "Diamond") space station. In the early 1960s, America's unmanned Corona series of reconnaissance sat-

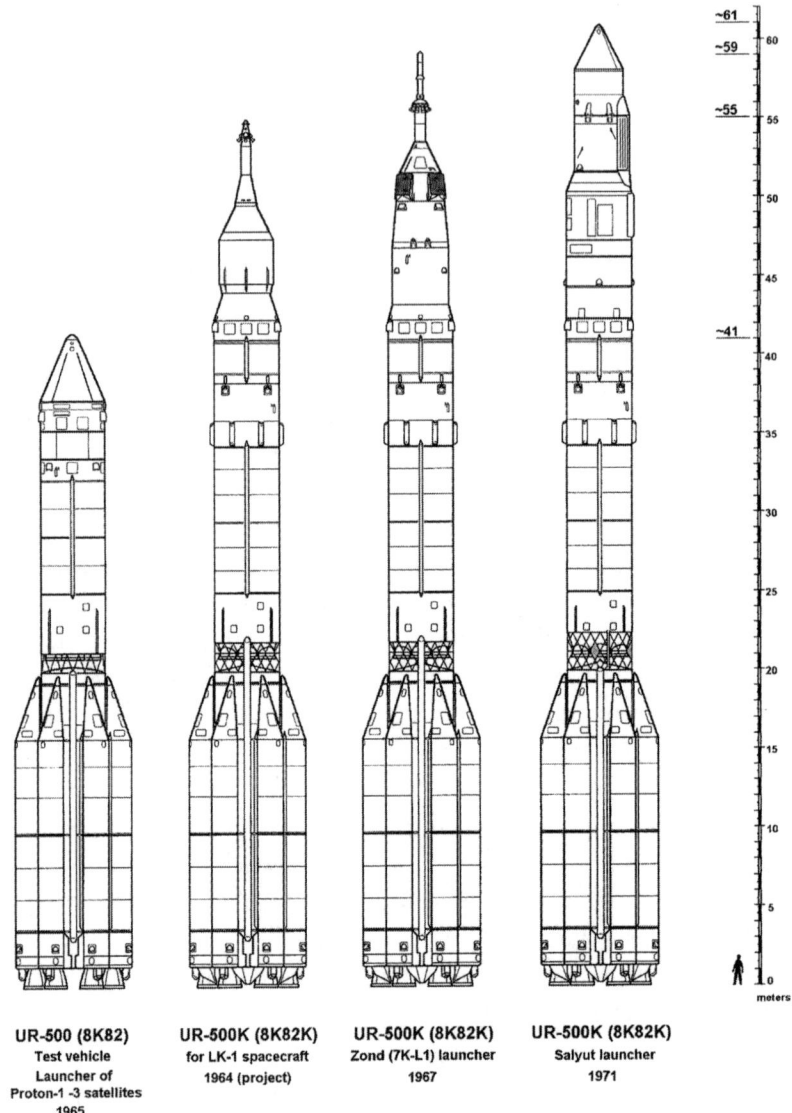

				~61	
				~59	60
				~55	55
					50
					45
				~41	40

UR-500 (8K82)
Test vehicle
Launcher of
Proton-1 -3 satellites
1965

UR-500K (8K82K)
for LK-1 spacecraft
1964 (project)

UR-500K (8K82K)
Zond (7K-L1) launcher
1967

UR-500K (8K82K)
Salyut launcher
1971

9. Chelomei's UR-500 Proton became a workhorse of heavy lift and is still used today.
Courtesy NASA.

ellites were already orbiting over the Soviet Union, attempting to photograph regions of the country that aircraft could no longer access since the May 1960 downing of a U-2 spy plane. Indeed, the Soviets had been working on a similar satellite capability, yet they lagged behind the Americans.

Almaz was conceived as a three-man space station, whose primary purpose was to gather intelligence on the West with the use of a large telescopic camera system and other reconnaissance equipment. Once in operation, the crew could aim the camera to focus on targets of interest during each ninety-minute orbit. They could shoot photographs, develop them, and then transmit those images back to the ground. Almaz received initial approval for early design work in 1962. The green light to fully develop it into an operational space station didn't come about until early October 1964 when Nikita Khrushchev gave his full approval to the program.

Looking at the Almaz design from the side, it resembled a spacecraft with a pointed nose section; two stair-stepped cylindrical sections, with the fatter section in back; two solar arrays; and a rear docking port. A little over 11 meters in length and a maximum diameter of 4.15 meters for the rear cylindrical section, Almaz was larger than any other manned spacecraft the Soviets had built to that time. In this configuration, three cosmonauts would ride the nearly twenty-ton vehicle into orbit atop a Proton booster, in a VA (*vozraschaemyi apparat*, Russian for "return craft") capsule bolted to the front of the space station. In keeping with Chelomei's modular approach, the VA capsule was also planned for use on OKB-52's LOK spacecraft, a vehicle being designed for a manned lunar-flyby mission. The capsule would act as an escape vehicle in the event that a problem occurred with the rocket during ascent. After achieving orbit, the crew transferred into the Almaz station via a hatch in the base of the capsule's heat shield, like the Gemini-B spacecraft intended for the MOL. At mission's end, the crew would return to Earth in the VA capsule. Externally, the VA resembled a slightly smaller version of an Apollo capsule. A long section on the front of the capsule, which superficially resembled a launch escape tower, contained the orbital maneuvering thrusters and the retro-rockets for deorbit and reentry. Each capsule was designed to be refurbished and flown on up to ten missions.

Cosmonauts would occupy the stepped-cylinder main body of the Almaz. Located in the smaller diameter section was a bedding area, seats, exercise equipment, a lavatory, and a small galley with a table for meals and drinks to be prepared. The three cosmonauts would man the station in rotating shifts of eight hours each around the clock. One crewmember would rest while the second manned the sensor equipment. The third crewmember would assist the second one during breaks from exercises needed to combat the long-term effects of weightlessness. The larger body housed the cameras and sensors for the photography of targets of interest on Earth. The largest of this equipment was the two-ton reconnaissance camera and telescope system code-named Agat-1 (Russian for "Agate," a semiprecious stone). Behind the main body were two solar arrays for electrical power generation and a docking port for other spacecraft.

From a workstation in the front of the main body, located between the Agat-1 camera system at the rear and the smaller living section at the front, a cosmonaut could monitor the reconnaissance sensors and view what the camera saw. With tracking systems built into the camera, it could hold focus on a target on the ground or ocean. This was not an easy task to accomplish since an orbiting spacecraft would only be within view of a target for maybe a minute or two at most, depending on whether the target was directly under the spacecraft or off to one side.

The camera's viewfinder could survey an area as large as one hundred kilometers, but by looking at a panoramic display screen on the control console, the cosmonaut operator could zoom in and focus on an area as tight as one hundred meters. According to Vladimir Polyachenko, head designer of the Almaz at OKB-52, when he was interviewed for Nova's "Astrospies" television program on PBS, "We could see details that were half a meter long from 250 kilometers [away] in outer space. For example, we could see the make of the car, if it's a Ford or Toyota." Some Western analysts with knowledge of Almaz have speculated that its camera's resolution was even greater than that, possibly approaching the MOL KH-10's three-inch resolution. But MOL's camera was housed in a much larger area than where the Agat system was contained, meaning the KH-10 likely had a higher magnification capability.

The Almaz was designed so it could fix on a target on the ground long enough to shoot a picture of interest and cancel out the motion blur of the spacecraft. Once the image was taken, development of the film could take place on orbit for further analysis. If the cosmonauts photographed anything of immediate interest, the images could be transmitted to the ground via a video camera. The minimum time needed to shoot a photo,

Almaz & TKS Spacecraft
(original concept)

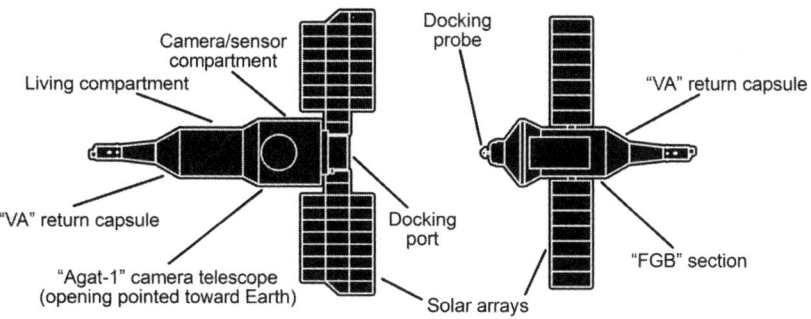

10. The Almaz system was Chelomei's answer to the MOL. Courtesy of the author.

develop the film, and transmit the image back to the ground in this manner was about thirty minutes.

Almaz was also equipped with its own defense system. The Soviets were very concerned that an American spacecraft might be sent up to intercept and rendezvous with the Almaz, perhaps attempt to board the craft, or at worst try to destroy it. To combat this perceived threat, a proposal was drawn up to mount a Nudelman-Rikhter 23 mm automatic cannon (the same type used in aircraft such as the MiG-15) on Almaz. The cannon system was mounted in a fixed position on the belly of the station, in line with its longitudinal axis. In order to aim the cannon, the entire station had to be rotated. Once the target was acquired visually through a periscope, the cannon could be fired by the Almaz crew or remotely from the ground to deliver a devastating blow to any intruding spacecraft.

While the Almaz station's configuration superficially resembled the MOL design, it isn't entirely known if its configuration was derived from the MOL or if it was designed totally independent of that program. During the "Astrospies" interview, when asked about how much the Soviets knew about the MOL, Polyachenko replied, "We had some information about the MOL program from open and closed channels. We had our sources."

One essential difference between the two programs was that while MOL was designed for thirty-day missions and intended to deorbit for burn up after the crew departed in the Gemini capsule, the Almaz was planned to remain in a parking orbit awaiting the next crew. A new three-person crew and supplies would come from the TKS (*transportnyi korabl snabzheniia,* Russian for "transport supply ship") spacecraft, a vehicle almost as big as the Almaz. Like Almaz, the TKS needed a Proton booster to reach orbit. The TKS was almost a self-supporting space station itself, fully intended to be more than just a ferry vehicle. The forward section comprised a VA capsule, complete with a hatch in the base of the heat shield like the one launched on the Almaz. Behind the capsule, in the FGB (a Russian acronym meaning "functional cargo block") section were contained additional fuel tanks, supplies, solar panels, and other equipment deemed necessary for additional Almaz missions.

In a sense, the TKS was part spacecraft and part logistics module. Once it was docked with the Almaz at an aft port on the space station (behind the main solar arrays), the resulting spacecraft would have about two-thirds more useable internal volume than the original Almaz, and the TKS would become an integral part of the complex. Its fuel supplies would be used for orbital reboosting, and its solar arrays would augment the original Almaz ones, as solar cells gradually lose their electrical generation capability over time. In the cargo area, updated equipment could be sent up for refits of the station. Film was provided for the Agat camera, while food, water, and other consumables would replenish the cosmonauts' vital supplies.

The TKS could also be launched unmanned, which meant that mission planners could use the vehicle to extend a mission if it were proceeding well with the original crew. The second return capsule could then be used to send developed pictures, recording cassettes, or other experiment packages home for more thorough analysis on the ground by Soviet intelligence agencies. As an added bonus, the FGB section of a TKS could remain operational while docked with Almaz, even after the VA capsule had returned home. With support from a TKS ferry, an Almaz station could potentially remain in orbit for a lot longer than thirty days, and multiple TKS launches could potentially allow Almaz missions to last over a year.

Two Major Shake-ups

In late 1964, things were going well for OKB-52. The bureau had produced cruise missiles for the Soviet navy, continued development work on the UR-100 and UR-200 ICBMs, and had just been granted approval to build the

UR-500 and Almaz. But things were transpiring behind the scenes to shake up the Soviet space industry as a whole and affect the direction the country would take in response to American plans for spaceflight.

The first event occurred on 14 October 1964 when Nikita Khrushchev was removed from power as the Soviet premier and First Secretary by members of the Presidium and Central Committee in a move led by Leonid Brezhnev. The reasons given for the change were "policy failures" during Khrushchev's tenure, but the main trigger for the ouster was the Cuban Missile Crisis two years earlier. At the end of the crisis, Khrushchev agreed to remove nuclear missiles from Cuba in exchange for the United States reciprocating from its missile bases in Turkey and Italy.

Khrushchev had developed many political enemies during his time in office, since military leaders were affronted by his planned cutbacks for the military (many of which had already taken place) and since political leaders did not like his compromises with Western nations. Khrushchev's advanced age and behaviors were also of concern to other Communist Party officials. While the premier was still sharp mentally, physically he began to tire more easily. Indeed, Khrushchev seemed to know his days were numbered and did nothing to impede the process of his ouster. Instead, he lamented that for the future of the Soviet Union, it was good that he should go quietly like this, rather than be carted off and summarily executed as Beria had been or perhaps poisoned as Stalin was rumored to have been, instead of dying from a stroke as official sources had stated.

In Khrushchev's place, Alexei Kosygin assumed the role of Soviet premier, while Brezhnev became First Secretary of the Central Committee in the collective leadership roles. The reason for the joint leadership was that a plenum of the Soviet Central Committee forbade any member after Khrushchev holding both positions, as they did not want one individual to have that much power again.

Khrushchev's dramatic removal from power affected Chelomei's ambitions, as it meant that the designer had lost his greatest political ally and was now unsure how the new leaders would accept him. By the middle part of 1964, it had been decided that Yangel's R-36 ICBM would become the new rocket in the strategic Soviet arsenal. While continued development work on the UR-200 was suspended, enough of them had been built that OKB-52 was allowed to use them in trial launches. Chelomei used this as a means of testing the waters, believing there was always a chance that such a decision favoring the R-36 might be reversed and potentially the UR-200 could still gain approval for operational deployment.

In previous years, Chelomei had been able to call Khrushchev directly to report the results of his test launches, and the premier accepted the news happily. Now, though, he had no idea how the new leadership might respond. After a successful launch of the UR-200 a little over a week after Khrushchev was removed from office, a nervous Chelomei placed a call to Kosygin. He hadn't bothered to phone Brezhnev, as Ustinov was a friend of the new First Secretary; Kosygin's official role as the chairman of the Council of Ministers still held sway in matters of national defense. Kosygin accepted Chelomei's report and told him that in the future he should only report to his own minister, Sergey Afanasyev, the newly appointed head of the Ministry of General Machine Building (MOM), and not call again. Chelomei reportedly held on to the phone receiver for a few moments after Kosygin had hung up, listening to the static.

In early 1966 another event took place. OKB-1 was considered to be somewhat overcommitted in the projects they had taken on, as work on both the N1 and the Soyuz spacecraft were falling behind schedule. However, there was still some optimism that chief designer Sergei Korolev would help see things through to a successful conclusion and that the Soviets would beat America to the moon. It was not to be, though. On 16 January 1966, only two days after his sixtieth birthday, Korolev underwent surgery to remove a polyp from his colon. The polyp turned out to be a cancerous tumor, and its removal led to heavy blood loss. Korolev's heart, already weakened as a result of his imprisonment in the gulags three decades earlier, had experienced at least one heart attack in 1960, and his workaholic lifestyle wasn't helping matters. Given the stress associated with major blood loss, his heart stopped beating completely, and the attending surgeons were barely able to get it restarted. Then they were unable to insert an airway into Korolev's lungs due to an improperly healed broken jaw, a byproduct of his harsh treatment at the hands of the NKVD. Korolev never regained consciousness and died a few hours later. The only consolation coming from his death was that the Soviet leadership subsequently revealed his name and achievements to the Soviet people and the rest of the world. In death, Sergei Pavlovich Korolev finally received belated official and public recognition for the incredible successes of Soviet spaceflight for which he was mostly responsible—an overdue acclaim that he was denied in life.

Following the death of Korolev, his deputy, Vasily Mishin, was placed in charge of OKB-1. While Mishin was a competent designer, his management style was quite different, lacking the sheer will and determination of Korolev. It was up to Mishin to manage the efforts of OKB-1's Soyuz spacecraft and the N1 booster, while at the same time overseeing how subcontracted elements from other bureaus would integrate properly into OKB-1's designs. Mishin tended to take a far more cautious approach to design and testing than his predecessor; many individuals, including several cosmonauts, feel this is how the Soviet Union lost the moon race.

In March 1965, at the direction of the MOM, OKB-1 was renamed, becoming the Central Design Bureau of General Machine Building (TSKBEM). At the same time, OKB-52 became the Central Design Bureau of Machine Building (TSKBM). No doubt the name changes were brought about to help confuse Western intelligence agencies. At the same time, Dmitry Ustinov became a secretary of the Central Committee of the Soviet Communist Party, responsible for management of defense and space issues.

Looking at the size of both design bureaus, TSKBEM had six hundred thousand engineers and workers at its disposal, while TSKBM had only eight thousand. That Chelomei's group would have any hope of being able to secure development work for future space projects, with Khrushchev no longer in charge and Ustinov's role in government management, seemed almost like a David-versus-Goliath task. But Chelomei's success in military circles continued to gain some allies. Sergey Afanasyev, the newly appointed minister in charge of the MOM, also seemed to hold the technical capabilities of Chelomei and his design bureau in high regard. Through government decrees, work continued on both the UR-500 Proton and the Almaz station, mainly because of continued U.S. efforts on the MOL project.

Changes were in store for Almaz. During early work to refine the Almaz design, it gained so much weight that it could not be launched with a manned VA capsule attached. As a consequence, the design was altered, allowing the crew to be delivered on a TKS spacecraft once the station had achieved orbit. However, given the work needed to develop the TKS with TSKBM's relative inexperience in designing manned systems, they rapidly fell behind schedule. A solution was found in pairing Almaz with a ferry version of the Soyuz spacecraft. TSKBEM had been working on a military space station based on the Soyuz design, since early 1964, only a few months after the United States announced the MOL project. The Soyuz-based station wouldn't have been as large as the Almaz, mainly because it was only intended to be launched by the R-7 booster. In order to man the station, a ferry version of the Soyuz was needed to take cosmonauts to it.

A ferry version of a craft can be very different from one designed to take an active part in a mission from beginning to end. To begin with, the vehicle has to be able to survive in a powered-down state for a lengthy period of time and potentially handle extreme temperatures with no active thermal control capability. Secondly, the longevity of the systems aboard the craft also have to be taken into account; if a vehicle remains in orbit longer than its certified design life, the risk of a critical systems failure increases. With the TKS spacecraft, the intent was for the FGB node of the craft to become an active part of the space station with its power-generation capabilities and systems operating the entire time. With the exception of its solar arrays being used to augment the power-generating capabilities of the station, a Soyuz ferry needed to remain in hibernation, not becoming active until the time came to return home.

Plans were therefore drawn up for an interim configuration of the Almaz, redesigned to dock with the Soyuz ferry. This would allow the Almaz to potentially be launched before the TKS spacecraft was even ready to fly. But there was some risk involved in that decision as it also meant that the Almaz program was now tied directly with the Soyuz program and would be affected by delays encountered during development of the Soyuz. It also meant that the first Almaz stations could not be resupplied and have their systems augmented by those of the more capable TKS spacecraft. It would be nearly impossible to resupply a large quantity of film for the Agat camera or keep the station's power-generation requirements in check with the smaller solar arrays found on the Soyuz. Essentially, the early Almaz stations would become single-mission vehicles, much like the MOL vehicles they had been designed to counter.

On the lunar-program side, additional changes were in store for both design bureaus. While TSKBEM continued work on the NI and associated hardware for a planned lunar landing, they were tasked with using Chelomei's Proton booster to launch a stripped-down version of Soyuz that

Almaz & Soyuz Spacecraft
(as built and flown)

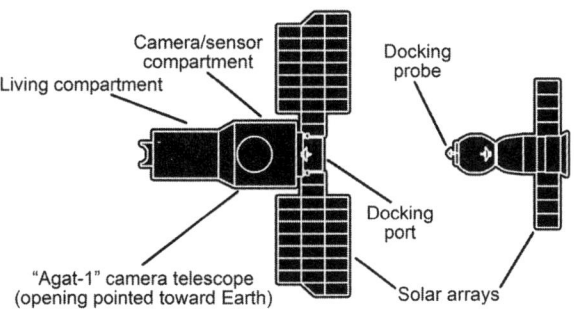

11. Almaz was altered to allow Soyuz spacecraft to dock with it. Courtesy of the author.

would conduct a manned flyby of the moon. This entailed a single loop around the moon with no attempt to enter lunar orbit, and the craft would head back to Earth on a free-return trajectory. This craft eventually became known to the West as the Zond (Russian for "Probe") spacecraft. Chelomei had already been given approval to develop his LOK spacecraft for a similar mission when Khrushchev was in power, but the LOK was ultimately canceled in favor of the Soyuz-based design. Again, it meant that the potential success of yet another program was tied directly with development of the Soyuz.

In November 1966 the first unmanned Soyuz test craft was launched as Cosmos 133. The vehicle achieved orbit as planned, but its control system malfunctioned, leading to a rapid loss of the craft's attitude-control fuel. Attempts were made over a two-day period to deorbit the craft. Finally, the ground controllers were successful, but when it was determined that the Soyuz would land in China, the order was given to destroy the craft using a self-destruct system.

A little over two weeks later, attempts were made to launch a second Soyuz spacecraft on an unmanned test flight. The countdown went flawlessly until the moment of ignition, when the engines shut down. As puzzled technicians arrived at the pad and the gantries were moved back into place, the escape rockets from the Soyuz launch abort system suddenly fired, carrying the spacecraft away from the booster to a perfect landing under parachutes. Flames belching from the escape rocket ignited the still-fueled third stage and set it on fire. Engineers and technicians, including Mishin, ran for cover before the rocket exploded about two minutes later. Thankfully, unlike the Nedelin launchpad disaster in which an exploding Soviet booster killed dozens of pad workers in the fall of 1960, there was only one fatality from this incident, although several others received seriously injuries. Since the Soyuz craft never made it into orbit, it was never assigned a Cosmos mission number, and the failure was never publicly acknowledged.

Results were more encouraging in February 1967 with the launch of Cosmos 140. The craft encountered attitude-control problems, but it remained in control without a complete loss of fuel. When the time came for retrofire, though, the craft reentered the atmosphere in a ballistic trajectory, and the resulting g-forces would have killed any crewmembers on board. The descent module survived the fiery extremes of reentry, and the parachutes deployed as planned. But then it smashed through an ice layer covering the Aral Sea, sinking down ten meters into water hundreds of kilometers off target. Divers were able to recover the craft for later study by engineers. The flight was considered to be of sufficient success that approval was given for a manned test flight. On 23 April 1967 *Soyuz 1* launched into orbit with cosmonaut Vladimir Komarov at the start of a particularly ambitious flight plan. One day later, *Soyuz 2* was to have been launched with a three-man crew of Valery Bykovsky, Yevgeny Khrunov, and Aleksei Yeliseyev to rendezvous and dock with its sister ship. Once docked, it was intended that Khrunov and Yeliseyev would conduct a transfer EVA over to *Soyuz 1*, enter the craft, and return to Earth with Komarov, leaving Bykovsky to return home alone aboard *Soyuz 2*. In this flight, the Soviets had planned to duplicate many of the mission goals NASA's now-completed Gemini program had already achieved. The space walk was of critical importance, as the planned lunar landing required one cosmonaut to conduct an EVA to transfer from a Soyuz craft to the lunar lander, since the early Soyuz docking system had no pressurized tunnel to allow for an internal crew transfer.

Komarov immediately ran into problems once *Soyuz 1* reached orbit when one of the two solar arrays jammed during deployment. As these arrays were the only electrical power generation for the Soyuz, it led to other problems. On the ground, the *Soyuz 2* crew were making preparations to fly with an added goal of trying to repair the stuck solar array on *Soyuz 1*, in addition to their other tasks. That night, thunderstorms near the Baikonur launch-pad caused lightning strikes that affected the R-7 rocket's electrical systems. Plans to launch the next morning were scrubbed.

Meanwhile, for Komarov, things had gone from bad to worse. By the thirteenth orbit, the automatic stabilization control system was completely dead, and the manual backup system was only partially effective. Orientation detectors on the craft were also malfunctioning due to the power-generation problems, making it hard to determine the correct attitude of the spacecraft. Plans were made to terminate the mission, and Komarov executed a manual retrofire on the eighteenth orbit. After the successful reentry, a drogue chute jettisoned and slowed the craft, but the main parachute did not deploy. Komarov manually fired the reserve chute, but it became tangled in the drogue chute. *Soyuz 1* crashed back to Earth at around ninety miles per hour, killing Komarov on impact. Retro-rockets, intended to cushion the shock of touchdown, were triggered on impact, starting a small blaze. By the time rescue helicopters arrived at the scene, little remained of *Soyuz 1* apart from some smoldering wreckage.

As it turned out, if *Soyuz 2* had flown, there would likely have been four fatalities, not just one. A postcrash investigation revealed that there was a serious defect in the application of the ablative-material coating in that craft's descent module. Normally the coating was applied with a set of covers installed over the descent module's parachute compartments. On *Soyuz 1* and *2* the covers were apparently left off, and some material was applied inside the openings for the parachutes. This material restricted the size of the openings, so when the main parachutes were packed inside, they were a much tighter fit than intended. This made it extremely difficult for them to deploy and open under normal operation.

The loss of *Soyuz 1* meant that there was a lengthy delay to all projects dependent on the Soyuz vehicle, including the Almaz. Work continued on the Zond craft, and it completed several unmanned flights in 1968, with a couple of successful flybys of the moon. Despite these successes, the Zond still was far from perfect, and Mishin was reluctant to risk a cosmonaut crew. The manned lunar-flyby mission was quietly canceled a few months after *Apollo 8* and its three-man crew flew to the moon, orbited ten times, and returned home in December 1968.

As for the N1 booster, Chelomei's predictions about it proved to be correct. Due to the slipping schedule and in order to keep costs down, the N1 was never static tested on the ground as the Saturn boosters had been. This meant that flight-testing was absolutely critical. The first N1 booster was stacked at the launchpad in May 1968, but cracks were found in the first stage structure, and the booster was rolled back to its preparation facility at Baikonur for repairs. History records that it didn't roll back quickly enough, as an American surveillance satellite shot photographs of it, revealing to the West what plans the Soviets had for the moon. The revelation of that first N1 booster was allegedly a major part of NASA's decision to send *Apollo 8* to the moon at the end of 1968.

Two unmanned launches of the N1 were attempted in 1969. The first launch went well early on as the vehicle cleared the tower. The first stage engines were commanded to throttle down during the period of maximum dynamic pressure (max q) at the twenty-five-second mark. At sixty-five seconds, they were commanded to throttle up again. But the result was too sudden, and unanticipated vibrations ruptured LOX pipes near a gas generator. A fire broke out, and all engines were shut down within five seconds. The launch abort system on top of the rocket successfully pulled the unmanned Soyuz craft away to safety, but the N1 continued on a ballistic path and fell back to Earth about forty-five kilometers downrange from the launchpad.

The outcome was much worse on the second launch attempt. On 3 July 1969 the second N1 lifted off the launchpad, reaching an altitude of about two hundred feet before what was believed to be a loose bolt was ingested by a fuel pump. The onboard control system detected the problem, but instead of commanding just one or two engines associated with that pump to cease operating, it shut down twenty-nine of the thirty motors. As a result, the almost fully fueled N1 toppled and came crashing back down on the pad. The ensuing explosion almost completely destroyed the pad and heavily damaged an adjacent pad where another N1 was stacked—presumably for a manned attempt if the flight had been successful. The massive damage set the Soviet lunar program back by about two years.

A little over a week after the second N1 launch ended in failure, *Apollo 11* successfully made it to the moon, with Neil Armstrong and Buzz Aldrin becoming the first humans from planet Earth to set foot on the lunar soil. While the Soviets had yet to give up entirely on the goal of landing a Soviet cosmonaut on the moon, they now began to look for alternative ways to achieve another space first.

One of the alternatives under consideration was a manned civilian space station. For that to happen, however, it would take a high degree of cooperation between two design bureaus that were considered rivals. Such cooperation would also be against the wishes of both Vasily Mishin and Vladimir Chelomei. Nevertheless, a plan was being formulated by some middle-level engineers and managers to try to do just that, in a way that was so un-Soviet-like that nobody could have expected or predicted it.

3. Salyut

In 1969 the Soviet space program was at a crossroads. On the manned-spacecraft side, the Soyuz spacecraft had chalked up a success as *Soyuz 3* flew with Georgi Beregovoy at the controls on 26 October 1968. It was intended to rendezvous and dock with an unmanned *Soyuz 2*, although it only succeeded with the rendezvous. Still, *Soyuz 3* managed to return to Earth successfully after a four-day mission, showing that the flaws that had contributed to Komarov's death had been properly addressed.

On 16 January 1969 *Soyuz 4* and *Soyuz 5* succeeded in completing the tasks that were first intended for Komarov's flight: docking and EVA transfer of two crewmembers. Yevgeny Khrunov and Aleksei Yeliseyev left their crewmember Boris Volynov in *Soyuz 5* and spacewalked across to *Soyuz 4*. The two men then entered *Soyuz 4* and returned home with its commander, Vladimir Shatalov.

Soyuz 4's return went well, but the same could not be said for *Soyuz 5*. After retrofire, the service module did not cleanly separate from the descent module, and *Soyuz 5* with cosmonaut Volynov on board began a front-first reentry. Volynov revealed in interviews conducted nearly three decades after the incident that he was hanging in the harness of his craft watching the front of the Soyuz cabin distort from the heat when the remaining struts holding the two modules together finally tore away. The descent module then righted itself and came through the rest of reentry just fine. Volynov was not out of danger, though. The main parachute initially didn't open fully, although it eventually did. The craft also hit the ground at a higher-than-normal speed since the braking rockets didn't fire. Volynov's seat harness broke at impact, and he broke some teeth as his face was smashed against the instrument panel. But he survived to fly in space again.

With rendezvous and docking still a big part of the planned lunar-landing mission (once the NI rocket's problems were solved), the Soviets needed all the experience they could get. So they attempted to set more records on their next mission. Georgi Shonin and Valery Kubasov would fly *Soyuz 6* and attempt to observe and film the docking between *Soyuz 7* and *Soyuz 8*. *Soyuz 7* would be crewed with Anatoly Filipchenko, Vladislav Volkov, and research cosmonaut Viktor Gorbatko. *Soyuz 8* would include *Soyuz 4* veteran pilot Vladimir Shatalov and EVA veteran Aleksei Yeliseyev. The plan again was for cosmonauts to spacewalk from one craft to the other after docking.

All three craft launched normally, but the Igla rendezvous system was unable to achieve a lock between *Soyuz 7* and *8*. Two attempts were made at manual docking. But without the Igla, Shatalov (flying the actively docking spacecraft) could not use the main engines to close the distance quickly and had to rely on thrusters only. It was very hard to judge distances manually due to sun glare out the portholes, so the two craft were never able to dock. *Soyuz 6* was able to get close enough to see *Soyuz 8* at one point, but it was not equipped with a docking probe or an Igla rendezvous system and could not make a close approach safely. It was a costly effort for what was ultimately reported in Soviet papers as a successful attempt to set records for the largest number of spacecraft and people flown simultaneously in space by one country.

Soyuz 6 had one secondary objective to achieve, the first testing of on-orbit welding techniques. The size and weight of the welding equipment in the orbital module was the main reason why *Soyuz 6* was not equipped with a docking capability. Early plans called for an engineer from the Electrical Welding Institute in Kiev to accompany the experiment in orbit, since they designed the equipment. But when the weight of the passenger and the equipment became too heavy to fly both, the task of its operation fell to Kubasov.

After the hatches were sealed and the OM (orbital module) was depressurized to conduct the tests, Kubasov activated the experiment. The Vulkan welding experiment tested out three types of arc and electron-beam welding over various types of metals from aluminum to stainless steel. The cosmonauts monitored the progress while the ground received the data. When done, the OM was repressurized, and Kubasov went in to inspect the work. A stray arc from the experiment almost ended up burning a hole in the OM, which could have vented the atmosphere into space. Upon seeing the damage, the crew immediately sealed the hatch again in case a breach was imminent. But after a few minutes of monitoring with no sign of a loss in cabin pressure, Kubasov went back and retrieved the samples. *Soyuz 6* safely returned to Earth later that day.

The lack of a successful docking put a dark cloud over the whole mission. Fingers were pointed back and forth between Gen. Nikolai Kamanin, who was in charge of the cosmonaut corps at Star City, and Mishin, the head of TSKBEM, while trying to determine whether the rendezvous failure was due to poor equipment design or poor training. No answer was found to the other's satisfaction, and this feud of sorts would continue for quite some time.

Meanwhile at Chelomei's TSKBM bureau, design and construction of the first Almaz station cores were progressing, albeit at a slower pace than what Afanasyev and Ustinov would have liked. When the dictate was made for the first Almaz stations to use the Soyuz ferry while design work continued on the TKS spacecraft, Chelomei resisted the idea. So when designs from TSKBEM came over showing what equipment was needed for incorporation of the Soyuz docking system, Chelomei would reduce the work down to a bare minimum. He did not want any unnecessary equipment from Mishin's bureau to mess up his station.

If the relationship between Chelomei and Sergei Korolev was considered cordial when Korolev was alive, Chelomei and Mishin's relationship was a much more contentious one, as both men seemed to despise one another. The problems did not go unnoticed by Ustinov. He commented that it seemed as if both designers were acting like children and using their bureaus as their own personal "principalities." Yet in order to fly Almaz at the soonest possible time, the two men needed to work together . . . or did they? Even if Chelomei's station were to fly successfully, there was another problem. Almaz was still a classified military project.

The Birth of *Salyut*

At this point, serious looks were made at NASA's plans for the space station, which would ultimately become known as *Skylab*. The Americans were making preparations to convert the S-IVB stage of their Saturn boosters into an orbital space habitat. After launching it into orbit, astronauts would use it to set endurance records of one to three months in space while conducting various experiments.

A small group of engineers at TSKBEM led by Boris Raushenbakh looked at whether or not a similar station conversion might be possible with an R-7 rocket, using it to beat the Americans and score yet another space first. But with the mass of such a station and the long-term nature of equipment required for record-setting missions, there were concerns that TSKBEM's engineers would not be able to come up with the larger on-orbit engines required for reboosting of such a heavy structure when its orbit began to decay.

Enter designer and deputy chief Boris Chertok into the mix. A senior man at the bureau since the earliest days of OKB-1, he was responsible for control and guidance issues for rockets and spacecraft. Chertok was also the number-four man in charge at TSKBEM behind Mishin, Sergei Okhapkin, and Konstantin Bushuyev. Chertok knew that engines designed for Soyuz weren't powerful enough for a station. But his friend Aleksei Isayev, chief designer of OKB-2 (renamed Himmash in 1966), might be able to provide what was needed.

Boris Chertok and Aleksei Isayev's friendship went back to just before the Second World War. Back then, Isayev was working on a rocket-powered aircraft design for use in attempts to break several speed and altitude records. Chertok was brought on somewhat early in the project to provide additional input. A few days after the design was formally submitted, Germany invaded Russia, and priority was given to building weapons of war instead of setting records. The design was adapted to become a rocket-powered fighter plane; although it underwent extensive flight-testing, it did not make it into operational use.

Even without an operational fighter to his credit, Isayev learned valuable lessons about designing rocket engines and fuel tanks. At the end of the war, Isayev and Chertok both accompanied Sergei Korolev to Germany to help recover useable materials from the V2 rocket program. Both men later took part in Korolev's early postwar rocketry work.

When asked by Chertok about a new engine design, Isayev replied that his bureau had already designed larger engines for Chelomei's Almaz space station. This revelation spawned another idea. To this small group of engineers, the idea seemed like a logical one—take the unfinished Almaz and adapt the proven systems of the Soyuz spacecraft to it, resulting in a space station that could potentially fly before the American *Skylab*.

As the proposal moved forward, Konstantin Feoktistov was brought into

12. Konstantin Feoktistov risked his career to get a civilian space station program approved for development.

the mix. Feoktistov was the design head on the Soyuz spacecraft. Before Soyuz, he was involved in modifying the Vostok spacecraft so that three men could fly it into orbit as the Voskhod. He was also one of the three cosmonauts who flew on the first Voskhod flight. Therefore, he was one of the few spacecraft engineers out there with both design and spaceflight experience.

With Feoktistov's involvement, his immediate boss and the number-three man at TSKBEM, Konstantin Bushuyev, was made aware of this "conspiracy" as it were. Bushuyev was not in favor of approaching Mishin about the project, as he knew Mishin would reject it outright. Chelomei would also resist any attempts to appropriate his Almaz for use by a rival design bureau, and MOM minister Sergey Afanasyev would likely reject the idea as well, since it might mean a delay in getting Almaz ready for flight.

While Chertok, Bushuyev, and Feoktistov were enthusiastic about getting backing for their proposal, they were at a loss for who would give the approval to proceed. Outside the hierarchy of TSKBEM, there was one man who could give their plan the proper endorsement, and that was Dmitry Ustinov. It was highly irregular for anyone under the level of a chief designer to have direct contact with Ustinov, unless they had been given prior approval to do so.

While the group wanted to schedule a meeting with Ustinov, there was reluctance to make the call due to potential political backlash if the project was rejected. Feoktistov decided to arrange the meeting himself, since he was already considered to be outspoken and something of a maverick anyway. Feoktistov was not a member of the Communist Party either, meaning his effort carried an even greater risk. Still, he made the phone call and, surprisingly, was successful in being granted a meeting with Ustinov to propose the idea.

In late 1969 the meeting with Ustinov took place. All the engineers involved in the space station design study were present, as was Sergei Okhapkin, who was responsible for development work on the N1 rocket for Mishin's bureau. All the senior leadership from TSKBEM were present except for Mishin himself. Mishin was on vacation at the time, so he was conveniently not invited. Chelomei was in the hospital for a medical condition and similarly unavailable as well. Afanasyev was present at the meeting, as were other high-ranking government members directly involved with Soviet space policy. Feoktistov himself reportedly made the formal pitch and laid out the benefits of the plan.

By using an Almaz station core along with the already-proven systems of the Soyuz spacecraft—such as the solar panels and the support and guidance systems—a new space station could potentially be ready for flight in about a year. Such a project would still involve a fair amount of work to integrate and test everything properly. But the engineering group was confident they could do it if given the approval.

Ustinov liked what he heard. While Almaz would be ready to fly by 1972, the schedule was tight, and program delays might cause the schedule to slip behind Skylab's. Almaz was also a military project, so it couldn't easily be unveiled to the world without potentially revealing its true purpose. Ustinov also liked the idea of being able to stick it to Chelomei as payback for Chelomei's courting of leaders to get what he wanted in years past. While Ustinov had to accept Chelomei's Almaz project, it didn't mean he had to necessarily back it to the exclusion of all else.

Ustinov asked Sergei Okhapkin if this project might impact work on the N1. The answer was no, as a totally different team of engineers was working on the N1. Afanasyev voiced no objection either as Ustinov wanted only one design bureau (TSKBEM) to work on coordinating this "civilian" station project so there would be minimal impact on Almaz after the initial hardware transfer. The task would fall on a group from TSKBEM to oversee the work. All the team would need were some unfinished Almaz space station cores so they could get started and a Proton rocket to launch it into orbit. Cosmonauts to fly the missions were plentiful, as the ranks were flush with candidates training for the Zond and lunar-landing missions.

The relative ease by which this back-channel proposal gained endorsement and approval came as a pleasant surprise. It was almost a perfect storm of timing and luck. It was green-lighted almost immediately with an official decree signed early in 1970. The Soviets craved a new space success, as opposed to following in the footsteps of where the United States had gone before. General Secretary Brezhnev was also looking for something to downplay the successes of the United States and NASA on the moon. So this project had the potential of giving him yet another prestigious Soviet spaceflight first.

Naturally, the reactions from Chelomei and Mishin when they heard about the project were not exactly positive. They spoke with one voice in saying that they did not want to partner up in the creation of a long-duration space station. But the Kremlin's decree was firm. Chelomei got assurances that the new station would not further delay work on the Almaz. At least Chelomei could get valuable test data from the flight. But the rival design bureau would be given credit for this project, and that did not sit well with him. Mishin, on the other hand, was very furious that his top engineers had gone behind his back to do this project, and he threatened to send anyone else not already connected with the station project "to hell" if they occupied themselves with work in support of it.

At this point, the station became known as the DOS (a Russian acronym meaning "Long-Duration Orbital Station"). The small design team at TSK-BEM now had a decree to do work and a plan, but they needed additional manpower to implement it. They found some from Chelomei's TSKBM factory in the town of Fili. The Fili factory had once been part of Vladimir Myasishchev's OKB-23 aircraft design bureau. The factory was forced to become part of Chelomei's group by a state decree a decade earlier. So the Fili group wanted to show up Chelomei as well, and they were more than willing to take part in this project as a matter of pride.

TSKBM engineers from Fili in connection with a team from TSKBEM would do the primary design work of the station as they merged systems together. The stations themselves would be assembled at TSKBM's Khrunichev Machine Building Plant alongside the Proton rockets also being assembled there. Other firms would contribute work on support systems and outfitting the scientific equipment on board.

The man placed in charge of the DOS station program for TSKBEM was Yuri Semyonov. He was an engineer and manager who to that point had been involved in work on the lunar-flyby program with the Zond spacecraft. Since cancellation of Zond was imminent, he was available. It has also been reported that Ustinov wanted Semyonov since he was the son-in-law of Andrei Kirilenko, fourth down in the Kremlin hierarchy. So Semyonov had high political connections and could likely get things done, should either Chelomei, Mishin, or anyone else try to place obstacles in the path of the DOS project. It wasn't long before Semyonov was tested.

While Almaz was still about two or three years away from flight, around a dozen Almaz cores had already been assembled. The meeting between Chelomei and Semyonov to arrange for the transfer of some cores was a tense one as Chelomei accused the junior engineer that he was stealing his work. Although Semyonov had a decree from the Kremlin to take the cores, it took a direct phone call from Sergey Afanasyev before Chelomei would allow four of the cores to be transferred.

Anatomy of the DOS

At a glance, the external configuration of the DOS design didn't look that different from Almaz. The biggest change was that instead of docking port for a Soyuz at the rear of the DOS, as was intended for Almaz, a docking port was placed in a new housing on the front of the station. In addition to the docking port, this new housing contained two solar arrays of the same type as used on Soyuz, a scientific telescope, equipment for the Igla rendezvous-and-docking system, the life-support system, and telemetry antennae. Consumables storage tanks were also mounted on the outside of this new section.

Through the main transfer compartment, cosmonauts would enter the smaller of the two stepped cylinders of the Almaz station core. Internally, the layout of these cylinders was quite a bit different from Almaz, since the DOS engineers didn't have to worry about placement of the massive Agat camera system. While the primary living section for Almaz was the smaller cylinder and the larger cylinder was the workspace for the surveillance mission, in the DOS the smaller cylinder became more of a mission-support workspace where the station's controls were located. A worktable and galley were kept in the smaller compartment, but other items such as the exercise equipment, a refrigerator for food storage, and sleeping bags were housed in the main work compartment in the larger section. Internally, the two compartments formed one rectangular-shaped space with no real difference in their internal dimensions. But thanks to its larger exterior dimensions, the main work compartment carried larger-volume storage lockers than the smaller cylinder. At the rear end of the main work compartment, separated by a small partition, was the hygiene and toilet station. Urine was collected by a vacuum system, while solid fecal waste was stored in separate holding tanks.

The interior of the DOS was designed with a strict sense of direction, meaning there was a floor, walls, and a ceiling; each was defined accordingly with a different color. It was believed that this would help with orientation of the cosmonauts, make it easier to locate stored equipment, and potentially help to stave off bouts of space sickness. Stored equipment was kept in lockers so that it would be less likely for damage to occur from a cosmonaut bumping into exposed hardware. Each internal panel was designed to be removable for maintenance. Self-maintenance of station systems was important when the nearest systems engineer was located hundreds of miles away down on Earth. Cleaning could be done with a vacuum cleaner, and the crew also had access to dry and wet tissues for their own personal hygiene, as well as antibacterial towels (similar to antiseptic wipes) for washing in place of a bath or shower.

Two seats were provided at the control section for the commander and a flight engineer. The station could be flown manually with controls that resembled those found on a Soyuz. Displays were provided to help with orientation, as was a periscope for additional visual sightings. Sometimes it was necessary for crewmembers to fly the DOS in order to carry out scientific observations, such as when manual aiming of one of the station's telescopes was needed. Orientation of the station could be controlled with gyrodynes, which could move the station slowly. Orientation thrusters could be used for quicker maneuvers or to stabilize the station during reboost burns. Thruster use would be minimized to ration the onboard fuel supply for as long as possible.

Each cosmonaut was allocated four meals per day in the form of two breakfasts, a lunch (which is the largest meal in many eastern European cultures), and dinner. They could prepare their meals on a table located aft of the control section, and a hot plate was provided for heating liquids, such as coffee or soups, while the crew was "seated" at the table. Total calorie intake for all four meals was about three thousand calories, if all the meal items were consumed. A water tank provided the water rations for all three cosmonauts, and each occupant was allowed up to two liters of water per day. The water source was connected with additional water tanks located in the main work compartment. The water supply was launched into orbit with the station; since Soviet spacecraft did not use fuel cells like NASA's Apollo spacecraft, there was no way to top off the tanks. Silver ions were added to the water as a biocide to make it safe for human consumption.

On the back of the station, in the spot originally intended to be the docking port of Almaz, was a modified Soyuz service module containing a second set of solar arrays. The biggest difference between this module and

DOS-1 Space Station (Salyut)

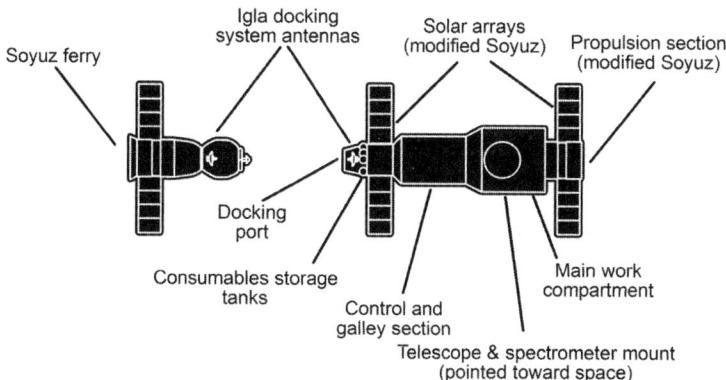

13. The first civilian station had a few differences from Almaz. Courtesy of the author.

the standard Soyuz unit was that it contained the larger engines needed for orbital maneuvering and reboosting of the station during its time in orbit. While the solar arrays used by the DOS were of a proven design, they were not nearly as powerful as the arrays intended for the Almaz, and electrical power would be at a premium. To supplement the electrical system, while the Soyuz ferry was docked with the station, its solar arrays would be hooked into the station's grid to provide additional energy. Yet even with the Soyuz, these early solar arrays would lose efficiency the longer the station stayed in orbit. The first DOS station could only support up to two missions with three crewmembers each for a month at a time before it would have to be abandoned.

Thermal regulation of the station's internal temperature was done with a set of internal and external heat loops linked to a set of heat radiators mounted on the main body of the station. Heat transfer was done with an ethylene-glycol mixture similar to car antifreeze pumped through the cooling loops. Internal temperature was regulated to a comfortable fifteen to twenty-five degrees Celsius (or fifty-nine to seventy-seven degrees Fahrenheit). Humidity could be maintained from 20 to 80 percent through the use of water-condensation plates.

While the DOS was occupied, the cosmonauts would be doing more than just floating around, as the station was packed with about one and a half tons of scientific equipment. Studies would take place around the clock, with the three crewmembers doing eight-hour shifts similar to what was planned for Almaz. So for each cosmonaut, that meant eight hours for work; eight hours for support tasks such as hygiene, meals, and exercise; and the final eight hours for sleep. The Soyuz could be used as a dormitory of sorts during this mission, so the cosmonaut at rest would sleep inside the ferry craft to minimize interruptions by the others.

A broad set of scientific experiments in multiple disciplines was planned for the DOS with studies in astronomy, Earth observation, life sciences, and medicine. Taking advantage of the opening located on one side of the Almaz core originally intended for the Agat camera, several different telescopes and spectrometers were mounted in a conical unit inside the station's main working compartment. It stretched almost from the floor to the ceiling. With the station being based on Almaz, many drawings and pictures showing the internal layout of the DOS had the conical compartment airbrushed out so as not to draw the attention of Western intelligence analysts. Unlike Almaz, the opening for the telescopes would be pointed toward the heavens instead of down toward Earth. Three additional instruments—an ultraviolet telescope, a gamma-ray telescope, and an astrophysical telescope—were placed in other parts of the station.

For ground control of the DOS missions, the Soviets would use a new tracking network originally intended for their Soyuz and lunar missions. The main control center was located in Yevpatoria, Ukraine, on the north shore of the Black Sea. Given that the center was located several hundred miles from Moscow almost due south, it meant that the only way to travel between the two cities in quick fashion was by airplane. The Soviet Union itself contained several tracking stations that could maintain direct communication with the station as it passed over the Soviet Union. But given that each orbit would place the spacecraft farther west, there were times of the day when the orbit would not cross Soviet borders. So for that, the Soviets developed a fleet of tracking ships that were built for two-way communications, to supplement a fleet of smaller vessels that could only receive transmissions. Communications between the space station, the tracking ships, and the control center in Yevpatoria could be routed through communications satellites.

Laying the Foundation for the DOS

While work progressed on designing and building the new space station, the focus of Soyuz flights changed from short-duration to long-duration spaceflights. The first was *Soyuz 9*, which was a precursor mission to the first DOS flight. The spacecraft was loaded with as many consumables as possible to support two cosmonauts for an eighteen-day endurance test. At the same time, several three-person crews began training for long-term stays aboard the new station. The cosmonauts already selected for missions to the Almaz stations continued their training separately because of Almaz's intelligence goals.

Cosmonauts Andrian Nikolayev and Vitali Sevastyanov were selected to fly the *Soyuz 9* mission. Nikolayev was no stranger to spaceflight, as he became the third cosmonaut to fly behind Gagarin and Titov when he commanded the *Vostok 3* spaceflight. Nikolayev's nickname among the cosmonauts was the Iron Man, because he held the record for the longest time a cosmonaut spent in isolation testing on the ground. The Iron Man seemed like a perfect fit for this mission. In addition, Nikolayev was married to the first woman to fly in space, Valentina Tereshkova.

Vitali Sevastyanov, on the other hand, was an engineer from TSKBEM who began working for Korolev after graduating from the Moscow Aviation Institute in 1959. He worked on the Vostok spacecraft and also assisted in the training of cosmonauts before he himself was selected for training in 1967. Sevastyanov was an avid chess player and would play the first chess match in space on *Soyuz 9* against an opponent on the ground. He was selected for this mission since his engineering background would come in handy to help deal with any unforeseen problems that cropped up with the spacecraft.

There was a lot more to flying a Soyuz for such a long duration than just loading it up with more provisions. A new life-support system and carbon dioxide scrubber that could withstand constant use for three times longer than previous flights had to be installed. Exercise equipment had to be fitted for the cosmonauts to help combat the effects of long-term weightlessness. In zero gravity, certain muscles begin to atrophy since they aren't acting against gravity as they would during normal activities on Earth. A new type of load-bearing suit was also to be tested on this flight. This suit, which would eventually become known as the penguin suit, contained several elastic bands that would force the muscles to work in ways they wouldn't normally do so otherwise. It was a form of passive exercise, in a sense. In preparation for the upcoming space station flights, the shock absorbers on the crew couches were modified to make the landing more bearable after weeks spent in zero gravity.

Soyuz 9 launched on 1 June 1970. The cosmonauts on board took up the call sign "Sokol" ("Falcon" in Russian), which was Nikolayev's call sign from *Vostok 3*. While on orbit, the cosmonauts were referred to as Sokol 1 and Sokol 2 instead of by their real names. Space research was the key to this flight, and the cosmonauts conducted biomedical research experiments, photography of Earth, navigation exercises with a sextant, and evaluation of the new penguin suit. The flight program was so packed with tasks that midway through the mission, the crew began to evade the planned exercise program. Even after they were scolded by mission control for doing so, they still did not exercise. When the crew returned to Earth on 19 June, they were barely able to stand. Reports say that it took ten days for the crew to recover to their pre-mission fitness level. But they endured. All indications were that the spacecraft performed well and that the mission could have potentially been extended by a couple of days. In the process, the crew set a new space endurance record of eighteen days, eclipsing the fourteen-day mark set by *Gemini 7* almost five years earlier.

Back at Star City, DOS mission training continued. The plan was to select two primary crews to fly to the first station to conduct missions of nearly a month in length, with the second crew backing up the first one on the initial mission. A third crew was assigned to act as a backup for the second crew and would train to fly to the second DOS station about a year later with a fourth crew training for the second DOS-2 mission.

The cosmonaut corps in those days was made up of a few of the old group of air force officers from the Vostok and Voskhod flights. In addition to that, there were some cosmonauts selected from the strategic rocketry forces and engineers who joined the program from Korolev's bureau. All were considered to be good candidates for spaceflight, but crew selection turned out to be a political tug-of-war between General Kamanin, the air force officer in charge of the cosmonaut training; Mishin; and several agencies with competing agendas.

Gen. Nikolai Petrovich Kamanin was not a simple bureaucrat. He was awarded Hero of the Soviet Union in 1934 when he successfully led an aircraft evacuation of the crew of the ss *Chelyuskin* after the ship was stranded and crushed by arctic pack ice. During World War II, Kamanin was a colonel in command of air divisions that fought in the battles of Stalingrad and Kursk, with Kamanin himself flying combat missions periodically. Tactically, he came up with some very innovative strategies and personally led an IL-2 Sturmovik aircraft attack on a massive German airfield in the occupied town of Lviv, Ukraine, with only light losses to his unit. For this action, Kamanin was awarded the Order of Kutuzov, one of the highest decorations in Soviet and later Russian military aviation. General Kamanin had been assigned to head the cosmonaut training office since the program's early days. This gave him broad powers over many portions of the Soviet space program and the military personnel involved with it, from approval of equipment on spacecraft to crew selection to spacecraft recovery at the end of the mission.

The feud between Kamanin and Mishin led to some back-and-forth in the selection process for the first DOS crews. Kamanin wanted at least two military officers to fly, and Mishin wanted his engineer-cosmonauts to fly. At the same time, Kamanin also objected to Mishin wanting to assign two cosmonauts with flight experience to each mission when so many others (particularly air force ones) had not yet flown even one mission.

Eventually, the crew nominations were whittled down to four groups of three cosmonauts, with the list being finalized in early 1971. The crew of *Soyuz 10*, flying the first mission to the new station, would be veteran cosmonauts Vladimir Shatalov and Aleksei Yeliseyev. Joining them would be engineering cosmonaut and flight engineer Nikolai Rukavishnikov. For *Soyuz 11*, first-EVA veteran Alexei Leonov would command the flight. He would be joined by Pyotr Kolodin and *Soyuz 6* cosmonaut Valery Kubasov. The third crew, planning to fly to the second DOS station, would be Georgi Dobrovolsky, Vladislav Volkov, and Viktor Patsayev, with Volkov being the only veteran on that crew, having flown on *Soyuz 8*. The fourth crew included rookies Alexei Gubarev and Anatoli Voronov and veteran Vitali Sevastyanov from *Soyuz 9*.

Salyut Flies!

On 15 April 1971 a Proton rocket with the first DOS station sat ready for launch. While smaller elements of the station were shrouded with aerodynamic covers to prevent damage during ascent, the large-diameter working compartment was visible and bore the name *Zarya* in big red letters on the side of its green external insulation. The word meaning "sunrise" in Russian, it was intended to be the name of the station, although the Kremlin expressed their dissatisfaction with it and pointed out that the Chinese were working on a rocket design with apparently the same name. "Zarya" was also a radio call sign used by the control center when in contact with manned spacecraft. So a new name had to be selected. After discussions, the name *Salyut* was suggested and agreed on unanimously. In Russian, the word means "salutation" or "greeting," and it also had a secondary meaning as a "salute" to the late Yuri Gagarin since his flight as the first man in space had occurred exactly ten years before the scheduled launch day. The salute in memory of Gagarin also seemed an appropriate tribute since the cosmonaut had died in a plane crash several years earlier and was still mourned by many. The timing of the launch in this new era in Soviet spaceflight seemed perfect. Discussions were made about repainting the name on the side of the station, until Boris Chertok pointed out that nobody would see the name anyway once the vehicle was in orbit. So the name *Zarya* remained.

On early Monday morning, 19 April 1971, the Proton rocket carrying *Salyut* rose from its launchpad at the Baikonur complex and headed for orbit. *Salyut* was injected into a 200 by 222 kilometer orbit at an inclination of 51.6 degrees. Everything seemed to go well, although telemetry indicated that the launch cover designed to protect the scientific instruments and telescopes located in the station's working compartment had not jettisoned. If that cover could not be released, it meant that a large portion of the planned telescope observations could not be carried out.

The crew of *Soyuz 10* was ready to launch on 22 April, but the countdown was aborted when an umbilical plug did not come free. Heavy rain before the launch and ice condensation from the liquid oxygen tanks had frozen the plug into place. The fully fueled rocket was kept on the pad, and plans were made to try again the next day. This time, the launch was successful, and *Soyuz 10* rocketed into orbit at 02:54. Moscow time. In keeping with a mission radio call sign protocol for Soyuz flights, the crew of *Soyuz 10* was known as "Granit" (Russian for "Granite").

Shatalov was the most experienced cosmonaut when it came to rendezvous and docking, and he would need all his experience for this flight. As planned, the Soyuz spacecraft would spend two days in a slightly higher orbit than *Salyut* as it waited for the station to catch up on its lower orbital path. An initial rendezvous burn was made on the fourth orbit to adjust the path of the Soyuz; from that time on, it was a matter of waiting for the distance to close.

Rendezvous in space can be explained in part with an analogy of two race cars on a two-groove circular race track. With both cars going the same speed, the car on the inner groove seems to go around the track faster than the car on the outer groove, as the outer car has to cover more distance. So if the car on the inner groove is behind the car on the outer groove, it doesn't take long to catch up with each successive lap or orbit. Eventually, the two cars close the distance with one another to the point where they are side by side. But unless an attempt is made by the car on the outer groove to drop to the inside to complete the rendezvous, or the one inside moves to the outside, eventually the distance between the two cars would get larger once again, and they would miss one another.

The analogy is a simple one, however; in reality, rendezvous in space is a bit more complicated than that, with speeds being much higher and distortions in the orbits causing a few other challenges. Plus, all of this is taking place in three dimensions, with the orbit paths potentially crossing to the left or right of one another. But ultimately it boils down to one spacecraft meeting another spacecraft at a precise point in orbit where the speed differential is not too high so they can station keep, or dock, with one another. The failure to rendezvous and dock between *Soyuz 7* and *8* showed just how challenging things can be.

On the second day, *Soyuz 10* closed to within sixteen kilometers of *Salyut*, a distance close enough for the Igla automatic rendezvous system to take over. It locked on, and the cosmonauts became passengers as they watched the distances close between Soyuz and *Salyut*. Back on the ground, things were in a state of excited chaos as engineers and guests crowded the control room at Yevpatoria. Mishin was there, along with Gen. Kerim Kerimov, a military engineer with responsibilities in managing the station program. Both men would frequently interrupt the controllers for Igla system updates, as there were concerns it might fail again. But there was no failure. The interruptions added to the stress level in the room during the approach, while things were much calmer in space as the cosmonauts monitored the approach from their vantage point.

It had been planned for *Soyuz 10* to dock with *Salyut* while over Soviet territory, but it had only closed the distance to about 800 meters when the two spacecraft passed out of range of the land-based tracking stations. At 150 meters from *Salyut*, Shatalov took over the approach manually for the final phase of docking. He approached *Salyut* at about 0.3 meters per second and slid the docking probe into the drogue on the station. With mechanical contact being made, the automatic docking process should have taken place, with the probe retracting to bring the two spacecraft together for a hard dock with all the capture latches secured. But something was wrong, as the probe seemed to get dragged sideways across the drogue during retraction and made a scraping noise in the process. *Soyuz 10* was stuck, only halfway docked.

When the spacecraft got within range of a tracking station on the next orbit, the crew relayed the problem to the ground. Everyone waited for the representatives of the design team who had built the probe to explain what could have gone wrong. Mishin exploded in anger as a result of this failure. Further debriefing of the crew after the mission provided enough clues for Boris Chertok at least to determine what had happened. As per procedure, the spacecraft's thruster control system was kept active when soft dock was made. So when the probe began its retraction sequence to bring the two craft together, the spacecraft's guidance system detected a change in orientation and tried to fire thrusters to correct itself while the Soyuz was hooked to *Salyut*. The resulting torque forces damaged the probe, and it was unable to retract any farther. In hindsight, if the thrusters were switched off at the right time, the docking would have happened normally. But there was no way to override the automatic docking sequence once the probe contacted the drogue. So once initial contact was made, a cosmonaut would have needed to act quickly to shut down the thrusters.

Attempts were made to see if posigrade thruster firings from the Soyuz could force the probe to retract fully, but to no avail. *Soyuz 10* was still unable to establish a hard dock. What was worse, though, was that since the Soyuz had not finished docking, it couldn't easily undock either, as the system would only work once the craft were fully attached. Attempts to try the normal undocking procedure didn't work, and the Soyuz just spun around on its bent probe when it tried to separate from *Salyut*.

In an emergency the cosmonauts could still come home by jettisoning either the docking mechanism or the orbital module from the rest of the spacecraft. But that would leave the docking port fouled, and no other spacecraft could dock with *Salyut*. *Soyuz 10* was also near the end of its consumables life, as it was only intended to fly for three days in space at most with the supplies it had on board. The crew had to come home soon.

An undocking solution was found, and procedures were read up to the crew. Inside the Soyuz orbital module, Rukavishnikov used cables to bypass the docking latch sensors to make the system think it was fully docked. When given a command from the control panel, this tricked the docking system into thinking it had been given a command to undock by *Salyut*. *Soyuz 10* was free.

Soyuz 10 spent the next few orbits visually inspecting the station's docking port to see if it was damaged and also inspecting the condition of *Salyut* itself. The station had survived its trip into orbit quite well, and the docking port was ready to receive another crew. But visual inspection confirmed that the protective cover over the main telescope had not come free. Those instruments were now useless. *Soyuz 10* safely returned home, performing the first night descent and landing by a Soyuz craft. Except for the lack of light outside, the descent and landing were uneventful, and recovery crews were on hand to greet the cosmonauts once they touched down.

At the end of the mission, state officials publicly applauded the mission as a success, achieving all its goals. In the announcements made after the launch of *Soyuz 10*, no mention was made that it would dock with *Salyut*, only that it would conduct close proximity operations with the new space station. The crew were welcomed as heroes, and they stuck to the story that entry and occupation of the *Salyut* station was never a goal of the mission, only rendezvous and docking with it, which were achieved . . . at least in part. It wasn't until years later that the real story of the failed hard dock emerged.

Of the three cosmonauts from *Soyuz 10*, two would never fly in space again. Vladimir Shatalov agreed to take over duties as the officer in charge of cosmonaut training, since General Kamanin's mandatory retirement was imminent. Aleksei Yeliseyev figured that three spaceflights in three years were enough for him. So he retired from flying as well, but he remained active in the space program until retiring in 1985. That left only Nikolai Rukavishnikov to fly again.

Crew Change

Once *Soyuz 10* returned home safely and it was determined that *Salyut* was still capable of accepting another crew, preparations were made for the flight of *Soyuz 11*. Work progressed on modifying the docking system and revising procedures so that portions of the docking sequence could be controlled manually. Now the Soyuz thruster control system could be turned off once a soft dock was established. There were calls for the head of the designer who came up with the failed docking probe by some managers. But cooler heads prevailed, and no disciplinary action was handed down since a board of inquiry led by Boris Chertok determined that it was an unanticipated oversight that led to the *Soyuz 10* docking failure as opposed to negligence on the part of anyone on the design team. By the end of May, a new beefed-up docking probe had been successfully tested.

At Star City, Leonov, Kubasov, and Kolodin continued their training as preparations were made for their flight to the station. With Leonov knowing that he was going to be commander of the second mission on *Salyut*, some of his colored pencils and paint brushes had already been launched aboard the station in April. Already an accomplished painter, Leonov had plans to take his form of artistic expression literally to new heights. Leonov's extracurricular activities met with some criticism from Mishin. Not long after *Salyut* entered orbit, there were indications that one of the ventilation systems had failed. Mishin made some disparaging comments that Leonov's art supplies must have broken loose and floated into the vents, jamming the mechanism.

Leonov let those comments slide, but he did not have much confidence in the chief designer as a manager. Leonov felt that Mishin's lack of Korolev's drive and his overly cautious approach had kept the first-EVA veteran from making the first manned lunar flyby in the Zond spacecraft before the Americans. While Leonov respected Mishin's technical expertise as an engineer, he still felt that the chief designer was a bad manager and not willing to take risks when it counted most. Plus, Leonov felt that Mishin also drank way too much.

As the first man to perform a space walk, Leonov was looked up to by other cosmonauts. There was a lot of prestige in having Leonov involved with a program. Flight engineer Valery Kubasov was a young civilian engineer who joined up as part of the first class of engineering cosmonauts in 1966. Prior to that, he worked for Korolev's bureau on spacecraft trajectories and guidance. Joining them was Pyotr Kolodin, an officer in the Soviet Armed Forces who had been in the cosmonaut program since 1962 as part of the second class of cosmonauts. He had performed several duty assignments in support of the early space missions.

All seemed to be going well until the final days before launch. The *Soyuz 11* crew had their physicals, and the capsule with the crew's custom-fit couches was placed on the rocket. The rocket had already been rolled out to the pad in anticipation of a launch within a couple of days. But doctors expressed concern about a dark spot found on an X-ray of Kubasov's right lung. It hadn't appeared in previous X-rays.

Further testing was ordered, and the shading was verified as being in the lung itself and not something caused by the X-ray film. A blood test also indicated that Kubasov had an elevated white cell count, which could mean an active infection. What could it be? Kubasov seemed to be in perfect health otherwise, as did the rest of the crew. Specialists were consulted, and it was decided that Kubasov should be scrubbed from the mission. The prevailing concern was that it might be tuberculosis. If Kubasov should fly to the station with it, in the enclosed environment of *Salyut* and the Soyuz, the others might become infected as well. But even with this dark spot, Kubasov seemed to still be as healthy as ever and showed no signs of pulmonary problems.

Managers went back and forth on what to do. Should Kubasov be scrubbed and Vladislav Volkov from the backup crew fly in his place? Or should the entire crew be stood down and replaced with the backup crew? The managers who argued in favor of scrubbing Leonov's crew for Dobrovolsky's figured that the crewmembers had trained together and had developed interpersonal relationships with one another. An upset to that balance might have unknown repercussions. A mission rule to do so had already been in place, but it had never been enforced this close to launch time. Kamanin was in favor of flying Volkov in place of Kubasov after consulting with others. With Volkov already a space veteran, he should have been able to handle the job just as capably as Kubasov; with Leonov also a space veteran, the mission should potentially have gone better than one flown by a crew with much less operational experience and only one mission veteran. But Mishin argued in favor of changing the whole crew and used the mission rule drawn up by the air force to argue his point. The final decision was made by the State Commission at a meeting on 4 June. The backup crew would fly instead of Leonov's crew.

When Leonov heard about the decision, naturally he exploded into rage and demanded that he be allowed to fly given that he had trained long and hard for this mission. Leonov tried to get Mishin to change his mind on the matter. In Leonov's half of the coauthored book *Two Sides of the Moon*, the veteran cosmonaut relayed what happened in the heated discussion he had with Mishin as the chief designer apparently said to Leonov, "Don't forget you shared a room with Kubasov, perhaps you both drank from the same glass. We can't take the risk of you becoming ill while in space." That argument does seem somewhat petty, though, since after the crew replacement was made, Leonov's entire crew made a couple of appearances with Dobrovolsky's in public gatherings. If Kubasov really had been that infectious, it is doubtful he would have been allowed in the same room.

The decision was hardest on Pyotr Kolodin, who had not flown in the seven years he had been a cosmonaut. Reportedly Kolodin said to Mishin, "To them, I am the 'white crew'—they're all pilots and I'm a missile man." The term "white crew" was in reference to support and launch crews as opposed to flight crews. Apparently, Kolodin had a suspicion that he would not fly and resigned himself to possibly never flying a mission in space ever. The perceived pecking order of cosmonauts had the best assignments going to the air force cosmonauts first and the civilian cosmonaut-engineers second. Those cosmonauts who joined from the ground forces ranks were the low men on the list.

Changes had to be made to the Soyuz during the last day of launch preparation, as the flight couches for Leonov's crew had to be removed and replaced with ones for Dobrovolsky's crew. The technicians did a superb job in getting the spacecraft ready to fly. Apparently, while Volkov was happy to be granted an earlier flight as the crew's sole space veteran, Dobrovolsky and Patsayev, according to Leonov, were said to be noticeably nervous about the whole thing, as they had only been selected as a crew just four months prior and hadn't intended to fly until a year later at least. Leonov also wrote that Patsayev met with him privately to apologize for how the whole situation unfolded.

Many weeks after *Soyuz 11*'s flight, Kubasov was finally given a clean bill of health. The medical report concluded that he must have inhaled and had an allergic reaction to a pesticide chemical that was being used on the grasslands at Star City near the cosmonaut apartments during one of his regular five-kilometer runs. Kubasov himself reported many years later that it was an allergy to a type of pollen and that the reason the dark spot hadn't been present on an earlier X-ray was that the previous exam took place during the late winter months, before the flowers in Star City's gardens came into bloom.

The Crew of *Soyuz 11*

Lt. Col. Georgi Dobrovolsky was born on 1 June 1928 in Odessa, Ukraine, near the Black Sea to Russian parents. Odessa was also the childhood home of Sergei Korolev, but the two did not meet until many years later when Zhora (as Dobrovolsky was known to his friends and family) joined the cosmonaut ranks. When Dobrovolsky was thirteen, the Germans invaded the Soviet Union and occupied Odessa. Dobrovolsky was an active resistance fighter until he was captured by the German ss when they found a revolver on him during a search. He was tortured and subjected to electric shocks to try to get him to reveal other resistance members. But he didn't break. After being sentenced to twenty-five years of hard labor, Dobrovolsky managed to escape a month later when one of his relatives bribed a guard. A few weeks later, Odessa was liberated by the Soviet Army. Dobrovolsky next tried to enlist in the army, but he was kept out due to his still extremely young age.

After the war, Dobrovolsky completed primary school. He wanted to become a merchant marine at the Odessa Nautical School, but his application was turned down. Instead, on the advice of a friend, he decided to enter a school that trained young men to become pilots before enrollment in the military. Dobrovolsky excelled at this and was accepted into the Soviet Air Force for military flight training. He rose quickly through the ranks, eventually becoming a squadron commander and an instructor pilot. He married a student mathematician named Lyudmila Steblyova in 1957, and together they had a daughter named Marina. Dobrovolsky took some correspondence classes, as he knew he was a bit weak in certain aspects of his education; he wanted to become an aeronautical engineer. He attended the Air Force Academy and graduated around the time that the first cosmonauts were flying into orbit. The Soviet Air Force Academy was not the same as the U.S. academy; the Soviet academy was a command-staff-level school that groomed officers in the Soviet Air Force for leadership roles and positions of command.

After encouragement from his squadron commander and consulting with his wife, Dobrovolsky submitted an application to become a cosmonaut. While the first class of cosmonauts was made up of relatively young pilots, the next class would be made up of older pilots with extensive flight experience. He was accepted and joined the cosmonaut ranks as part of the second class of cosmonauts in 1963. While he was training to become a cosmonaut, his wife had their second child, a daughter whom they named Natalya.

Vladislav Volkov, by comparison, had a much different early life. He was born on 23 November 1935 in Moscow. His father was an aeronautical engineer, and his mother worked in an aircraft factory during the war. He also had an uncle who was a pilot. It seemed inevitable that Volkov would have a career in aviation, but he wasn't sure what path that would take. At his uncle's urging, Volkov entered school to become an aeronautical engineer and studied at the Bauman Moscow Aviation Institute. While attending school, he met Lyudmila Birykova. They married in 1957, and she gave birth to their only child, a son named Vladimir.

Upon graduation, Volkov ended up working at Department Number 4 of Korolev's OKB-I design bureau. His organizational skills meant that he was assigned as a deputy to the leading engineers on both the Vostok and Voskhod spacecraft projects. At the same time, he also enrolled in flight school and obtained a sport pilot's license, which allowed him to fly light airplanes.

According to a few of Volkov's friends and colleagues, Volkov was a very driven individual and tended to be jealous of others if he didn't get what he wanted. He was very fit as he played both soccer and ice hockey. If he ever got the ball in soccer matches, he wouldn't pass. He would advance the ball to try to score a goal or miss.

When plans were made for the three-person Voskhod flights, Volkov was under consideration to join the cosmonaut program as one of OKB-1s cosmonaut-engineers. He felt he had a good shot thanks to his health, his flight experience, and his engineering background, but he didn't get the assignment. When he complained bitterly to Korolev, the chief designer responded calmly, "You are still too young. There is time. It is impossible to send everybody on spaceships. Somebody has to design them." After briefly considering a career change, Volkov remained at his job; in 1966 after Korolev's death, Mishin put forth his name as one of twelve cosmonaut-engineer candidates. After passing the air force medical screening, Volkov trained as a cosmonaut flight engineer.

Volkov was also the first accredited journalist to fly in space, thanks to articles he had written previously for the Soviet Army's newspaper *Red Star*. He kept a personal diary of everything that happened on the *Soyuz 7* flight and impressed his fellow crewmates in the process. Before launching on *Soyuz 11*, he had sent proof pages that he had written for his biography to a publisher, as there were plans to publish this book once he returned home.

Compared to Vladislav Volkov, Viktor Patsayev was almost a directly opposite individual. Patsayev was also a civilian engineer, but he was quiet and reserved. He liked to listen and analyze. He would typically avoid conflicts, and he would try to prove his points only with indisputable facts, rather than with his own opinions. While Volkov was driven to do things by his force of will, Patsayev typically liked to do things mainly so he could figure out how they worked. Spaceflight and how devices worked in it seemed to fascinate him.

Viktor Patsayev was born in Aktyubinsk, Kazakhstan, on 19 June 1933 near the Russian border. His father was director of the local bakery, but he was also a Soviet army reservist. Young Patsayev loved books from a very young age, as he would read almost anything he could get his hands on. Patsayev wanted to attend school at age five, when most children wouldn't attend until age eight, but his parents didn't allow him to do so until age seven. Even being one year younger than his classmates, Patsayev's exam scores were considered excellent.

When the Germans invaded the Soviet Union, his father was called to active duty and was killed in action, defending Moscow. This greatly affected a Viktor Patsayev who was not quite eight years old, and he seemed to mature beyond his years as a result. He still continued his quest for knowledge. Patsayev's mother remarried after the war, and he attended school to complete seventh and eighth grade with two step-brothers.

Patsayev would read history books and mathematics and natural sciences text books. He also developed a love of science fiction as he read the novels of Konstantin Tsiolkovsky and even books by Jack London. In high school, he and a few friends built their first telescope and would spend hours looking at the heavens. It is believed that this is the time when Patsayev became interested in spaceflight. He also taught himself German, and years later he became fluent in English as well.

By many accounts today, Viktor Patsayev would be considered a genius in a Western society. But he wasn't just a bookworm; he was a very active athlete. He was also a competition sharpshooter and archer. After high school, Patsayev wanted to become a geologist. But while his scores on entrance exams to get into the Moscow Geology Institute were considered good, they were not quite good enough, and he was denied entry. Instead, he enrolled in the Penza Industrial Institute. At Penza, Patsayev applied for and was accepted into a class that would study the (then) new technology of computers. He graduated as a mechanical engineer in 1955 with honors.

Patsayev spent the next few years designing scientific instruments that were launched into the upper atmosphere by balloons and sounding rockets. During this period, he married his wife, Vera, who was a researcher at the Central Scientific Research Institute for Machine Building (TSNIIMash). They had two children—a son named Dmitry, born in 1957, and a daughter, Svetlana, born in 1962. It was during this period that Viktor Patsayev met Sergei Korolev. Patsayev read articles written by Korolev and approached the designer with a desire to transfer from his current job to OKB-1. While Patsayev was very competent at his job designing instruments for sounding rockets, he wasn't very happy where he was working; Korolev's work seemed like a better prospect. Korolev was impressed by the young engineer's experience, and a position was found for Patsayev at OKB-1 a little over a year after the two men first met one another. At OKB-1 the young engineer worked on many critical components for Soviet spacecraft, including the life-support systems. After a few years working in this capacity, both he and Vladislav Volkov were assigned to the recovery teams that would retrieve cosmonauts returning from space.

In 1967 Mishin was recruiting OKB-1 engineers to participate in the

14. The *Soyuz 11* crew of Patsayev (*top*), Dobrovolsky (*center*), and Volkov (*right*) dressed in Soyuz flight gear for the era.

Soviet lunar-landing program. After passing the medical exams, Patsayev was accepted as part of the second group of cosmonaut-engineers and was selected for lunar space-mission training. During training, Patsayev's quiet, analytical approach to his schooling caught the attention of his classmates and instructors alike. He was determined to do a good job but didn't make any moves to call attention to himself. Instead he let his work do the job.

In 1970 Viktor Patsayev accepted a transfer to the DOS project, where he became assigned to the third crew of cosmonauts along with Dobrovolsky and Volkov. While Viktor Patsayev was the third member of the crew, he didn't let that bother him, as he reportedly said in one interview, "Your position on the crew—flight engineer, researcher, physician, or commander—isn't important. In order to work well together, we have to believe in and respect one another, and we must celebrate the achievements of our crewmates. That is the foundation of a crew." Considering Viktor Patsayev apparently valued honesty and sincerity throughout his life, this statement likely sums up his character the best.

Flight of the Yantars

Two days after the crew change was made, on 6 June 1971 the crew of *Soyuz 11* journeyed to the launchpad. As during the previous Soyuz flights, they wore only lightweight clothing. This was common Soviet practice, as pressure suits had not been used for launches since *Voskhod 2*. Even in the Soyuz missions where space walks were planned, the cosmonauts conducting those space walks would not don their suits until preparing for the EVAs themselves. For this flight, the crew was given the radio call sign "Yantar" (Russian for the semiprecious stone "Amber"). So Dobrovolsky was Yantar 1, Volkov was Yantar 2, and Patsayev was Yantar 3. On orbit, they would be referred to only by those designations during radio transmissions with the ground.

At 07:55 *Soyuz 11* rose from its launchpad and headed into orbit with a flawless liftoff. The crew was in good spirits as the bad memories of the abrupt crew change gave way to the excitement of their mission and the duties they had to perform. Just before liftoff, Volkov prompted the crew to wave farewell to the controllers on the ground via their onboard camera, which they did. The spirit of a nation would ride with *Soyuz 11* into orbit, as this flight of the Yantars would be the first mission covered extensively in the Soviet press while it was taking place.

Upon reaching orbit, *Soyuz 11* began its one day trek to let *Salyut* in a slightly lower orbit catch up to it. Due to the lessons learned from the *Soyuz 10* mission, not only were changes made to the docking system, but extra propellant for the thrusters was loaded for added safety margins during the rendezvous. An additional day of provisions was on board so that potential docking troubles wouldn't necessarily cause a mission abort.

During the early morning hours of 7 June, *Soyuz 11* sighted *Salyut*. At a distance of sixteen kilometers, the Igla rendezvous system locked on to *Salyut*, and the Soyuz began to close with it. As before, the intention was to dock with the station while the spacecraft was over the tracking stations of the Soviet Union. But like *Soyuz 10*, the final approach didn't occur until after the craft had moved out of range of the tracking stations. It would be another twenty-three minutes before contact was reestablished to know if docking was successful.

At the end of the blackout period, live television signals and telemetry from the Soyuz revealed that the crew of *Soyuz 11* had successfully hard docked with *Salyut*. When docking occurred, it was so smooth that the crew reportedly did not feel the contact, and the modified control system did not produce any of the wild oscillations experienced by *Soyuz 10*. They switched off the thrusters, retracted the probe, and achieved a firm hard dock.

On the next orbit, the seals with the station were verified. Finally, the order was given, and at 10:45 Moscow time on 7 June, Viktor Patsayev entered *Salyut*. There was a stale smell in the air; this was due to the failure of six of the station's eight ventilation fans a few hours after the station entered orbit. Leonov was in the control center at this time. Remembering the insulting remark that Mishin made about his art supplies, he apparently asked a question to be relayed up to the crew to see if his art supplies had anything to do with failure of the fans. The crew replied that was a big negative as Leonov's pencils and brushes remained where they had been stashed before launch in their proper location. It turned out that some loose material left inside before launch did get trapped in the fans. The smell was caused by burned insulation on two of the fans, but they continued to operate normally once the blockages were cleared. And all fans were quickly restored to service. Once everything was established and working properly, the crew was ordered to have a meal and get some rest inside the Soyuz. During the rest period, the station cabin would be cleansed of its stale air.

Congratulatory telegrams were received in the control center from all over the Soviet Union on the success of the mission so far. The station was equipped with onboard television cameras so that engineers on the ground could see the activity going on in Earth orbit. Like an American spaceflight to the moon, the Yantars would report on their experiences in space, and snippets of these video transmissions would be released to the Soviet news agencies on an almost-daily basis. The Soviet people seemed captivated by the events, having never had extensive coverage of any previous mission, not even the American flights to the moon for that matter.

Even with the loss of some instruments due to the stuck cover, the *Soyuz 11* crew still had plenty of scientific experiments to keep them occupied. Exercise made up a fair portion of the day for each crewmember, as the doctors on the ground had come up with a grueling exercise regime to try to minimize the effects of weightlessness and avoid a repeat of the long recovery time of the *Soyuz 9* crew. Each crewmember was to spend two hours a day tethered to a treadmill while wearing a chest expander for hard exercise and then spend another thirty minutes of "light walking" on it before retiring for scheduled rest periods. Some "sports days" were also booked into the flight program, where the cosmonauts would perform more strenuous exercise to simulate competition sports.

At certain times during the flight, they were also authorized to wear the new TNKs (a Russian acronym meaning "training loading suits"), or penguin suits, for additional exercise loads on the body. These suits had been refined from the original *Soyuz 9* suit. Whereas the earlier version was attached to a wall in the orbital module, these new suits could be worn anywhere in the station. It turned out that the suits were so comfortable to wear that the crew asked for and received permission to wear them for longer periods. A negative-pressure body apparatus was also used in conjunction with the penguin suit to help counteract the changes in fluid distribution that take place in space. Each cosmonaut's medical state was monitored by a special belt they wore around their chest.

Each cosmonaut took extensive notes in their diaries during the flight. Astronomical observations with a gamma-ray telescope and an ultraviolet telescope in the transfer compartment began on the sixth day of the flight. These observations could only take place during the thirty-five-minute period when the station was on the night side and the stars could be seen. To aim the telescopes, Dobrovolsky would rotate the station while Patsayev directed him where to point it.

The cosmonauts had problems with low lighting. To help provide necessary power to critical pieces of equipment from the Soyuz-based solar arrays, designers deliberately limited illumination levels in some parts of the station. As a result, if a piece of equipment had to be operated in one of these areas, it was often difficult to see it to make the proper settings and record the results.

Growth experiments with plants and tadpoles were conducted on orbit to check the effects of weightlessness. The frog eggs were launched aboard the Soyuz and hatched in orbit. When the tadpoles reached a certain level of maturity, the embryos were frozen and returned to Earth for scientists to study the effects of weightlessness on their cellular development. For the plants, chinese cabbage and bulb onion seeds were grown in a small hydroponics lab. Blood samples from each crewmember were also regularly collected for later analysis. A portable radiation meter was used to check for radiation exposure as well.

The flight's routine continued for ten days without interruption, with the crew doing their regular shift of space science and observations in an around-the-clock schedule. They also transmitted regular "Cosmovision" television broadcasts to the ground in which they would conduct small tours to different parts of the station and describe their activities, using the onboard television cameras.

There was some concern that breaking the circadian rhythm cycle of the human body for such a long time—with one crewmember at rest and two awake and with each on a different cycle—might cause some nervous stress, but this cycle was maintained throughout the flight until toward the end. Concerning how the crew behaved toward one another, things were reported to have gone well for the most part, although Volkov did try to override Dobrovolsky's authority as mission commander, which apparently led to some heated exchanges on orbit between the two of them. Patsayev, on the other hand, seemed to keep up with his work in an almost-obsessive fashion. He would quietly conduct his experiments, collect his findings, and take notes in his diary. Patsayev also continued to shave regularly, while both Dobrovolsky and Volkov began to grow beards.

On day eleven the routine was interrupted by a serious incident. Volkov smelled smoke; when he went to investigate, he noticed smoke coming from a wall panel. When the next communication pass started, he reported seeing a "curtain" in the aft part of the station. The word "curtain" was a code word for fire, but the flight controllers on the ground didn't understand and asked him to repeat. Volkov replied in plain Russian, "There is a fire on board!" He also indicated that the crew was entering the Soyuz and that they needed the procedures for an emergency undocking because in their haste they did not grab the procedures book. His concerns were justified, as fire can be a very insidious thing in airplanes and spacecraft. While only smoke might be visible, there could be a raging inferno going on inside the panel.

Immediately, cosmonauts Nikolayev and Yeliseyev, who were on the communications loop at the control center, did their best to calm down the crew. They figured the fire was likely being caused by a piece of scientific apparatus, and they instructed the crew to switch off all power to the scientific instruments, find the specific source of the smoke, and then retreat to the Soyuz if all else failed. They relayed those instructions up to the *Salyut* crew, but the station passed out of range before the crew could report the results.

When the station next orbited close enough to a tracking station for an update, the crew reported that they were still aboard the *Salyut*. The smoke was no longer being produced. But there was still smoke in the cabin, and they all had headaches, probably from carbon monoxide exposure. As suspected, the fire had been caused by a short in one of the scientific instruments. But the condition of the station was stable otherwise. Yeliseyev continued to reassure the crew and to help put their minds further at ease, reading up the instructions for emergency undocking. That way if the situation got any worse, the crew could still evacuate. He also instructed the crew to turn on the filter fans and let them cleanse the air inside the station.

All seemed to return to normal, and the crew were instructed to rest at that point and resume normal operations the next day. There were some concerns as to the mental state of the crew. The person who reported the fire when it occurred and reported again when the fire was out was Volkov, not Dobrovolsky as it should have been in the proper chain of command. When the situation calmed down, Dobrovolsky resumed control. After some additional communications sessions between Volkov and engineers on the ground, it had come out that Volkov tried to resolve the situation himself rather than relying on his crewmates. Mishin communicated directly with his young cosmonaut-engineer and tried to reassure him. His instructions were simple: respect the chain of command and carry out the orders of the commander. Volkov replied that the crew would decide what to do together. It isn't entirely known if indeed the crew wanted to evacuate the station over Volkov's wishes or if it was only Volkov himself who wanted to evacuate. None of the crewmembers' diaries seem to tell the whole story. But Dobrovolsky's diary did contain one curious entry. It said, "If this is harmony, what is divergence?" in apparent reference to the crew's relationship after the fire.

The source of the fire was traced to a seized cooling fan. When the fan jammed, the motor continued to try to drive it until the stator began to pump out dense smoke. The crew restored the other systems one by one back to operation and continued with the mission. In the coming days, the routine would return to normal, although there were still some concerns about Volkov's behavior. He had developed a bit of a short temper and was prone to making simple mistakes while being slow to admit he made them unless pressed on the issue by ground controllers.

As the days wound on, mission managers on the ground discussed the possibility of extending the mission to thirty days. But it was decided instead that the cosmonauts would undock on the evening of the twenty-fourth day as originally planned and come home on the early morning of the twenty-fifth day just before sunrise. This would beat the old record set by *Soyuz 9* by the required 10 percent needed to recognize it as an official record with international agencies.

Salyut continued to perform well with no major problems, as did the crew. As for the experiments, they had fallen a little behind schedule since the unusual nature of the work and sleep schedules and the disruption of the fire had thrown things into slight disarray. But the cosmonauts were still continuing to work and collect data. When they returned, they would have a treasure trove of material for the scientists on the ground to look over.

During the twenty-first and twenty-second days of the flight, the Soviet space program suffered two setbacks. First, Aleksei Isayev, head of the Himmash (OKB-2) design bureau, suffered a lethal heart attack and died at only age sixty-three. Then the next day, on 27 June, the third test flight of the N1 moon rocket ended in failure. After liftoff, the rocket experienced a spin along its axis that could not be corrected by the roll-control motors. The spin caused the support structure between the second and third stages to fail. The upper stages toppled and exploded while the out-of-control yet still-powered lower stages shot past and crashed into the desert about thirty kilometers away, resulting in a large crater. It had been hoped that the *Salyut* crew would be able to observe the launch from orbit as part of a rocket launch detection experiment, until delays on the ground prevented it.

During the final days of the mission, the sleep cycle of the three crewmembers was shifted so that they would all be awake and alert for undocking day and the return to Earth. Efforts were made to close down and store the experiments while transmitting the last findings to the control center. The station was given a thorough cleaning, and it was returned to automatic function. Tests were made on the Soyuz's systems as it was brought out of hibernation. Everything seemed to be in working order, and it was time to return home.

On the final dock day, the crew entered the Soyuz and sealed the hatches between the two craft. As a final procedure before undocking, they also had to seal the hatch between the orbital module and the descent module. But there was a problem, as a warning light indicated the hatch was still open even when properly shut. The problem was traced to one of several buttons that surrounded the hatch. When the hatch is closed, these buttons get pushed and the light goes out, kind of like an overhead dome light in a car or a refrigerator. But one of the buttons was barely being pushed when the hatch was fully seated. So the decision was made to put a piece of tape over the button in order to bypass it and turn out the indicator light. With that done, *Soyuz 11* undocked from *Salyut* automatically. With two orbits left to go before reentry, all the crew had to do was sit back and enjoy the ride.

Two hours before reentry, General Kamanin made a call to the spacecraft with a final weather report and instructions for what to do after reentry while awaiting pickup. This would be the last flight he would oversee as the director of cosmonaut training and the recovery forces, as he was retiring and turning his duties over to Shatalov. The conversation went well. Kamanin concluded his remarks with, "I wish you a soft landing. See you soon on Earth." Dobrovolsky acknowledged the instructions; told Kamanin that the crew was "excellent"; and concluded his reply with, "We thank you for your help and good wishes."

Silence

At the proper time, *Soyuz 11* was ready to come home. Over the Atlantic Ocean between North and South America, just to the north of the equator, *Soyuz 11* started the retrofire sequence at 22:35:24 GMT as the main motors of the Soyuz service module fired for about three minutes to slow the craft's orbital velocity so that it would descend into Earth's atmosphere. Once retrofire was complete, the typical reentry sequence for the Soyuz would have it pitching upward in a near-vertical orientation and jettisoning both the orbital and service modules from the descent module about five to ten minutes later. The orientation was needed in order to ensure that all three modules would reenter Earth's atmosphere on different trajectories with no chance of a collision. The thrusters on the descent module would then orient the capsule's blunt end forward into the reentry path and keep it aligned properly. Roll-control thrusters on the capsule would also help to guide the craft on a controlled reentry trajectory where the cosmonauts would only experience about three times the force of gravity on their bodies. If this system failed, the capsule would reenter on a steeper ballistic trajectory, and the cosmonauts might experience gravitational forces of over 9 g's. A ballistic reentry would make for an uncomfortable ride, but the custom-fitted crew couches were designed to minimize the effects.

The jettison sequence took place at about 22:47:28 GMT. With the propulsion module severed from the descent module, all communications ceased except for signals on a VHF transmitter located inside the descent module. Good telemetry signals were received from the descent module's VHF antenna when it passed within range of the western-most land-based Soviet tracking stations, but there was no voice contact. The signals were lost when the capsule entered reentry blackout at 22:54 GMT. Soviet tracking radars picked up the descent module making a normal reentry a few minutes later. The drogue and main parachutes deployed properly, and *Soyuz 11* made a perfect touchdown on the steppes of Kazakhstan at about 23:16 GMT on 30 June 1971. Upon touchdown, the descent module flipped over on its side, as typically happens. A recovery helicopter visually monitored the descent of the craft and was on-site right away to render assistance.

Minutes went by with no radio response from either the crew or the recovery forces. Things were eerily quiet. Thirty minutes after touchdown, it was reported to General Kamanin by radio from an officer on the scene that the situation was "a most tragic one." The three cosmonauts—Georgi "Zhora" Dobrovolsky, Vladislav "Vadim" Volkov, and Viktor Patsayev—were dead in their couches.

When the recovery forces opened the hatch to the descent module, they found the cosmonauts with their eyes closed as if they were asleep. There were signs of dark-blue bruising on their faces and trails of blood from their noses and mouths. They had no signs of life, but Dobrovolsky's body was reportedly still warm. The bodies were removed from the capsule as doctors and medics at the site tried to administer CPR on the three cosmonauts. A camera crew sent to film the recovery with movie and still cameras documented the futile attempts to revive the three cosmonauts while officers looked on in shock and disbelief.

An investigation was immediately launched while the announcement of their deaths was made to the Soviet people. General Kamanin and Shatalov were flown to the landing site along with cosmonauts Alexei Leonov and Vitali Sevastyanov to inspect the capsule. According to the results, all the systems seemed to function properly, and everything seemed to be in place except for one valve that had its handle position moved by about ten millimeters. Further testing was done on-site when the recovery crew closed all the valves and pumped up the capsule to a high internal pressure. No leaks were detected. The capsule was then returned to Moscow for further analysis.

After the bodies were autopsied, the results were conclusive. All three crewmembers had died from decompression and oxygen starvation. Once they had lost consciousness, the trapped gases in their bodies had ruptured their eardrums, and their blood boiled. Bruising along their faces and extremities was caused by capillaries rupturing in a near vacuum. An onboard black box designed to record several parameters of data, from the spacecraft's automatic control systems to the crew's heart rates and other metabolic functions, was also checked for clues.

Through the medical analysis and data from the black box, it was determined that by the time medics had gotten to the cosmonauts, they had already been dead for nearly thirty minutes. They had died from rapid decompression and had been exposed to a vacuum for as long as eleven minutes before the pressure began to come up inside the capsule as it descended into Earth's atmosphere.

On 1 July the cosmonauts' bodies lay in state for eight hours at the Central House for the Soviet Army in Moscow as tens of thousands of mourners filed past to pay their respects. Other cosmonauts formed an honor guard to watch over their fallen comrades during the proceedings. The bodies were then cremated in preparation for their burial in the Kremlin wall alongside Sergei Korolev, Vladimir Komarov, and Yuri Gagarin.

As a gesture of good will and solidarity among space travelers, President Nixon asked astronaut Thomas Stafford to be the NASA representative at the funeral for the *Soyuz 11* cosmonauts. When he got the call to go to Moscow, Stafford was on his way to Belgrade, Yugoslavia, to take part in a space exhibition. During the previous funerals for Komarov and Gagarin, NASA had offered to send astronauts along as representatives, but the Soviets kindly turned them down. For this one, though, they accepted Stafford with open arms and made him a pall bearer for a portion of the procession as the ashes of the three cosmonauts were taken to the Kremlin.

Investigation and Aftermath

Just what had killed the crew of *Soyuz 11*? The investigation ultimately focused on two pressurization valves in the descent module. Each valve was equipped with both a manual and an automatic shutter. From launch to landing, one valve would have been manually closed, the other manually opened. And both would have had the automatic shutters closed. When the main parachute opened on the Soyuz, small explosive charges were designed to blow open the automatic shutters in both valves, and the one open valve would equalize air pressure between the capsule and the outside. The Soyuz descent module was only intended for use up to about thirty minutes by itself, so no internal tank of oxygen was provided. When the capsule landed, the crew might be incapacitated from their reentry ordeal, especially if they were returning in a weakened state. So the open valve would keep fresh air coming into the capsule and prevent suffocation until rescuers arrived. The second valve was provided in case the descent module landed in water. In that case, whichever valve was not submerged could be opened and the other one kept closed to prevent water leaking into the capsule.

As to why the valve failed, it was speculated in Western publications that the Soyuz capsule's explosive bolts, designed to separate the orbital module from the descent module, had fired all at once instead of in a properly timed sequence, as was done in the Apollo craft, and that the resulting jolt jostled the valve open while the capsule was still in space. But this conclusion was incorrect as the Soyuz explosive bolts were designed to fire at once on a common circuit. The resulting jolt should have been no more than what the valve should have normally seen. When the valves were analyzed, it was found that they were not torqued to the proper specifications called for in the design and were way below the proper tolerances. So when the explosive bolts fired, the automatic shutter check ball inside the manually open valve was dislodged, and the air leaked out into space. Further testing on other Soyuz descent modules, both flown and still under construction, revealed that none of them had valves at the proper torque settings either. *Soyuz 11*'s valves had the lowest torque numbers of them all.

With the size of the valve in question, once it had opened in what essentially was still a vacuum, it would have depressurized the capsule completely within thirty seconds to a minute. It was figured that the crew would have maybe fifteen seconds at most of useable consciousness before becoming incapacitated. They would likely have heard a whistling sound when the valve opened and probably also felt their eardrums pop as the pressure began to reduce. Dobrovolsky unlatched his seatbelts to check the descent module's hatch since it was the source of the undocking problem earlier. The other two crewmembers shut off the alarms and radios to listen for the source of the leak themselves. Eventually, they apparently focused on one of the two valves and began to close it, although they ran into one final problem. According to *Soyuz 11*'s factory and preflight documentation, the manual knob of valve number one should have been closed and valve number two should have been open. But it turns out the sequence was reversed, and valve number one was open instead of valve two. Valve one was the source of the leak. Patsayev, who was located closest to valve two, would have run into a roadblock, trying to close an already-closed valve. However, it looks as though the leaking valve was located and Dobrovolsky began to close it, as indicated by the valve handle's position on landing and his proximity to it. But the crew simply ran out of time.

Under normal conditions, the valve would take thirty-five seconds to close. Even if they had focused immediately on the valves instead of the hatch, isolated the correct valve, and closed it in record time, the pressure in the cabin would have fallen below a level to maintain life in less than thirty seconds. Without a tank of oxygen to repressurize the cabin, the crew would have been dead anyway. If the proper valve had been closed and the other one open as originally specified, there might not have been an accident, as the valve that failed was not supposed to be open. So ultimately the *Soyuz 11* crew died because of a couple of careless mistakes they likely had no control over.

Mishin to his dying days maintained that the crew could have given themselves more time if one of the cosmonauts had put his finger over the valve's opening to keep the air in while the valve was closed manually. Others were a bit more realistic in their assessment that more than likely there was nothing the crew could have done to save themselves once the valve was opened to space. Leonov, in his biography, said that he communicated instructions up to the crew to close the valve, around the time they undocked from *Salyut*, since he felt that the idea of having one valve manually opened as the checklist indicated was not smart and that the crew could always open it after they were descending on the parachutes. Leonov also maintains that if he had been flying the mission, he would have closed both valves himself before the orbital module was jettisoned. Whether or not Leonov's crew would have survived if they had flown will ultimately never be known.

Even in the early days of the investigation, there were calls to launch another crew to *Salyut* as normal since the station was operating properly. But all Soyuz flights were put on hold while the investigation determined what happened. The practice of flying crews on the Soyuz without pressure suits had been criticized by General Kamanin years before, as he felt it was an unnecessary risk. Korolev, Mishin, and Soyuz spacecraft designer Feoktistov all figured that flying with pressure suits was a waste since crewmembers on submarines don't wear wet suits in anticipation of a leak. Plus, the Soyuz spacecraft was already at the limits of its design weight. Pressure suits and their associated equipment would incur a substantial weight penalty, and the physical size of the descent module meant that only two suited cosmonauts could fly instead of three nonsuited ones.

Ustinov had heard enough, though; he made the official mandate: cosmonauts from then on would fly into orbit and back wearing pressure suits. Modified Sokol lightweight pressure suits used in high-altitude aircraft were added, along with a system that could rapidly repressurize the craft in the event of another pressurization failure, even if the crew were not wearing the suits.

Since *Salyut*'s design life would be exhausted by the time Soyuz was ready to fly again, the decision was made to deorbit the station in October of 1971. It reentered and burned up after flying successfully for 175 days. Even in its unmanned state, the station had performed well and likely would have continued to do so if a second crew had flown to it.

4. The Apollo Applications Project

In 1965 things were advancing quite well on NASA's road to the moon. The Gemini program was hitting its stride in providing answers for rendezvous, docking, EVA, and living in space up to two weeks. Development and testing work was also progressing on the Apollo spacecraft at the contractor sites. The schedule was tight, but it looked like the goal of reaching the moon by the end of the 1960s was within reach.

But there was a growing problem. NASA knew what its near-term goals were, but it didn't have any set goals beyond the lunar missions. Support for the next step was split among leaders at NASA who wanted to focus on just getting Apollo to the moon and those in Congress who wanted NASA to do another program to help sustain America's leadership role in space. Certain Congressional leaders feared that if NASA did not have a goal after Apollo, it would make the funding fights on Capitol Hill more challenging. At the same time, there was also a rather vocal opposition to NASA from some Congressional leaders who felt that spaceflight was a waste and that the funding would be better spent elsewhere.

When NASA's charter was drawn up during the Eisenhower administration, the funding for NASA would come from Public Health and Welfare as opposed to its own direct funding on the same tier as a Department of Defense agency, such as the U.S. Air Force. Part of the reason for this is that President Eisenhower wanted NASA to be a civilian agency with an open space program instead of a military one with potentially secret goals. The situation continues to be one of the primary reasons why funding debates in Congress typically come down to the space program versus welfare programs, even today.

There was also a concern that scientists were being left out in the cold by NASA. In the perception of some scientists, Apollo's goal of reaching the moon seemed to exclude the question of what should be done when astronauts get there. The perception among the scientific community was that scientific inquiry seemed to be tacked on to manned spaceflights as an afterthought. On the other side, many of the early astronauts expressed their dissatisfaction with scientists, since many felt that doing experiments tended to get in the way of flying the mission. Some would try their best to minimize the research if they felt that it interfered with the flight-testing goals.

On the unmanned-mission side, research satellites and probes began providing useful data on Earth and the solar system. There were calls among members of the scientific community that after the moon landings, NASA should devote more time and finances toward unmanned missions and research satellites, arguing that those missions would provide more scientific data than what a manned flight to the moon ultimately could. With these concerns in mind, NASA began looking into a post-Apollo mission for its astronauts. The result would become known as the Apollo Applications Program (AAP).

The principal backer of AAP was NASA associate administrator for the Office of Manned Spaceflight George Mueller (pronounced *Miller*). Mueller was an electrical engineer with a doctorate in physics. He had twenty-three years of experience in research and teaching positions, along with a good eye for both management and practical engineering solutions. Mueller was a very capable manager, helping to organize the Office of Manned Spaceflight in Washington DC and establishing the pecking order of the other NASA centers based on what roles they would play in the development of hardware for manned spaceflight.

George Mueller had already wielded his authority in mandating the use of "all-up" testing in the Saturn V rocket program. Prior to that point, rocket stages had been test-flown individually to verify their systems. He sensed that the Apollo program was most certainly going to fall behind schedule if the original rocket-testing process were used. So going against the wishes of the NASA centers, Mueller came up with the radical solution that all the Saturn V rocket stages should be stacked and tested live during one launch. The gamble would prove to be a successful one, as the two unmanned test flights of the Saturn V performed well with relatively few problems in 1967 and 1968.

Mueller made some additional decisions regarding the Apollo program, as he wanted to see the capabilities of the spacecraft expanded and made "multimission" capable with the ability to do more than just fly to the moon

15. George Mueller charted the course of NASA's post-Apollo future. Courtesy NASA.

and land on it. As production of the hardware got better, performance would improve. With these improvements, missions could be extended and more capabilities could be added.

NASA astronaut selection was also looking toward the future. In June 1965 the fourth class of astronauts selected was made up exclusively of scientists. Many in NASA and Congress considered these appointments to simply be a political stunt to appease the National Academy of Sciences in their concerns that NASA should be doing more for science, but several of the new astronaut candidates had backgrounds and experience that were considered potentially important to NASA's post-Apollo future.

Of the six scientist-astronauts selected, only Harrison Schmitt had a scientific background that had a direct benefit to Apollo's lunar-mission goals: geology. The others were either physicists or physicians. Almost all were required to take undergraduate pilot training once they joined NASA to qualify for jet flight, so it would be about a year before they could contribute in specific assignments as astronauts.

The AAP office first opened its doors in August 1965. George Mueller's intention for the program was to keep NASA's work in spaceflight continuing at the conclusion of Apollo's lunar program. Complex programs typically require a minimum of five years to get from drawing board to flight hardware. This time is needed for hardware design, systems testing, integration, and possible flight-testing before an operational mission can fly. Off-the-shelf hardware can reduce some of the work, but not by much since equipment still needs to be modified and tested to see if it works in its new role.

If work did not start on what NASA would do next soon, there was a real chance that a large time gap would form between the last Apollo mission and the next manned spaceflight; the likely result would be program layoffs and budget cuts. Mueller knew that a great team of civil servants and contract workers had been established in the creation of Apollo, and he wanted to see that workforce kept intact after Apollo's conclusion.

But there was a catch with AAP. Since Apollo's goal of landing a man on the moon was driving NASA's budget, it meant there was practically no money left over for a follow-on program. There was also the question of what the mission of AAP would be. Building a space station seemed like a logical goal, but the exact roadmap to pursue for a station design was unclear. If AAP began work on a station, which path should it choose? Should it be all new or use off-the-shelf hardware as much as possible?

Origins of the Wet Lab

At the Marshall Spaceflight Center (MSFC) in Huntsville, Alabama, center director Wernher von Braun was also concerned about the future. Located on the grounds of the U.S. Army's Redstone Arsenal, the job of this NASA center was to design and develop the Saturn family of boosters for NASA's lunar program. It was the Saturn that got von Braun's team transferred from the army to NASA, leading to the MSFC's creation in 1960.

By 1965, work was progressing steadily on the Saturn V as the MSFC was responsible for designing and testing the S-IC stage of the Saturn V and its F-1 engines. There were still many problems to overcome, but things were progressing steadily. Work was also finishing up on the Saturn IB booster being built for the Earth-orbit testing of the Apollo spacecraft. All this work was possible thanks to the original Saturn I two-stage rocket, which up to that point had been used for early testing of Apollo hardware and the launching of three Pegasus micrometeoroid detection satellites. With some tweaks to the Saturn I design, it might be possible to do other things with it.

Engineers at Marshall had conducted studies to see if a spent S-IV rocket stage (the upper stage of the Saturn I) left in orbit could be converted into a space laboratory as a workshop of sorts. The Pegasus program already had proved that an S-IV-based spacecraft was possible. The Pegasus satellite payload remained attached to the spent rocket stage, and the resulting vehicle looked almost like a space station itself.

The idea was a novel one. A Saturn I would launch an S-IV into orbit. The S-IV's fuel tank would provide fuel to the rocket engines as normal. But at the front of the rocket stage would be an airlock assembly with a docking port for an Apollo spacecraft. The exterior of the rocket stage would be reinforced with insulation shielding to protect the interior. An Apollo CSM would be launched next, and it would dock with the spent S-IV stage. The astronauts would conduct space walks to drain the remaining liquid hydrogen (LHX) out of the S-IV's large fuel tank and then repressurize it with gaseous oxygen to turn it into a space habitat for experiments. The idea of converting a spent rocket stage into a living space became known as the "wet lab" concept.

Additional proposals and studies were done with the S-IVB rocket stage of the improved Saturn IB design. The Saturn IB was also facing the prospect of being a short-lived program, as there were no plans for its use once the unmanned and manned Earth-orbit testing for the Apollo program had been completed. The S-IVB offered even more potential for conversion into a space laboratory, as its internal volume was greater than that of the original S-IV stage. It was also more lightweight than the S-IV, meaning the S-IVB could potentially haul more equipment into orbit.

Von Braun considered these proposals very carefully. His primary concern was to find a way to expand the role of the Marshall Spaceflight Center, and he wanted to do it for two reasons. First, he knew that without any follow-up rocket-stage development, contracts after Apollo and the role of the MSFC and its rocketry team would likely be broken up and sent to other NASA centers once the lunar program had concluded. Like George Mueller, but on a smaller scale, he wanted to keep the Marshall team intact. Secondly, von Braun also knew that in order for Marshall to survive, it would have to diversify its role in manned spaceflight and do much more than just design and develop rocket boosters.

Douglas Aviation in California, the prime contractor for the S-IVB rocket stage, had also done similar proposals for a wet lab workshop, as they had been hoping to expand their role in manned spaceflight for many years. They saw space as the future of their company. They had bid on and won many contracts for both the DOD and NASA, including the U.S. Air Force MOL project; yet they had enjoyed only limited success. The MOL program was also classified, so they couldn't share much of its development with any NASA programs. The relationship between Douglas and the Marshall Spaceflight Center was a healthy one, so each group bounced many ideas off one another to help flesh out a plan for the shape of what NASA's first space station should look like.

Eventually, a merging of the minds took place between Mueller, von Braun, and Douglas Aviation. Personnel from the Manned Spacecraft Center in Houston (MSC, what would eventually become known as the Johnson Space Center, or JSC, after President Johnson's death in 1973) also provided their input into the project and were encouraged by what they saw. MSC put together its own team of engineers to study the wet lab proposals, but there were many roadblocks that these proposals would have to overcome before becoming an operational program.

The biggest factor to any AAP station proposal coming from Marshall, MSC, or Douglas moving forward was that it could not interfere with work being done on the lunar program, regarding either engineering or budget. Everything had to be conducted in parallel, because for all intents and purposes, it was a separate program in much the same way that Gemini and Apollo were parallel programs of one another. With luck, AAP missions might take place concurrently with the later Apollo missions, but the lunar program still took priority in mission planning and budget circles. Some provisions were made in NASA's 1967 budget to fund AAP studies, but in the end, this only came to $42 million, which was little more than a drop in the very large bucket of what was needed.

The Apollo Telescope Mount, a Mission for AAP

As for what AAP would do with a lab once it got into orbit, a key piece of the puzzle began to fall into place in 1966. Observations of the sun from telescopes on the ground charted that the next peak of solar activity would take place in 1969–70. It would be nice if a solar observation telescope of some sort were ready to fly in orbit by then. Solar observations from ground-based telescopes share many of the same problems with telescopes observing other areas of the sky, thanks to the filtering properties of the atmosphere.

Getting a telescope and other instruments to fly above the atmosphere could pay off big with scientific data about the star at the center of our solar system. Having a crew on orbit to operate the telescope could allow for data to be collected and analyzed in real time, as opposed to a remotely operated system where data would have to be beamed back and analyzed before new instructions were radioed up. Changes can take place on the sun very rapidly, and a manned system could adjust more quickly as needed.

NASA's Office of Space Science and Applications (OSSA) was moving forward with plans to fly such a telescope, but there were questions of what form it should take. Homer Newell, the head of the OSSA, began talks with Mueller in 1966 to see if such a telescope could fly as part of AAP. The OSSA was in favor of flying a small telescope in the service module of an Apollo spacecraft. The space intended for it was already being developed for later Apollo missions, and it would eventually become known as the SIM (scientific instrument module) bay. Mueller wanted to develop a more capable instrument, perhaps built into a modified lunar module (LM) with the idea being that a bigger telescope could accomplish potentially greater results. In August 1966, NASA deputy administrator Robert Seamans authorized the project to move forward and went with Mueller's idea.

The Apollo Telescope Mount (ATM) plan seemed like an elegant one. It would be built in a heavily modified LM descent module. The LM would not fly to or land on the moon. Instead, it would be flown in Earth orbit. The landing legs of the LM would be replaced with a large solar array that would generate power for the ATM. Additional sensors and detection equipment would also be fitted. Astronauts would fly the ATM from the LM's ascent stage, using the onboard maneuvering thrusters. Astronauts would aim the ATM at specific regions of the sun during each orbit to collect data on points of interest. When done, an Apollo command and service module (CSM) would redock with the ATM, and the astronauts would reenter the CSM and come home. The ATM would fly free in orbit either to burn up on reentry or wait for another crew, depending on how it was outfitted. There was also the option of using the ATM for unmanned solar observations after the crew departed.

There were some drawbacks to the design, though. First of all, a Saturn IB could launch either an Apollo CSM or LM (in original configuration or ATM configuration), but typically not both, unless the weight was kept down to a bare minimum. Secondly, Houston had concerns about using the LM in this manner because of the risk of astronauts being stranded in orbit with no way to return home, since it had no heat shield. Plus, with such a lightweight spacecraft, there were concerns that movement of astronauts inside the LM ascent stage would jostle the instruments and potentially corrupt the data. There was also a concern that the work required to turn a LM into an ATM would run the risk of putting Grumman Aerospace, the LM contractor, even further behind in their delivery schedule of the first LMs for flight-testing.

Even with these roadblocks, work continued on the ATM, enough for a full-size mock-up to be built. Acknowledging the problems himself, Mueller proposed that it might be better to dock the ATM with the proposed wet lab orbiting workshop. The ATM still could fly free to conduct its observations, but it would remain tethered so that the astronauts would at least have a lifeline to get back to the workshop if a problem developed. In the end, the free-flying ATM concept was kept in reserve while other AAP mission ideas were considered.

Things were moving along well until 27 January 1967. On that day, a fire at Cape Canaveral in the *Apollo 1* spacecraft killed the crew of Gus Grissom, Ed White, and Roger Chaffee during a dress rehearsal a few weeks before launch. How and why the fire took place would have broad repercussions for the future of American manned spaceflight.

A pure-oxygen system was used in Apollo, since it was easier to develop than a two-gas oxygen-and-nitrogen system. Earth's atmosphere is made up of about 30 percent oxygen and 70 percent nitrogen, along with a few trace gases. During launch and ascent, the cabin pressure would bleed down to only 5 psi. At this pressure, the astronauts would still get the same amount of oxygen as they would receive on Earth. The use of 5 psi also meant that it was easier to design space suits that could operate during EVAs without blowing up like a balloon. The same practice had been used for both Mercury and Gemini with no problems. No one at NASA suspected that it might be a problem during Apollo.

On the launchpad, in order to seal the cabin properly once the hatch had been closed, the capsule was pressurized to 15 psi, which is a little over atmospheric pressure at sea level (14.7 psi). At 15 psi in pure oxygen, materials that don't normally burn will do so. When the fire broke out, it spread rapidly, and the internal pressure of the cabin went up even higher. The astronauts were unable to get out because the entry hatch of the block 1 Apollo command module was designed to open inward, not outward.

With the loss of the *Apollo 1* crew, the entire manned space program was put on an indefinite hold while the fire was investigated. For a large portion of 1967, all the NASA centers devoted many hours to investigating the cause of the fire and taking steps to prevent it from ever happening again. AAP's goals were set aside temporarily. To help prevent a similar disaster from happing in the future, the spacecraft cabin would be pressurized at launch with an oxygen-nitrogen mixture, while astronaut crews sealed in their space suits would continue to breathe pure oxygen. As the cabin environment bled down to 5 psi on ascent, pure oxygen would replace the nitrogen for the rest of the mission.

AAP Becomes a Space Program

In 1968 AAP was given its own line item in NASA's budget, which was the first major step to becoming an actual program. Astronauts were also being assigned to AAP. However, there were storm clouds on the budget horizon as opposition in Congress to the Apollo program had been brewing for years. Money had been funneled to NASA for a program that had yet to fly astronauts into space, let alone to the moon. The loss of the *Apollo 1* astronauts didn't help matters either. NASA's budget for fiscal year 1968 was due for a major slash.

Support for AAP was not going to come from NASA's chief administrator, either. The man in charge of NASA at the time, James Webb, was not in favor of AAP. He let the preliminary work progress, but he also saw it ultimately as an obstacle. Webb's primary responsibility for NASA, as he saw it, was to safeguard the lunar program. That meant that any program that intended to use Apollo hardware was going to take a backseat to the lunar program in terms of financial and hardware support. Webb was rather vocal in expressing his disdain for AAP and never put forth any efforts to champion it to Congress. Even when AAP got funding, Webb treated it as a hedge fund to help keep development on the lunar program going without financial interruption.

When the 1968 budget came out, there were cuts across the board by the House of Representatives. The Senate considered many of those cuts to be a bit too deep and reinstated some of the funding, but the damage was still done. A primary component of making AAP a robust follow-up program to Apollo was keeping the Saturn IB and Saturn V production lines going. Without more rocket boosters, AAP was shaping up to be a short program. Unfortunately, budget cuts forced the Saturn IB production line to shut down. So AAP would have to make do with already-built Saturn IBs to launch the wet lab and fly three manned missions to it, with a fifth rocket acting as a backup. The Saturn V remained in production, but it was exclusively intended for the lunar program and not AAP.

AAP itself didn't get all the funding needed to pursue its initial goals either, as it only received three-fifths of the budget that was requested. So grandiose plans featuring many spaceflights in support of an orbital workshop, a free-flying ATM, and other Earth science–based experiments were scaled back to a more modest proposal that could be done with fewer flights. Rather than becoming a major program in its own right, AAP was instead becoming a bridge from Apollo to whatever came next, whenever it might be. At least AAP was now an official program. With this funding and a mission mandate, work could now get underway to start building and testing hardware with a goal of flying the Orbital Workshop (OWS), a module that functioned as NASA's first space station.

Going Dry

By late 1968 the OWS looked very much like the station that would eventually become known as *Skylab*. It was an S-IVB rocket stage designed to power itself into Earth orbit with a single J-2 engine. Strapped to the sides of the stage were a pair of large solar arrays. Once the OWS reached orbit,

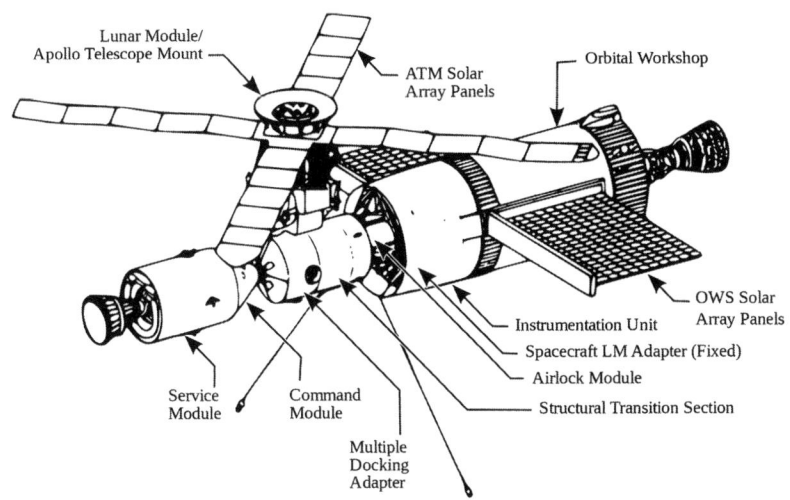

Lunar Module/
Apollo Telescope Mount

ATM Solar
Array Panels

Orbital Workshop

OWS Solar
Array Panels

Instrumentation Unit

Spacecraft LM Adapter (Fixed)

Airlock Module

Structural Transition Section

Service
Module

Command
Module

Multiple
Docking
Adapter

16. A 1968 drawing of the s-iv b wet workshop with lunar module–based Apollo Telescope Mount (the *Skylab* that almost was). Courtesy NASA.

the arrays would unfurl to generate the electricity needed. Mounted at the front of the s-iv b was a large airlock with multiple docking ports. An Apollo csm would dock at the front port, and a lunar module reconfigured as an atm would dock at one of the radial ports. The atm's heavily modified lm descent stage had four solar arrays arranged like a Dutch windmill. These arrays would generate power for the ows when docked and for the lm when the atm flew free (albeit in a tethered state). It all looked good on paper, but how would things shake out in testing?

In 1968 von Braun began work on a secret project at Marshall in support of the ows. This was a neutral buoyancy tank (nbt). In this deep freshwater pool, astronauts would be weighed down so that they would neither float nor sink, and the experience would be similar to weightlessness. Unlike the ocean-based facility used by the mol, the space suits worn would be pressurized with air. Although water is more resistant to movement than space, the laws of physics in the tank still apply. Every action has an equal and opposite reaction. So if an astronaut is not anchored properly and uses a wrench to turn a bolt, the wrench will turn him the other direction.

Building this facility was a violation of proper expenditure procedures within NASA, but von Braun did it anyway since he knew that if the wet lab proposal were going to work, his engineers had to test the concepts in a practical matter. George Mueller was made aware of the tank on a visit to Huntsville in early 1969, and he didn't think it was a bad idea at all. Thanks to the NBT, engineers found that refitting the S-IVB from a rocket stage into a workshop in space was turning out to be a lot harder than originally anticipated. Von Braun and his managers felt that a can-do attitude would make the wet lab work. But when they began their first NBT runs, they ran into several difficulties.

After going into the NBT himself with von Braun, to investigate and experience the problems firsthand, George Mueller was no longer a champion of the wet lab concept. He didn't abandon the idea entirely, and neither did von Braun, as both men still believed that it could be made to work. But specialized procedures, tools, and techniques take time and money to develop, and AAP didn't have very much of either resource.

So Mueller began calling for a dry lab approach. Technically, the dry lab would be very much the same as the wet lab. But it would be fully outfitted as a space laboratory on the ground and furnished with all the provisions it needed before it flew. When a dry lab was launched, it would be ready to begin scientific work once the first crew arrived. There was just one problem with this approach. Part of the reason why so much effort was put into the wet lab was because the only rocket booster available to launch it was the Saturn IB. The wet lab needed to act as a rocket stage since the Saturn IB's first stage alone was not powerful enough to get a fully furnished lab into orbit.

Logically, the best rocket for lofting a dry lab would be the Saturn V, but all the Saturn Vs were spoken for. Since it was not in the budget, another Saturn V could not be built either. But even with this setback, efforts were stated to alter the OWS design into a dry lab at both the NASA center and contractor levels, just in case a Saturn V were made available. The development of a dry workshop would also alter the ATM. If a Saturn V were available to launch a dry lab, the ATM could be integrated into the OWS completely, as one spacecraft, since the Saturn V had plenty of excess lifting capacity to handle both assemblies.

In July 1969 the *Apollo 11* mission successfully landed on the moon. Since the first lunar landing was a top priority, NASA had been holding many of its resources in reserve until that goal was achieved. With the lunar-landing goal met, some of those resources could be used for other things. After *Apollo 11*'s successful flight, NASA transferred one Saturn V rocket over to AAP to fly the OWS.

The decision had been a few months in the making already when it was publicly announced. James Webb retired from NASA at the end of 1968. Webb's deputy, Thomas Paine, took his place in the head office. Paine was much more receptive to the AAP project, and he also recognized that NASA needed a post-lunar-mission goal. So work secretly began behind the scenes to put a Saturn V at AAP's disposal. As a result of the Saturn V transfer, the weight constraints and necessity for a wet lab were no longer obstacles. The OWS would fly dry.

AAP Becomes Skylab

The official name change from AAP to Skylab occurred in early 1970. There had been dissatisfaction among many in NASA with the AAP acronym, as it had become the butt of jokes with terms like "Almost a Program" and "Apples, Apricots, and Pears." NASA administrator Thomas Paine formed a committee for name suggestions, as he felt a good name could score some public relations points and spark general interest in the program. Many names were considered, but eventually NASA selected the name Skylab, submitted by Lt. Col. Donald Steelman, a U.S. Air Force officer on duty with NASA.

Some changes were afoot on the management side, as George Mueller left NASA in late 1969 to take a job with General Dynamics. Dale D. Myers from North American Aviation (soon to become North American Rockwell) joined NASA and took over management of AAP when Mueller left. Wernher von Braun was also transferred away from MSFC to Washington DC to become NASA deputy associate administrator for planning. Von Braun's time in Washington was short, retiring from NASA in 1972 before dying of cancer in 1977 at age sixty-five.

In the astronaut office, *Apollo 7* veteran Walter Cunningham became chief of the Skylab branch Flight Crew Operations Directorate. A lot of people credit Cunningham's input with helping to turn Skylab from an amalgamation of experiment proposals and ideas into a cohesive set of hardware that could do the job once it arrived on orbit. Other astronauts had held this particular post during the early days of AAP, but the prevailing attitude among the astronaut corps at the time was that the OWS was so far down the line that not a lot of work should be focused on it, when the lunar program was the big show.

In his autobiography, *The All-American Boys*, Walt Cunningham considers his work on Skylab for the two years he was part of the program to be his real contribution to manned spaceflight. Cunningham had already been known as a good manager, and he had worked closely with scientists while working at the RAND Corporation in the early 1960s. Because of this experience, Cunningham seemed to provide the correct skill set that Skylab needed when it transitioned from a drawing board program to hardware.

Cunningham was aided in his tasks by Story Musgrave, who was selected in 1967 as part of a second class of scientist-astronauts. The group of eleven astronauts again had broad scientific backgrounds with only one or two having any skills that might directly contribute to the Apollo lunar program. NASA had plenty of astronauts already. So the second group called themselves, in a humorously ironic fashion, the XS-11, or "excess eleven" as it became more commonly known. The glut of scientists didn't last long, though. By the end of 1969, five members from each scientist-astronaut class had either been asked to leave or resigned while deciding that the astronaut program was not their cup of tea. That left only twelve, and Cunningham had a few of them at his disposal in addition to Musgrave. Other scientist-astronauts such as Owen Garriott and Joe Kerwin were also involved heavily in *Skylab* development, and both men would get flight assignments as *Skylab* crewmembers.

Cunningham also got some additional manpower when the air force's MOL project was canceled. He inherited the seven MOL astronauts who were hired by NASA and put them to work where their skills and knowledge of similar systems were invaluable. The scientist-astronauts became the direct interface with the scientists hoping to fly their experiments on the workshop, while the MOL astronauts would help with configuration and hardware design. These efforts helped to get the work of two NASA centers and five major hardware contractors coordinated properly.

Skylab was initially assigned a launch date in December 1972, and efforts ultimately began moving toward that goal as the hardware began to come together. But the date would slip due to further budget cuts and a lack of enthusiasm by the Nixon administration. Administrator Paine would eventually resign from NASA in mid-1970 as his own goals for the agency didn't match the White House's. When further budget cuts came down, acting administrator George Low took up the job in Paine's place until James Fletcher was appointed NASA administrator later that same year.

The latest budget casualties were *Apollo 20* and the Saturn V production line. There also wasn't enough money to fly *Apollo 18* and *19* along with *Skylab* either. Low figured that of the programs under the budget ax, Skylab at that time had the potential of returning more useful science than a couple of more missions to the moon. So the hard decision was made to cut the lunar missions after *Apollo 17* and keep Skylab on track. Having two unused Saturn Vs from the canceled lunar missions might have allowed for a second OWS to fly, but funding to fly a second Skylab was never made available. A second Skylab workshop was built, but it was kept as an engineering backup only and never flown. It eventually became an exhibit at the Smithsonian's National Air and Space Museum.

Anatomy of *Skylab*'s Orbital Workshop

If the twenty-ton *Salyut* was considered big, *Skylab* was truly massive, as the space station weighed nearly one hundred tons when it finally flew. The largest structure was made up of the converted S-IVB, which was slightly longer than a contemporary tractor-trailer truck. The exterior of the S-IVB was covered with a micrometeoroid shield. This shield would serve a secondary role in passively controlling the thermal temperature of the station's interior. Primary power for the station would come from a large pair of solar arrays located on each side of the station, which would fold out once the station reached orbit.

At the front of the S-IVB was the airlock section, which contained two Apollo docking ports and a separate airlock for space walks. Located on top of the docking adaptor and airlock section was the ATM, housing the instruments needed for observation of the sun and the section's four dedicated solar arrays. Many of the experiments located in the ATM used film, so periodically the astronauts would be required to perform space walks to retrieve film canisters and replace them with fresh ones.

Attitude control of *Skylab* would come from a set of nitrogen-powered thrusters for fast movements and control-moment gyros in the ATM for sta-

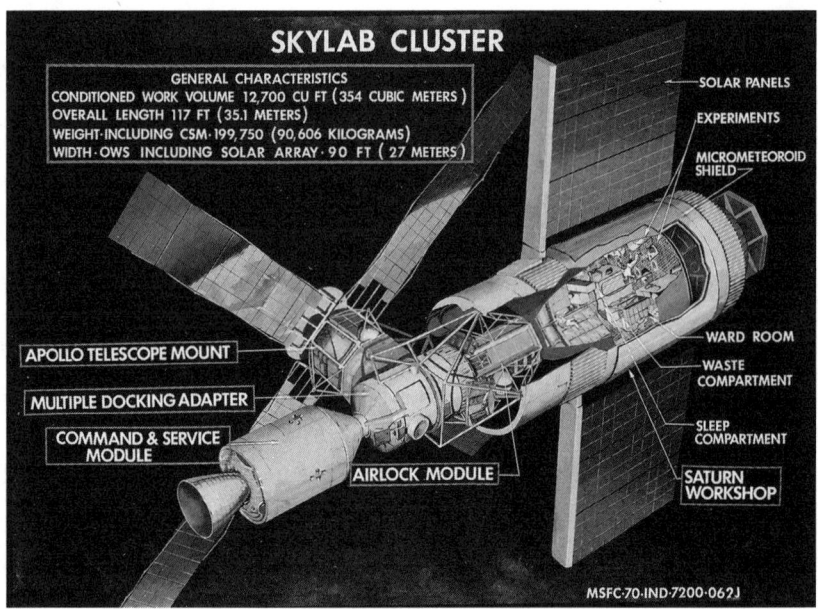

17. Illustration of *Skylab*. Courtesy NASA.

tion keeping and slower orientation changes. The control-moment gyros were designed along a similar principle to the gyrodynes used in *Salyut* but were larger in size. For most maneuvers, only the control-moment gyros would be used, in order to help save thruster fuel.

Like the Apollo spacecraft, *Skylab* would be pressurized to 5 psi in orbit. But there were medical concerns about breathing pure oxygen over a long duration, so a change was made to a two-gas system of 72 percent oxygen and 28 percent nitrogen. The internal pressure was kept low, since the Apollo spacecraft itself couldn't handle internal pressures higher than 8 psi and since a higher pressure would require astronauts to prebreathe pure oxygen to flush nitrogen from their blood prior to EVAs in order to help prevent the bends. Since the internal atmosphere in the lab was still only one-third of the pressure at sea level on Earth, sound didn't carry very well over long distances. So thirteen intercom system panels installed in various areas of the lab became an excellent investment once the station became operational.

Astronauts would launch into orbit on a modified Apollo CSM setup for long-duration missions. To do this, engineers removed some of the craft's fuel cells and associated cryogenic tanks, replacing them with storage batteries. Once docked, the remaining fuel-cell stacks would deliver their power to the workshop until their consumables were used up. A storage tank was installed in the service module to collect the waste water from the fuel cells. This was done so that water vapor from venting would not contaminate the delicate instruments of the workshop. Before coming home, the CSM would get its batteries topped off by *Skylab*'s power supply.

The supply of fuel to the service module's main engine was reduced since it was only intended for use in the deorbit burn, while the fuel supply for the spacecraft's reaction control system was almost doubled for rendezvous and docking maneuvers. One side of the command module was painted white to help with thermal control, since that side of the spacecraft would be constantly pointed toward the sun while docked with *Skylab*.

The CSM would dock with the workshop on a docking port in line with the station's longitudinal axis. A second docking port was provided on the bottom of the airlock module. This port would be used in an emergency and could accommodate a second CSM if needed. If a problem kept a docked CSM from returning home safely, NASA had a dedicated rescue CSM at its disposal. Two astronauts would fly the ship into orbit to retrieve the three-man crew on the station. The five astronauts would then ride home with three seated normally while the other two sat in couches in the spacecraft's lower equipment bay.

Skylab's docking port and airlock section also housed the control system for the ATM. Seated at a workstation, an astronaut could steer *Skylab* to aim its telescopes as needed at either the sun or other targets of opportunity. The airlock located in the "floor" of the docking compartment meant that the entire station didn't have to be depressurized during EVAs, but usually the third crewmember would stand watch inside the docked CSM in case an emergency arose.

For EVAs, *Skylab* astronauts wore slightly modified versions of the lunar suits designed for the Apollo program. But rather than using a self-contained, portable life-support system backpack, the *Skylab* suits would get their oxygen supply from a tether to the lab via the airlock. A safety lifeline tether was also provided. Breathing oxygen would be supplied through a control unit located on the astronaut's waist, and a secondary oxygen supply tank would be located on the astronaut's right hip.

The primary habitable volume of the station was in the S-IVB stage. The "floor" of the workshop was oriented vertically, while the airlock compartment's "floor" was oriented relative to the longitudinal axis of the station. The ows was divided into two main levels. The bottom-most level, known as the crew quarters deck, contained the living quarters for the astronauts. Each private room contained a sleeping bunk and blankets strapped to one wall and storage lockers for each crewmember's personal belongings along with a tape player. Music from each tape player could be routed to other parts of the station via the intercom system. Since *Skylab* had no washing machine, the supplied clothing and towels for each crewmember were thrown away once they got too dirty.

A ward room with galley table was provided next to a window with a view of Earth. The ward room housed the food freezer along with other supplies. The central galley table had both hot and cold water hoses for rehydration of drinks. To save valuable space on the Apollo csms, nearly all of *Skylab*'s food supply was launched into orbit with the workshop itself.

A fully enclosed bathroom with a wall-mounted vacuum toilet was also housed on this level. Solid and liquid waste were collected in separate bags. Each bag of fecal material was dehydrated and baked in special dryer ovens to remove moisture and help maximize storage space. Only small samples of urine were collected for analysis, and these samples were frozen for storage and the return trip. The remaining urine was collected in a disposable containment bag that was changed daily and discarded with the rest of the trash. A personal hygiene station was located next to the toilet, and here astronauts could use fresh water to wash their hands as if using a sink on Earth.

The center of the first floor housed a trash receptacle opening. Astronauts would place trash inside the container, close the inner hatch, and open the outer hatch to suck the trash into the s-ivb's empty liquid oxygen tank. The trash would remain there until the station was deorbited. The first level also included a shower in one corner. Crewmembers would enter the shower, pull the flexible curtain up to seal it with a ring on the ceiling, and spray water on themselves with a hose equipped with a spray fixture. A vacuum hose was used to remove the excess water droplets.

Half of the first level also housed much of the biomedical equipment used during the mission, including a negative-pressure device and the spinning chair used to test vestibular function. Control panels for electrical power and internal temperature were also located on the first level.

The second level was separated from the first by an open-structure floor webbing with a triangular pattern. This was a throwback to the wet lab design, as the webbing was designed not to impede the flow of liquid hydrogen while the s-iv b was operating as a fuel tank. Astronauts could get from the first to the second level of the lab through a central opening. As a side benefit, engineers designed a triangular foot restraint that could grip into the floor to anchor the astronauts if needed.

The second level housed some of the exercise equipment, such as a bicycle ergometer, a table for weight exercise, and a tape player for entertainment. It was also equipped with additional storage lockers. The twenty-foot-high ceiling for the second level was the inside top of the lhx tank, and the large volume provided by this twenty-one-by-twenty-foot area was wide open enough for all three astronauts to float around freely without bumping into anything. A complete ring of storage lockers were located along the wall midway up, and these became a zero-g running track for Owen Garriott. Other astronauts used them as gymnastics mats.

Astronauts found that they could locomote their bodies to a wall as needed and that air currents from the station's ventilation system would eventually move a crewmember to a wall if one tried to remain stationary in the middle of this cavernous area. So a firehouse pole provided by *Skylab*'s designers in the core of the station was usually stowed after a few days, once the crew got the hang of moving around in zero gravity.

The Science of *Skylab*

Since Skylab was the first dedicated mission of long-term manned spaceflight where the crew didn't have to spend most of its time actively flying the vehicle, science gathering was the priority. But the astronauts would not just be making observations and collecting data with their instruments; they would be experimental subjects as well. *Skylab* had an arsenal of biomedical equipment at its disposal.

Exercise would be a big key to fighting the effects of weightlessness; for that, astronauts had a few devices at their disposal. An isokinetic dynamometer weight table was provided. It was designed to provide simulated weight using electrical resistance. For cardiovascular exercise, a bicycle ergometer was used; scientists could monitor how well an astronaut performed by the amount of electrical current he generated from pedaling.

After the first two crews came home, it was determined that a treadmill might be more effective than the bicycle ergometer due to balance issues that cropped up during recovery. So a low-tech treadmill, designed by astronaut Bob Thornton and nicknamed Thornton's Revenge, was flown with the third crew. It was simply a sheet of slick Teflon material that the crewmember was strapped to in a standing position with a pair of elastic cords that gave a simulated weight force. Wearing socks, the astronaut could run in place on the sheet. It provided body loads similar to a regular treadmill, without moving parts.

The spinning chair tested vestibular function by spinning a person in one place while he moved his head. This was done to see if the activity could bring about an onset of space adaptation syndrome, a condition that tends to incapacitate astronauts with dizziness and nausea as their inner ear and eyes have a hard time adapting to the new environment. There is no way to predict with 100 percent certainty whether somebody will be prone to such a condition, and a lot of effort was expended during the Skylab program to try to understand it better.

Food was an important part of the Skylab mission since the astronauts' caloric and nutritional intakes would be monitored closely. The food preparations for Skylab were and still are the most elaborate of any space mission to date. Space missions flown to this point made do with camping-style food and dehydrated pouches that had to be reconstituted with water. But while the astronauts considered this form of cuisine tolerable for short-duration trips, there were concerns of low morale if a crew had to eat the same things on longer missions.

On *Skylab* the meal choices were expanded with canned foods that didn't require rehydration. Even cold-stored foods, such as ice cream and canned fruits, were available thanks to *Skylab*'s freezer. The only items that needed rehydrating were the drinks. Each astronaut had a food tray with slots for warming some of the food cans, and these were attached to the galley table during mealtimes. Each astronaut had a set of magnetic eating utensils that could be anchored to the food tray when not in use. Astronauts would regularly rotate who performed the kitchen duty to prepare the food at mealtimes.

Except for the small identification labels, most of the cans looked almost identical to one another. This caused a slight problem on the third manned

Skylab mission, as astronaut Bill Pogue accidentally heated up a can of ice cream intended for crewmate Ed Gibson. But rather than getting mad, Gibson simply put the can back into the freezer for a bit. When he pulled it back out, the ice cream had refrozen into a hollow ball. Ed then stuck some strawberries in the middle and enjoyed the first ice-cream sundae in space.

The meals selected for Skylab were a balancing act between nutrition, scientific data requirements, and the enjoyment of the crew. Each nutrient in the food had to be accounted for to determine what substances were being excreted in an individual crewmember's urine and feces. To establish a baseline of data both before and after the flight, astronauts began their special diets a week or two before launch and continued a week after returning.

Blood and tissue samples would be taken on orbit by each of the crewmembers, so even the astronauts without a medical background had to be trained in drawing blood from one another. Some blood studies could take place on orbit, but for the most part, the samples were preserved for the trip home. The astronauts weren't the only subjects of life science studies; pocket mice and fruit flies were also flown to the station for the crew to monitor their development. Two spiders were also flown and monitored to determine if development of spiderwebs would be different in a zero-g environment than on the ground.

The spiders were flown as one of the first experiments suggested by school students. Other student experiments were more sophisticated, and some were in areas that professional scientists hadn't even thought of before. Skylab was the beginning of many fruitful years of student-suggested scientific experiments flying in space. That relationship continued through the shuttle era, and student experiments also take place on the International Space Station (ISS) today.

One of the most important elements of Skylab's science gathering was the ATM. What started as little more than a retrofitted LM had become a sophisticated solar observatory that mounted eight different telescopes in a single housing. This arrangement would permit the collection of data from the visible-light to the ultraviolet spectrum and on X-ray wavelengths. Some instruments on the ATM were designed to study the sun as a whole, while others were designed to focus on specific regions or targets of opportunity. From the ATM control panel in the airlock module, an astronaut could see activity on the sun with up to four television monitors and focus on specific regions to collect data, depending on what feature was being studied. The entire workstation for the ATM looked about as complicated as a flight engineer panel on an old commercial jet airliner. In addition to the equipment used to monitor the cameras, it also contained controls for steering the station.

Instrumentation mounted in the ATM included two hydrogen-alpha telescopes designed to record light wavelengths emitted by hydrogen atoms in various forms. These telescopes could see things that were not observable in other light wavelengths. Although the sun is essentially a big ball of gas, it still has surface features; hydrogen-alpha filters produce the distinct deep-yellow images of the sun where these features are visible.

A white-light coronagraph was also provided. The corona is a cloud of gases around the sun that normally can't be observed unless light from the sun itself is blocked. A solar eclipse can also occlude the sun's disk to allow a corona to be seen, but only for short periods. While solar telescopes can use an occlusion disk to block the sun, the atmosphere can distort the effectiveness of the disk. Today, many ground-based telescopes can counteract atmospheric distortion with computers and special lenses to view the sun's corona, but in the early 1970s such technology was not yet available. *Skylab* would allow for the first long-term studies of the sun's corona.

Other ATM instruments included an extreme ultraviolet spectrograph and spectroheliograph in addition to other equipment that could detect ultraviolet wavelengths. An X-ray spectrograph and camera were also carried. Ultraviolet light and X-rays, which can be harmful to living organisms, are mostly filtered out by the atmosphere. Spectrographs are used in astronomy to help determine the chemical composition of stars and other heavenly bodies. Different atomic and molecular compounds produce different colors when their visible-light wavelengths are separated and analyzed. Since the sun is so hot and bright, it obscures almost everything in the visible-light wavelength. But ultraviolet light and X-rays can be analyzed to collect the same data. The spectrograph was designed to analyze parts of the sun's surface, while the spectroheliograph was designed to do the same thing to features found in the sun's chromosphere, a thin atmospheric region of the sun near its surface.

For Earth-observation studies, additional spectrographs, radiometers, scanners, and photography equipment were mounted in the bottom of the *Skylab* docking module near the second docking port. Scientific data collection was the key here as opposed to photographing specific points on the ground like MOL and Almaz. But astronauts also had access to hand-held photographic, television, and movie cameras.

Getting Hardware Ready for Flight

It was decided that a ground-based dress rehearsal of sorts should be carried out to help shake out any problems, since much of the equipment would be flying in space for the first time. So engineers devised a test called the Skylab Medical Experiments Altitude Test, or SMEAT. The test essentially involved locking three people up in a sealed environment for fifty-six days to conduct a simulated mission. Like an actual Skylab mission, the atmosphere in the SMEAT environment would be pressurized to 5 psi. The astronauts would evaluate the equipment, the mission plan, the food, and the test procedures to uncover any unforeseen problems. Additionally, the medical data collected would be used as a baseline comparison with the Skylab missions themselves.

Three Skylab support astronauts were selected to take part in SMEAT. The first two were MOL veterans Bob Crippen and Karol "Bo" Bobko. Dr. William Thornton, an XS-11 class member with experience in engineering and as a U.S. Air Force flight surgeon, was the third member. As per Skylab protocol, they would begin their preflight food diets a couple of weeks before "launch" and conduct a battery of medical tests on one another. The SMEAT experiment was conducted in the late summer of 1972.

The crewmembers did their jobs very well and uncovered issues in most everything tested. William Thornton became somewhat legendary during SMEAT. His physical build and extremely good shape made him perfect for uncovering problems with the bicycle ergometer, as he practically tested it to destruction, according to some accounts. Thornton's medical background also helped him to lobby the medical specialists that their one-size-fits-all approach to astronaut diets wasn't exactly a good idea, as people with different metabolisms can require a different caloric intake from one another. Thornton also proved that the one-size-fits-all urine bags didn't do the job either, as his urine output typically would exceed the bag's capacity. As far as dress rehearsals go, SMEAT was a valuable experience for almost all involved, and many credit its results as the reason why no major issues cropped up with *Skylab*'s biomedical equipment during its operational use.

At the Kennedy Space Center (KSC), rocket stages and hardware for *Skylab* began to come together in the vehicle assembly building alongside those for *Apollo 17*, as the NASA facility was the busiest it had been since 1968–69. Accommodations for *Skylab*'s Saturn IB boosters would mean that changes were in store for one of the launchpads.

During Apollo, Saturn IBs had flown from launchpads 34 and 37 of the Cape Canaveral Air Force Station. After *Apollo 7*'s successful launch, both pads were decommissioned, and useable equipment was repurposed for use at KSC's Launch Complex 39. Since the Saturn V was a much taller booster than the Saturn IB, the launch umbilical towers designed for it could not be used with the smaller rocket without modifications. The solution came in the form of a stilt pad structure built on the base of the launch tower so that the Apollo CSM and the S-IVB rocket stage of a Saturn IB would sit at the same height as a Saturn V. This stilt pad unofficially became known as the "milk stool."

The Saturn V pad used for *Skylab* was modified to accommodate testing and checkout of the lab prior to launch. Some supplies could only be loaded relatively late in the countdown, so a side door in the lab itself allowed for access at the pad. And modifications were made to the support structures to allow technicians to carry out the needed work inside. The side door was only used for launch preparations.

The *Skylab* crews were selected in 1971. The first crew was commanded by Charles "Pete" Conrad, a Gemini and Apollo veteran. Joining him were two rookies: scientist-astronaut Joe Kerwin, MD, and pilot Paul Weitz. The second *Skylab* crew consisted of veteran Alan Bean, rookie scientist-astronaut Owen Garriott, and rookie pilot Jack Lousma. The third crew was made up entirely of space rookies. Commanding the mission would be Gerald P. Carr. Joining him were scientist-astronaut Edward Gibson and pilot-astronaut William Pogue.

The crews were selected a little differently from crews of previous programs. During Apollo, chief astronaut Deke Slayton usually had the final say in crew selections, and he tended to favor test pilots instead of scientists. For Skylab, appeals were made to have two scientists fly on each mission, but in the end, it came down to a crew selection of only one scientist per crew. This meant that several qualified scientist-astronauts would not get to fly until the space shuttle began operations a decade later.

Launch Day

The first stacked *Skylab* Saturn 1B arrived at launchpad 39A in late February. On 16 April 1973, the *Skylab* workshop's Saturn V booster on its mobile launch platform began its long trek along the crawler way to pad 39B. Seeing two rockets stacked on both pads made for quite a sight at KSC. Two firing rooms at KSC's Launch Control Center would control the counts for both rockets simultaneously. The *Skylab* workshop would enter orbit first. If all went well with deployment of the workshop, the first *Skylab* crew would launch the next day to dock with the laboratory and set up operations. Pete Conrad's crew began their pre-mission diets and spent a couple of weeks in medical quarantine in anticipation of their mission. Everything seemed ready.

As with the Apollo flights, a crowd of spectators gathered at KSC to watch *Skylab* get off the ground. The usual VIPs were invited to the launch, but Pete Conrad and his all-navy crew also extended invitations to the 591 U.S. military POWs from the Vietnam War who had recently been freed from North Vietnam's Hanoi Hilton (some as recently as April 1973). Most were combat pilots, and some of these men had been held in captivity since the Vietnam War had begun. The POWs who were captured before July 1969 were not made aware that the United States landed a man on the moon until five months after it had happened.

Most of the astronauts had friends who flew combat over Vietnam. Conrad and his crew did not want to let the sacrifice of the POWs go forgotten or unrewarded. Among the former POWs who took up the invitation was air force colonel James Lamar, who was captured in May 1966 after bailing out of his heavily damaged F-105 Thunderchief. He watched the launch from ABC television's press facility at the cape.

On 14 May 1973 at 13:30 Houston time, the last operational Saturn V lifted off from the pad and rose into the cloudy Florida sky. Overcast conditions meant that spectators and cameras lost sight of the vehicle about one minute into the flight as it passed through a layer of clouds. Nobody was able to view what happened next, but the characteristic Saturn V rumble could once again be heard many miles away. After seeing the launch, Colonel Lamar was asked what he thought of the experience, and he replied, "That has got to be the most awe-inspiring sight I have seen in my life. That is something. Tremendous!"

All indications showed that the Saturn was performing brilliantly, and most of the data coming back from *Skylab* indicated the same. But in mission control at the recently renamed Johnson Space Center (JSC) in Houston, there was an indicator light showing that *Skylab*'s micrometeoroid shield had deployed a little over a minute into the flight. Flight controllers dismissed it as an erroneous reading, because otherwise the booster and the lab were following their preprogrammed course and performing well. Ten minutes after liftoff, *Skylab* was in orbit.

Where EGILs Dare

The *Skylab* controllers sent a series of commands to activate the station and deploy its arrays, as the vehicle passed within range of the tracking stations and communications aircraft on the first orbit. Data coming back indicated that one solar array was not giving power at all, as if it were missing; the second one was only giving a trickle of power, as though it were jammed partially shut. Even worse, the temperature inside the lab was slowly climbing past its nominal levels.

Analysis quickly determined that *Skylab* had indeed lost its micrometeoroid shield and that one solar array was likely gone as well. The shield also doubled as a thermal cover for the workshop. When the station reached orbit, the shield was designed to spring open like a larger tube held away from the surface of the OWS with a series of standoffs to keep it constantly shaded. But without the shield, the sun's rays would bake the workshop's structure and cook the insides. Even if the interior wasn't made totally uninhabitable, increased temperatures could degrade battery life, cause interior padding to off-gas toxic fumes, and potentially cause food to spoil. Without help soon, the Skylab mission was in danger of ending before a crew could occupy the station. The launch of the first crew, planned for the next day, was postponed indefinitely while engineers discussed both short- and long-term solutions.

For Skylab, the Apollo EECOM (Electrical, Environmental, and Consumables Manager) designation had been replaced with EGIL (Electrical, General, Instrumentation, and Life Support). Essentially, both jobs performed the same function, as the EECOM and EGIL were responsible for monitoring the items that keep equipment running and astronauts alive. But the EECOM would perform that job for the CSM, while EGIL would do it for the lab itself. Many of the Skylab EGILs were former Apollo EECOMs.

The quick fix for the temperature problem came from EGIL flight controller John Aaron. Aaron was already a legend at mission control for his quick thinking at critical times on several Apollo missions. To cool the lab's internal temperature, Aaron instructed the guidance flight controller to pitch *Skylab* down a bit, rotating the workshop's exposed surface partially out of the path of sunlight. Aaron would then monitor the power output of the ATM solar arrays to see if the lab had enough electrical power for its batteries. It was a highly unorthodox procedure and went against Mission Control protocol. But the flight director allowed it, and finally *Skylab* stabilized in an attitude that would keep the internal temperature from climbing too high. It would buy time for engineers to come up with more-permanent solutions.

Pruning Sheers, Parasol, and a Sunshade

When the problems were more fully understood, a plan of action began to take shape all across the NASA centers, while investigators tried to find out what caused the micrometeoroid shield to be torn away. Investigations would later determine that the shield had been poorly designed. While the lab was ascending through the atmosphere, a trapped air pocket began to expand; the resulting gap, in combination with aeropressure loads from the rocket's flight, caused the shield to tear away. When the shield went, it took one of the solar wings completely with it, and twisted metal jammed the other one shut. If the stuck wing could be freed, it would give *Skylab* 50 percent more electrical power generation than it had with just the ATM solar arrays.

Parallel efforts took place at the NASA centers to deal with both the solar array and the heating problem. Skylab backup mission commander Rusty Schweickart (a veteran of *Apollo 9*) and Joe Kerwin both took turns in Marshall's NBT to test procedures and equipment for freeing the trapped solar array. The plan was easier said than done, because while *Skylab* had been designed with EVA handholds around the ATM for the retrieval of film canisters, no such provisions were found around the rear of the workshop since there were no plans to conduct EVAs there. The closest EVA handholds were quite far from the site of the damage, so an astronaut would have extreme difficulty getting in position to cut any metal that might be holding the array in place.

It was also considered too dangerous to make any cuts with an astronaut close by, as the potential stored energy in a strap or attachment bracket could turn it into a projectile once it was cut. If an astronaut got hit, it could be a very bad day. Finally, engineers settled on a variation of a set of pruning sheers mounted on a telescoping pole, similar to the type used by workers to safely cut small branches on tall trees or electrical lines. A worker used a rope-and-pulley system to actuate the jaws of the cutter. The telescoping handle for the cutter was broken down into segments small enough to store inside the CSM cabin. The full-length pole would then be stacked together during the EVA.

For the thermal problems, engineers at both Marshall and JSC came up with solutions. The JSC team devised a folding parasol, or what essentially is a giant umbrella, much like what is used to keep picnic tables cool during outdoor gatherings. The parasol would be deployed through the scientific airlock on top of the station, as the airlock's opening was in the area exposed by the missing shield. Unfurling the parasol would be akin to threading a folded umbrella through the tiny opening of a car's sunroof and popping it open. The airlock and parasol deployment mechanisms were designed in such a way as to keep the interior of the station pressurized, while the shade itself was deployed from a modified scientific airlock experiment canister. The whole thing looked like a Rube Goldberg contraption, but if it worked, it would help save *Skylab*. The only minor problem is that experiments intended for deployment from the scientific airlock could not be used, as the airlock would be out of commission for the remainder of *Skylab*'s service life. Some of those experiments would eventually be placed outside the lab in racks and later retrieved during EVAs to service the ATM's film canisters.

Marshall came up with a twin-pole sunshade design. It resembled a giant awning. Plans for deployment called for it to be erected from two telescoping poles that were joined by a base plate to form a giant V shape. The baseplate was mounted on the truss structure used to support the ATM. Two reefing lines on the edges of the cover would pull the sides of the shade nice and taught over the exposed side of the OWS. These reefing lines would then be passed through parts of the docking module's support structure; the free ends were secured to the left and right sides of the ATM itself. The sunshade was capable of covering more of the lab's structure than the parasol but required a space walk to deploy it.

18. Astronauts Kerwin, Conrad, and Weitz (*left to right*) made up *Skylab*'s first crew. Courtesy NASA.

The decision was made to fly both the parasol and the sunshade, along with the cutter. The parasol would be used first, and the sunshade would be deployed later, after the stuck array was addressed. Work on the shades and the cutter took place almost around the clock. The final piece for the parasol was delivered by a NASA jet to KSC only a few hours before the launch. The crew of the first Skylab manned mission had nearly every resource of NASA at their disposal to save the laboratory. Long hours were put in by engineers, astronauts, and contractors, and everyone worked together toward a common goal. The stakes were very high.

"We Fix Anything!"

On 25 May 1973, almost two weeks after the crippled *Skylab* entered orbit, the program's first manned crew was ready for launch aboard *Skylab 2*. At 07:00 Houston time, the first manned Saturn IB to launch from pad 39B roared off its milk stool. A few seconds after liftoff as the rocket cleared the launch tower, Pete Conrad proclaimed, "[Clear] tower and Houston, *Sky-*

lab two! We fix anything! We've got a pitch and a roll program." And the rocket began its preprogrammed arc to achieve orbit. A few seconds later, Conrad proclaimed, "Boy, is that a smooth ride!" It was rather cloudy at KSC, and some low rain clouds began to move in over the viewing areas right before launch. So it didn't take long for the rocket to disappear into a cloud layer. Thankfully—unlike Conrad's last trip into space, on *Apollo 12*—no lightning strikes were reported that day.

Rendezvous took place almost on schedule; as the CSM closed with the lab, the crew conducted a fly-around inspection and were able to survey the damage for themselves. As predicted, the micrometeoroid shield was completely gone, and so was the left-side solar array. The right array was jammed shut and held in place by a one-centimeter-wide retaining strap that was firmly stuck to the rivets around it. Television pictures transmitted to the ground gave mission controllers their first look at the damage. A first attempt at trying to free the jammed solar array was made with Paul Weitz conducting a standup EVA outside the CSM while Joe Kerwin held his legs. From the pilot's seat of the CSM, Conrad tried to station keep with the lab while Paul tried to free the stuck array. It ended up becoming an exercise in frustration, though, as the only thing the astronauts really succeeded in doing was adding multiple four-letter obscenities to the flight-communications loop. The decision was made to abandon the task after a couple of tries.

Docking didn't go well, either, as the CSM's docking probe did not properly mate with the drogue assembly on the station. Backup procedures were tried, but to no avail. The last procedure that the crew tried before calling it quits was to climb back into their EVA suits, depressurize the cabin, open the docking tunnel, and remove the back plate from the probe mechanism to bypass some electrical connections that were causing the problem. Then, after the probe and drogue were aligned with one another visually, the CSM's thrusters were used to try to engage the docking latches. It worked, and the CSM was firmly docked with *Skylab*. The crew retired for a good night's sleep in the CSM at the end of a twenty-two-hour day.

The next day, the crew took air samples from *Skylab* to see if it was safe and entered the workshop. Their first order of business was to deploy the parasol, which they did. While it didn't quite cover all the affected area, the internal temperature began to come down over the next three days before finally stabilizing at a slightly high yet still-comfortable 26 degrees Celsius

19. *Skylab*'s missing micrometeoroid shield and jammed solar wing are visible in this photo. Courtesy NASA.

(78.8 degrees Fahrenheit). Initially, the crew would spend much of their time in the CSM or the airlock module, where it was cooler. But as the temperature came down, normal *Skylab* operations began, and the crew settled in.

Eventually, though, the jammed solar array had to be unfurled, as degraded battery performance from heat exposure and the station's power requirements meant that *Skylab*'s power-generation capability was running on the ragged edge. In the NBT at Marshall, Rusty Schweickart had been running simulations on ways to cut the strap on the jammed solar array, and a procedure was devised where one crewmember would attach a set of lines to part of the solar array's housing. These lines would be used to pull the erection boom free once the strap was cut. The second astronaut would anchor himself to the lab halfway up the side of the OWS and cut the strap with the pole device. The strap would still be about eight meters away from the astronaut with the pole, so it would require a bit of work for the cutter operator to anchor his position and manipulate the pole without coming off the lab. If the task did not succeed, although the first *Skylab* crew could still complete its mission with the power available, there was doubt that the second and third crews would be able to do so. Success or failure of this repair would determine *Skylab*'s future.

For the job, Conrad anchored the pulling lines while Kerwin operated the cutter. After a few tries, Kerwin was unable to get in a stable stance to cut the strap as he used one arm to hold onto an antenna on the lab and the second one to manipulate the pole. Conrad had trouble anchoring the lines also, as the holes in the erection beam's structure were a bit smaller than the ones used in the NBT simulations. Finally, Kerwin shortened his tether line by looping it around the antenna's bracket, almost like what a repelling climber does on a mountain; the resulting support from a taught tether line gave him a three-point stance that freed both of his arms for use in operating the cutter. Kerwin cut the strap, but the erection beam still didn't move. Conrad managed to wedge his body under the beam, and he pushed with all his might while Kerwin pulled on the lines as best as he could. It worked. The beam unfolded, and both men went tumbling off the surface of the OWS, with only their safety tethers and air lines keeping them from drifting away. The beam snapped into place, and the solar panels unfurled just as they were designed to.

After that, things settled down rather nicely as the mission returned to an almost-normal routine. The crew conducted their normal experiments and solar observations. Conrad and Weitz conducted one more EVA to perform a regularly scheduled changing of the ATM film. Undocking was uneventful, and a final fly-around inspection of the lab revealed that everything seemed okay for the next crew to occupy the station. The parasol had performed well and was still doing its job, so there was no need to erect the twin-pole sunshade on the first mission. It would fall to the second crew to do that. At the end of a twenty-eight-day mission, the crew of *Skylab 2* returned home successfully.

20. *Skylab* as it appeared after deployment of the parasol and freeing of the trapped solar wing. Courtesy NASA.

21. *Skylab*'s second crew of Garriott, Lousma, and Bean pose next to the ATM control panel inside a training mock-up. Courtesy NASA.

The Second Crew

The second crew rocketed into orbit on 28 July 1973, a little over a month after the first crew returned to Earth. Rendezvous and docking for the second crew didn't go quite as planned, though, as one of the CSM's reaction control thruster quads went off-line due to a fuel leak. The craft still had three more thruster quads, so docking could still take place. But with a loss of one-fourth of the ship's braking-thruster capability, it took a little more finesse to finish the rendezvous and docking maneuvers. Still, Alan Bean managed to execute a nearly perfect docking, and the *Skylab 3* crew set up shop in the station.

A few days after docking, a second thruster on the CSM began to show signs of a fuel leak. Two sets of the CSM's four thruster quads were enough to keep the spacecraft under control when returning home. But if a third one went out, the spacecraft would be unable to maneuver as effectively in orbit for either the undocking or the deorbit burn. So launch preparations were accelerated on the final Skylab mission's Saturn IB booster and the rescue CSM, in case it was needed. The fuel problem eventually stabilized with only two reaction control system quads having to be isolated from the spacecraft's fuel supply. The rescue craft was ultimately not needed.

The second crew opened up shop and began normal operations. The one major event that they had to accomplish early on was erection of the twin-pole sunshade. This was done on the first space walk. Bean's crew had practiced the procedure in the NBT, so it pretty much went off without any major problems. Additional EVAs were carried out for normal ATM maintenance along with the placement and retrieval of some scientific experiments designed by astronaut Don Lind originally intended for the scientific airlock.

Alan Bean's crew managed to double the endurance record set by Pete Conrad's crew while also doubling the scientific output. Upon return to Earth, the *Skylab 3* crew had spent fifty-nine days in Earth orbit. Healthwise, they seemed to be in pretty good shape thanks to the exercise regime. However, none of the crew's bodies made any red blood cells during the flight, and this worried the doctors a little. Some extra time was spent readapting to Earth gravity after the flight, since nobody had been in space that long before. Owen Garriott also did a little extracurricular "adapting" of his own, as he made arrangements to have a milkshake delivered to him on the recovery ship to give him brief relief from the mission's diet.

The Final Crew

The third manned crew was given a go-ahead to try to set a new three-month endurance record. But there was a minor problem with supplies. While *Skylab* still had plenty of oxygen and water to support a third crew, food was an issue, as the original provisions only called for a slightly larger supply of food than what the second crew had. Even with rationing, it could not be extended to over eighty days. To help with this, *Skylab 4* was sent into orbit with some energy bars, similar to the type used by campers and outdoorsmen. The bars would be cycled in as part of the normal food-plan schedule, with them being eaten every third day; they were included as part of the pre-mission diet as well. The bars did the job, keeping up with the crew's vitamin, nutrition, and calorie intake requirements, but they didn't do much for the taste buds.

Skylab 4 launched into Earth orbit on 16 November 1973. Unlike the two previous launches, which had been under cloudy conditions, the Saturn IB carrying the *Skylab 4* crew lifted off into a clear-blue Florida sky and was visible for miles. The mission was off to a great start. Docking occurred without a hitch at the end of the day, and the crew was greeted by three "occupants," as Al Bean's crew had played a prank on the newcomers. Before leaving, Bean's crew had stuffed three sets of *Skylab* clothing with bags to make three human-shaped dummies and strapped these "crewmembers" to the station's exercise and medical equipment. So *Skylab* now had six human-sized occupants. Gerald Carr's crew took the joke in stride as they got to work.

The flight surgeons on the ground scolded the crew for one minor breach of procedure early on. One of the astronauts experienced a bout of nausea, and the other crewmembers tried to keep it hidden by flushing the vomit-filled bag down *Skylab*'s trash receptacle. The crew made mention of it in the tape relay instead of during an on-air transmission in the hope it wouldn't be discovered, but it was. Normal procedure called for the bag of vomit to be returned to Earth with the rest of the bodily waste at the end of the mission.

This incident paled in comparison to what came next. Due to the combined workloads that the first and second crews were able to accomplish by the end of their missions, flight controllers and researchers packed the

22. *Skylab*'s final crew of Carr, Gibson, and Pogue pose for their portrait.
The display model behind them reflects the success of the previous on-orbit repair jobs.
Courtesy NASA.

mission plan for the third crew with more experiments and tasks than ever before, believing that they should have no trouble getting up to speed with previous experience to draw on. But there were problems almost immediately. The crew fell behind on the time line, and attempts to catch up did not work out too well. Mission control backed off the pace somewhat, but after a few days, the problems began to crop up again. Finally, the crew said enough to the whole situation and took an unscheduled day off, followed by a slow ramping up of their experiments step-by-step.

While the incident did result in a few bruised egos on the ground, the lessons learned were good ones. But the press had a field day with the incident by declaring it a "mutiny," while mission managers downplayed the whole thing. Skylab was NASA's first experience with long-duration space operations, and the learning went both ways with the astronauts on orbit and the flight controllers on the ground. While hardware testing tends to take priority, sometimes how man interacts with it is not as clearly understood, especially by people who have never flown in space. Ground controllers realized they should have backed off much more when problems began to crop up early on. Everybody involved was trying to get back on schedule and working too hard at it. But trying harder led to more mistakes being made. So the crew kept falling behind, until a breaking point was reached.

After the incident, the work schedule was a bit more relaxed as the astronauts did their assigned tasks at their own pace. By the end of the mission, the astronauts were getting more work done on a daily basis than the previous crews, and the third mission delivered a mountain of scientific data. The lessons learned by the end of the mission were valuable ones. Another bit of good news for the medical people was that the crew's bodies began making fresh red blood cells in orbit. Without that, future long-duration space missions might not have been possible.

Highlights of the *Skylab 4* mission include the first on-orbit filming of a solar flare formation as Ed Gibson tracked the flare from the moment he first detected it on the sun's surface. Pictures and drawings were also made of comet Kohoutek, the first comet observed by men in space. Both shirt-sleeve and space suit tests were also conducted of the Astronaut Maneuvering Stability Unit (AMSU) in the large volume of the OWS. The AMSU was a nitrogen thruster–propelled backpack intended for controlled, untethered spaceflight by an astronaut. The AMSU performed brilliantly in the tests, although it was never utilized during an actual EVA. Some elements of its design found their way into the Manned Maneuvering Unit (MMU), which flew in space a decade later with astronaut Bruce McCandless at the controls.

The third *Skylab* crew returned to Earth on 8 February 1974 after spending eighty-four days in space. As rookie crews go, they did a phenomenal job. But none of the astronauts would fly in space again, as all retired from NASA before the space shuttle began flying. There had been hopes that a fourth *Skylab* crew could be flown, as there was enough hardware in the pipeline to potentially support such a mission. But NASA wanted to focus its finances on developing the space shuttle and to use the remaining Apollo hardware for a docking mission with the Soviets. *Skylab*'s orbit was high enough that it was originally predicted to stay aloft until the early 1980s, when the space shuttle began its first flights. While *Skylab* was out of food, it still had enough oxygen and water on board to support continued operations, and the control systems were still operational for the most part. But occupation by future crews wasn't meant to be.

Skylab's orbit decayed faster than expected as increased solar flare activity caused Earth's atmosphere to expand. It was predicted that *Skylab* would reenter and burn up by 1979. Plans were made to use the shuttle to boost *Skylab*'s orbit. Astronauts Fred Haise and Jack Lousma trained for a possible *Skylab* rescue mission to utilize a special rocket stage remotely docked to *Skylab* from the shuttle in order to either reboost the station or deorbit it safely, but delays in the shuttle program caused the mission to be cancelled.

Flight controllers reactivated the lab as preparations were made to at least try to steer *Skylab* toward a reentry over water rather than having it come down completely uncontrolled. Mission control commanded the lab's attitude to change depending on if they wanted minimal or maximum air drag. The station was also commanded to tumble during its very last orbits to help it break up into smaller fragments early in its reentry and lessen the likelihood of large debris making it to Earth's surface intact.

Skylab ultimately came back to Earth on 11 July 1979 after spending 2,249 days in orbit. Most of the debris that survived reentry ended up landing harmlessly in the Indian Ocean. But some pieces ended up landing in sparsely populated Western Australia near the city of Perth. The chances of getting injured by debris falling from orbit are very remote, and nobody in the debris path was injured. Some large pieces of debris were recovered, and they can be found in museums. But other debris may still be lying undiscovered in Australia's western outback to this day.

The Skylab program taught a lot of valuable lessons. But it would be decades before NASA could completely exploit what it had learned and apply those lessons to a future space station. Some of Skylab's lessons would also be temporarily forgotten as NASA managers turned their attention toward other goals and interests. For the foreseeable future, the Soviets would have a monopoly on long-duration spaceflight, while NASA focused on the space shuttle.

5. The Rocky Road to Salyut Success

Problems, Problems . . .

After the *Soyuz 11* tragedy, the Soviets worked hard at getting another DOS station into Earth orbit, but it would take awhile as Ustinov's mandate to fly Soyuz only with pressure suits required time to implement. In addition to the upgrades required for using pressure suits, engineers used the delay to make some additional changes to the Soyuz vehicle in order to improve it for long-term storage. With the way the Soyuz assembly line was set up, equipment changes could be made pretty easily in order to configure the craft for a *Salyut* flight or for a stand-alone mission.

The most visible items removed from the redesigned Soyuz ferry were the solar arrays. In their place, engineers fitted a set of storage batteries. As a result, the craft could only operate in a powered-up state for about three days before having to come home, but the lightened load meant that the vehicle could launch into a higher orbit. A higher orbit meant that a Salyut could potentially operate longer with less of a need for orbital reboosts.

Everything seemed to be ready by mid-1972 to fly another station. Leonov and Kubasov (finally given a clean bill of health by doctors) were assigned as its first crew and trained to fly a mission that they had planned to fly before, but this time, they would fly as a pair instead of a trio. Pyotr Kolodin's prediction that he would never fly in space ultimately came true, as there was no room for a third crewmember in a spacecraft now capable of only fitting two people with pressure suits. Although he remained a cosmonaut for many years and was assigned as backup to other missions, Kolodin never got a chance to fly in space and eventually retired from the program.

The DOS-2 station was identical to the first one and had the name *Salyut 2* painted on its side, yet it would not be known by that name. The Proton rocket carrying the DOS station lifted off on 29 July 1972, but the launch vehicle experienced a second-stage failure about three minutes into the flight and the station came crashing back to Earth. Since it hadn't even reached orbit, the Soviets made no acknowledgment that it had flown and didn't even assign DOS-2 a Cosmos mission number. The failure of the Proton also meant that the relationship between Mishin and Chelomei was becoming more strained than ever.

Mishin wasn't having much better luck with his rockets either, though, as the fourth attempt to launch the N1 moon rocket again ended in failure on 23 November 1972. The rocket disintegrated in flight due to pogo vibrations. While repercussions were not immediately forthcoming against Mishin for yet another failure, the Kremlin began to take a hard look at the man in charge of TSKBEM.

The next flight of a Soviet station would come almost five months after the fourth N1 failure, as Chelomei's Almaz station was finally ready to fly in early 1973. The Soviets hoped that they would have this second station in orbit by the time the Americans were ready to launch *Skylab*. If the Almaz successfully made it into orbit, it would be called *Salyut 2* to disguise its military purposes. To differentiate the Almaz stations from the DOS stations, they were referred to by the acronym OPS, which stood for "orbital piloted station." The first Almaz would be known internally as OPS-1.

On 3 April 1973 a Proton carrying the OPS-1 launched successfully. Twenty-four hours after it achieved stable orbit, the TASS news agency announced to the world that it was called *Salyut 2*. Initial plans were to launch a crew to the station the next day. But prelaunch preparation of the Soyuz launch vehicle uncovered a problem, and the rocket would not be ready for flight until early May. *Salyut 2* operated normally in an unmanned state for almost two weeks until 15 April, when controllers noticed that the cabin began to depressurize and the station began tumbling. Over time, the other onboard systems failed, and *Salyut 2* was dead in space.

Initially, investigators felt that it was either a design or construction fault that caused the station to depressurize. Others theorized an onboard fire had caused the catastrophe. However, later analysis determined that *Salyut 2* may have been hit by debris from remains of its Proton launch vehicle's third stage. This determination was made possible by tracking data supplied from NASA to members of a Soviet team making plans for a possible joint mission with the Americans during a scheduled visit to Houston.

Three days after *Salyut 2*'s launch, NASA was tracking an object believed to be the Proton third stage, when it suddenly disappeared and they detected twenty-one smaller objects in its place. With further analysis, it was determined that some of these new objects could have crossed *Salyut 2*'s orbit, and a direct hit would have been devastating. If the station had been occupied, the crew would likely have been dead before they could have evacuated.

On 11 May 1973, three days before the launch of *Skylab*, the Soviets launched another Salyut station as part of the DOS program. This was DOS-3, and the new vehicle was much more capable than the first two DOS stations. Additional scientific equipment was also placed on board, making it heavier than its predecessors. The Proton rocket wouldn't be able to loft it into an orbit as high as the first Salyut, and the station would need to raise its own orbit with onboard engines before a crew could be launched to occupy it.

The control system for orientation would rely on a set of sensors capable of measuring the station's orientation relative to Earth's ionosphere. Because the sensors were very sensitive, only small corrective burns could be used, or the sensors would be overwhelmed with conflicting data. But due to a mix up at the control center, the thrusters were commanded to fire at maximum instead of minimum power. The guidance system overcorrected and tried to reorient itself in a back-and-forth tug-of-war, resulting in the new station ending up tumbling in space with its fuel reserves exhausted. Since the station had reached orbit and was tracked by Western observers, the Soviets designated it Cosmos 557 to disguise the failure.

The investigation into the loss of the DOS-3 station concluded that it was a lack of coordination between the design bureau and the control center that resulted in its loss. It was the third station lost within a two-year period. The use of ionic sensors in the design was perplexing since such equipment had never been flown on a test vehicle before, let alone an operational one. The Kremlin was not happy with this string of failures, especially when one of them was caused by an apparent comedy of errors.

Steps were taken to help ensure that this sort of thing did not happen again. A few engineers at TSKBEM were transferred from their postings, and the flight director in Yevpatoria during DOS-3's failure was dismissed. A big part of the problem with Soviet flight control was related to the fact that it was decentralized. A command would be passed from the control center to the army-operated tracking-and-control stations to feed up to the spacecraft, which would then read back the results. It is kind of like a commander of a ship issuing an order that gets repeated down the line until carried out. But with a spacecraft's orbit, the communications passes were very brief, and time was lost in relaying commands. This relay resulted in needless delays, when timing was critical.

Former cosmonaut Aleksei Yeliseyev was appointed as the new flight director, and his task was to streamline the control process by transferring command and control responsibility from the army stations to a centralized location. For the civilian missions, a new control center in Kaliningrad was built, and it would be modeled on NASA's mission control center in Houston. Yevpatoria would continue to be used for military Salyut missions and as a backup for civilian flights. It would take a few months to build the new center, but it would be ready when the next DOS was. The center was known as the Moscow Mission Control Center, but it is more commonly known today by the acronym TSUP.

More scrutiny was in store for the leadership of TSKBEM as the Kremlin took a hard look at Mishin's role in the loss of the DOS-3 station. While the formal report that led to the engineering shake-up didn't specifically name Mishin, the general feeling was that the chief designer's management style was largely to blame for the failures encountered by the bureau since his appointment. Granted, he had a lot on his plate with the lunar program, the Soyuz redesigns, and the Salyut program since he took over from Korolev. But a lot of stories began to circulate, and they were not praiseworthy of Mishin.

There were other concerns, as well, that were not related to just the operational failures. Before the DOS-2 launch failure in 1972, Mishin and Chelomei got together and drafted an agreement with one another. After the first four DOS stations had flown, Mishin wanted to turn over control of the entire Salyut program to Chelomei's bureau. Mishin still considered it a distraction to his engineers, so he wanted to rid his bureau of the DOS program. That way, TSKBEM could focus all its efforts on the N1 booster redesign and flight-testing. This went against proper procedure, as Mishin had not consulted his deputies on the matter before drafting the agreement. Many of Mishin's deputies who had been involved in DOS from day one drafted a strongly worded letter to the Central Committee expressing their extreme displeasure when they found out about this agreement.

As a result of the letter, the Kremlin sent Ustinov to meet with Mishin. Ustinov "encouraged" Mishin to drop the NI program completely, focus all of TSKBEM's efforts on the DOS program, and develop a more sophisticated space station capable of longer-duration flights. If Mishin didn't agree to this idea, his days as chief designer of Korolev's bureau were numbered. Mishin agreed to this bit of arm-twisting and kept his job, while Ustinov succeeded in getting the agreement with Chelomei nullified, thereby striking yet another blow to his longtime adversary.

After over two years of redesign and delay with no manned missions, a manned Soyuz spacecraft finally left the ground successfully on 27 September 1973. *Soyuz 12* was the first manned flight test of the new Soyuz ferry. Due to the short duration of its test mission, the Soviet news agency TASS announced the mission prior to the launch, lest Western observers take note of the brief mission and classify it as a failure. The cosmonauts selected for this mission were Vasili Lazarev and Oleg Makarov. *Soyuz 12* performed all its objectives without any real problems over a two-day period. The new Soyuz ferry would be ready when the next Salyut was.

Soyuz 13 flew in December 1973 as a further test of the revised Soyuz. But this particular vehicle contained solar arrays to power a scientific mission lasting one week. *Soyuz 13* carried the Orion 2 Space Observatory and other telescopes in its orbital module. The cosmonauts selected for the flight were air force cosmonaut Pyotr Klimuk and scientist-cosmonaut Valentin Lebedev, who was solely responsible for operation of the telescopes. *Soyuz 13* was touted in the Soviet media as the first flight of a dedicated space observatory, which conveniently glossed over the fact that the first Salyut was itself a space observatory until most of its instrumentation was crippled by the launch-cover failure. Celestial observation was not *Soyuz 13*'s only mission, though, as the cosmonauts also performed medical tests on one another and conducted protein-growth experiments.

The *Soyuz 13* mission was the first flight controlled by the new center in Kaliningrad. The mission went very well, and no major problems were uncovered with either the spacecraft or the TSUP. This mission also marked the first time that both Soviets and Americans were in orbit at the same time, as the *Skylab 4* astronauts were deep into their eighty-day mission during the *Soyuz 13* flight. No attempts were made to contact each other's spacecraft, as neither side had the proper communications equipment to do so. But both missions did observe some of the same celestial bodies with their instruments, including comet Kohoutek, which made its closest approach to the sun on 28 December 1973.

The *Soyuz 13* mission would be the last flight with Mishin in charge. Although Mishin had seemingly secured his position by agreeing to focus on DOS station development, the Soviet leadership decided that he should be replaced anyway. It is likely that the decision to do this was delayed until after the Soviet space program had some successes under its belt so as not to make it appear to Western observers as though Mishin was being removed due to failures. On 22 May 1974 Mishin was removed from his post as chief designer, and Valentin Glushko was appointed in his place. Mishin wasn't present for the management change. He was in a hospital when Alfansayev and Glushko both made an unannounced visit to the design bureau's offices to announce the change to its staff.

This management change likely struck many in the TSKBEM ranks as a bit of a sad irony, given that the design bureau once belonged to Korolev, whose contempt for Glushko was well known. Regardless, Korolev's former bureau finally had the strong management it had been missing since the original chief designer's death. The change came as no surprise to many in the know, given the problems of the past eight years. Glushko already had his own design bureau for building rocket engines, and he would merge TSKBEM with it. The new bureau would be called NPO (a Russian acronym meaning "Research and Production Association") Energia. The primary near-term goal for NPO Energia was to continue the space station successes that started with the first Salyut. As a long-term goal, Glushko would also focus his efforts on a new heavy-lift booster that would totally replace the fatally flawed N1 design. This new booster would make use of cryogenic propellants (LHX and LOX), something that Glushko publicly considered too risky when the N1 specifications were drawn up a decade earlier. The new booster would also be capable of launching a manned space shuttle, similar to what NASA had begun developing.

Mishin would ultimately be appointed as a professor of space rocketry technology at the Moscow Aviation Institute and went on to help to educate the next generation of engineers in the theory and mechanics of rocketry and spaceflight. It seems as though teaching life suited him. While he displayed a brash attitude and stubbornness with politicians, other engi-

23. Valentin Glushko merged Korolev's design bureau with his own.

neers, and officials, he was much more tolerant and supportive of his students, conveying his years of experience to them willingly. Mishin went on to publish several teaching papers and directed the school's Department of Space Systems and Rocket Design. The school's laboratory facility was named in his honor after his death in 2001.

Salyut 3, Almaz Flies

Meanwhile, over at TSKBM, final preparations were underway to launch the second OPS Almaz station. It would be almost identical to the OPS-1 station (officially designated as *Salyut 2*) that had been lost the previous year and was designed according to the layout that Chelomei had come up with when he first had the project green-lighted. There were some new creature comforts loaded onto this station that the first DOS, *Salyut*, didn't have. For starters, OPS-2 had actual sleeping berths on board, so cosmonauts didn't need to sleep inside the Soyuz. The round-the-clock work-and-sleep schedule originally intended for the Almaz was abandoned, because there were only two crewmembers instead of three on orbit. Both cosmonauts would be awake at the same time, so the concerns of one cosmonaut waking another during a mission were minimized. The exercise-equipment suite was improved with a decent treadmill. It would be put to good use, as the plans were for each astronaut to devote about two hours a day to exercise.

More critical to long-duration spaceflight, though, was a system for reclaiming and recycling water, which could collect water from the humidity in the air as well as on surfaces and condense it back into a state for use in electronics cooling and consumption by the cosmonauts. While it was not a totally closed-loop system, it was a start; even though missions still required regular infusions of water, the new equipment helped to minimize what was needed, freeing up valuable weight and room for other cargo on the Soyuz.

The station also had an onboard shower with a flexible, plastic ringed cover capable of totally enclosing a cosmonaut using it. It was similar to what flew on *Skylab*. The *Skylab* shower was an interesting experiment but not very effective, and no shower has since flown on American vehicles. The Soviets seemed to have better success with the shower and incorporated one in all its stations from *Salyut 3* to *Mir*.

The OPS-2 station, known to the world as *Salyut 3*, rocketed into orbit on 25 June 1974 and had a nominal deployment. The Soyuz rocket planned

Almaz OPS-2 (Salyut 3)
docked configuration

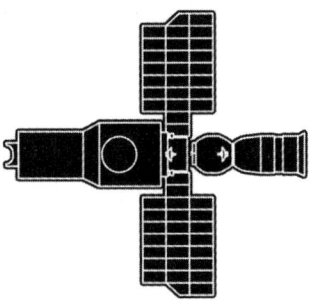

24. *Salyut 3* shown docked with a Soyuz ferry. *Salyut 5* had the same configuration.
Courtesy of the author.

to fly the first crew was also ready to go on time, and it flew into orbit on 3 July. Commanding the mission was *Vostok 4* veteran Pavel Popovich. Joining him on this flight was Yuri Artyukin, a rookie to space but not to the cosmonaut ranks. Artyukin was in training to fly the *Voskhod 3* long-duration space mission when the Voskhod flights were cancelled in favor of Soyuz spacecraft development. Artyukin had only just celebrated his forty-fourth birthday when he launched to *Salyut 3* and was slightly older than Popovich.

Since *Salyut 3* was a military space mission, it was controlled from the Yevpatoria control center in the Ukraine instead of the Kaliningrad facility. Western observers deduced that it was not a civilian Salyut since it used military communications bands instead of the normal civilian ones. Reportedly, the cosmonauts took part in some Earth-photography research experiments to help locate natural resources within the Soviet borders. Some tests of the station's Agat camera system also apparently took place, as some reference targets were set up at the Baikonur Cosmodrome for the station to photograph during orbital passes overhead. All indications are that the Agat system worked well, as did the onboard film-development system and the television system to wire the photographs back to the ground. *Salyut 3* also included a special reentry canister on the front of the station. At the end of an Almaz mission, this canister would be used to bring home up to 120 kilograms of developed film and other items for further analysis.

Ultimately, *Soyuz 14*'s stay on *Salyut 3* only lasted about two weeks, and the crew returned home after over fifteen days in orbit. Due to the short duration of the mission and the two-hour-a-day exercise regime, both cosmonauts were able to climb from their spacecraft without any assistance from the recovery forces. While the mission was not well publicized, it was an important milestone in Soviet spaceflight, because for the first time, a crew had launched to, occupied, and returned from a space station successfully. The new equipment worked as advertised, and Chelomei was proud of what his design bureau had achieved, even if it couldn't be publicly acknowledged. According to reports, *Salyut 3* had six more months of provisions on board to support additional crews.

Indications are that *Salyut 3*'s first occupation was a shakedown test of equipment as opposed to a fully operational Almaz mission. It would be about a month and a half before the next *Salyut 3* crew was scheduled to fly. The *Soyuz 15* crew consisted of Gennadi Sarafanov and Lev Dyomin. Both crewmembers were rookies, and little is known about them. Dyomin was the oldest Soviet cosmonaut to fly, to date, as he was forty-eight years old and held an engineering degree from the Soviet Air Force Academy in addition to his rank of colonel in the Soviet Air Force. By comparison, Sarafanov was only thirty-two years old.

The rocket carrying *Soyuz 15* lifted off on 26 August 1974 and entered orbit with no problems. Rendezvous occurred as normal, but when the spacecraft closed to within 350 meters of *Salyut 3*, problems began to develop. The Igla rendezvous system should have made a series of low-power thruster firings to continue a slow approach to the station, but a glitch in the system made it think the station was still twenty kilometers away. The Igla commanded the Soyuz to perform a long burn instead. Sarafanov aborted the approach within a distance of forty meters since the closing speeds were far too high for a safe docking, and some accounts say the two spacecraft just barely missed one another as they passed like two aerobatic planes performing an air-show routine. One additional attempt was made at a manual approach and docking. But the Soyuz exhausted its fuel reserves, and the mission was aborted after only two days in orbit.

A post flight investigation revealed some serious problems with the Igla, and it was felt that the entire system should be overhauled and checked before the next flight. The Igla on *Soyuz 15* reportedly was acting like it was mis-wired as it was doing long burns when it should have done short ones and vice versa. The cosmonauts should have been absolved of blame. But instead, they were given reprimands, and neither man ever flew in space again. Due to the investigation, no Soyuz ferry was available for a further attempt to dock with *Salyut 3*. A month later, the *Salyut 3* film return canister was successfully ejected and made a normal reentry and recovery within Soviet territory.

Engineers also conducted a test firing of the station's 23 mm cannon. While the gun could be fired with men on board, it was decided that it would be best not to risk it since the resulting jolt might cause hidden damage to the station. Plus, the hard mounting of the gun meant that the station's orientation had to be changed to aim it at a specific target. The gun fired apparently with no problems, but it isn't known if the gun fired on a target or not.

After recovering the film canister and firing the gun, engineers on the ground used the remaining few months of *Salyut 3*'s life in orbit to see how well its systems would hold up over time. *Salyut 3* was ultimately commanded to deorbit and reentered Earth's atmosphere over the Pacific Ocean on 24 January 1975. There would not be a long wait for the next manned station, though, as a month before *Salyut 3*'s fiery destruction, a new DOS station finally made it into orbit without problems. This was *Salyut 4*.

Anatomy of *Salyut 4*

Salyut 4 (DOS-4) was almost identical to the failed DOS-3 station. The biggest external difference from DOS-1 was the new solar array configuration. DOS-3 and *Salyut 4* were fitted with three large arrays capable of independently tracking the sun regardless of the station's orientation. With these new arrays, which were based on designs planned for Chelomei's TKS spacecraft, the station could orient itself in almost any direction while continuing to generate power. Yet the power output of the new station was not much better than that of the first Salyut with a solar array–equipped Soyuz ferry docked with it. Onboard power was still at a premium. For waste heat and regulation of internal temperature, *Salyut 4* made use of a more sophisticated coolant system than DOS-1, and it could operate in a wider range of temperatures and for a longer period of time.

Internally, *Salyut 4* was much the same as DOS-1, although some improvements were made. The new guidance and navigation systems allowed for independent flight with less input from ground control, while also reducing the workload of the cosmonauts on board. The systems also helped to reduce the fuel consumption, extending the onboard supplies. *Salyut 4* was the first Soviet station to be equipped with a telex printer to send up instructions from the ground without the need for a cosmonaut to write down commands sent by voice. *Skylab* made use of a similar system; in both cases, the printers paid dividends, as instructions could be sent up while crewmembers were either asleep or engaged in other activities.

With the success of *Skylab*'s ATM in its solar observations, the Soviets sought to achieve similar feats in solar astronomy. In the space originally designed for the Agat camera system of the Almaz, *Salyut 4* made use of a purpose-designed 25 cm focal-length solar telescope. Observation of Earth's atmosphere would also be a big part of the *Salyut 4* mission, so several spectrometers and a photometer were fitted to provide data on the composition of Earth's upper atmosphere. The data collected from these sensors during *Salyut 4*'s mission proved quite valuable in the study of Earth's ozone layer when discovery of the ozone hole over Antarctica was made in 1985.

Additional onboard sensors for upper-atmosphere study were fitted to measure the density, composition, and temperature of gases encountered by *Salyut 4*. It was felt that this data would help in the design of future spacecraft by minimizing radio interference caused by encounters with ionized gas, as the space station can become electrically charged (similar to producing a static shock by walking over a shag carpet in the dry winter months). For celestial observations, *Salyut 4* was equipped with an X-ray spectrometer and telescope system. *Salyut 4* was also the first spacecraft of any type to fly a light nuclear-isotope spectrometer, which had the capability to measure both the chemical and the radioactive composition of cosmic rays.

Salyut 4 also contained a full suite of exercise and testing equipment to gather data on the cosmonauts and how they adapted to weightless conditions. This included a rubberized-leggings system called Tchibis (a Russian acronym meaning "negative pressure," which was used by the system). The leggings were designed so that air could be pumped out in order to help combat the effects of fluid redistributing in the head and upper torso. Unlike previous negative–body pressure devices, these leggings were portable, as opposed to being fixed in one spot, but they limited the cosmonauts' mobility. About the only activities that the cosmonauts could perform while wearing the leggings were upper-body exercises.

Salyut 4 Space Station (DOS-4)

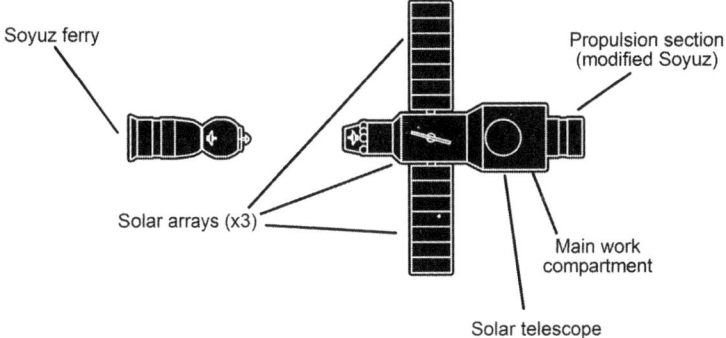

25. *Salyut 4* featured a few differences from previous stations. Courtesy of the author.

Looking at the manifest of *Salyut 4*'s scientific experiments, it doesn't seem to have been loaded with as much equipment as the first Salyut. The main reason for this is that DOS-1 was the Soviets' first pass at a civilian station, and they took a jack-of-all-trades approach to see what would work with off-the-shelf equipment. *Salyut 4* would be the first station to take advantage of purpose-designed equipment with a specific set of scientific goals in mind. Since *Salyut 4* had several years of design before it finally flew, the equipment was refined to a higher level of usefulness and durability than before.

Salyut 4 was still intended for a relatively short-duration mission. The station's on-orbit life was directly tied to the design life of its support systems and its onboard consumables. Even though *Salyut 4* included the water-recycling system first flown on *Salyut 3*, this new station could only be loaded with a finite amount of supplies, and the Soyuz spacecraft couldn't carry much more than the cosmonaut crew. Once the consumables ran out, that would be the end of the station's mission.

Launch and First Crew Occupation

Salyut 4 successfully launched on 26 December 1974. Its required maneuvers to place it in an orbit one hundred kilometers higher than *Salyut 1* were successfully achieved without incident. *Soyuz 16* was assigned to fly a stand-alone mission to test out equipment for the Apollo-Soyuz Test Project, so *Soyuz 17* would fly the first crew to the station. The *Soyuz 17* crew was made up of cosmonauts Alexei Gubarev and Georgi Grechko. Alexei Gubarev joined the cosmonaut ranks in 1963. This would be his first spaceflight. Georgi Grechko was also flying his first mission into space. He was an engineer who worked for Korolev during the early OKB-1 days; like *Soyuz 11* flight engineer Viktor Patsayev, he was originally selected as a cosmonaut to train for the lunar program.

Soyuz 17 lifted off on 11 January 1975 and achieved a successful docking with *Salyut 4* the next day. There were high hopes pinned on the mission, but *Soyuz 17* would not try to break the *Skylab* crew endurance record. Instead, the crew would set out to break the Soviet endurance record set by the *Soyuz 11* cosmonauts. The positive nature of the planned mission seemed to have carried over to the ground crew during launch preparations, because once the cosmonauts entered the station, they found a sign taped to the station's "floor" that said, "Wipe your feet."

For the most part, things went well on this flight as the regular work schedule and equipment usage apparently caused no problems. Early in the flight, though, the crew did have to fix a critical piece of hardware. Prior to *Soyuz 17*'s docking, the station's OST-1 solar telescope had been controlled automatically from the ground. Sometime during that automatic period, the control system for the telescope's secondary aiming mirror malfunctioned and damaged the angle indicator for it. The cosmonauts had no direct way to sight the secondary mirror; without a way to determine the secondary mirror's steering angle, the telescope, purpose designed for the station, would be useless.

The cosmonauts came up with a pretty ingenious work-around, though; they found that if they listened closely, they could hear the servomotor driving the secondary mirror and gauge approximately where it was by listening for how long the motor ran when they commanded it to move from one side to the other. Eventually they moved it into a neutral steering position and fixed it there permanently so that the telescope could still be used. The only drawback was that without the steerable secondary mirror, they would be unable to move the telescope's observing field independently of the station. Instead, they would have to move the station itself to aim the telescope.

For exercise, cosmonauts again had use of a treadmill along with a bicycle ergometer similar to the type that flew on *Skylab*. To combat the effects of weightlessness by forcing muscles in the body to erect a normal standing body posture, a refinement of the penguin suit was flown in addition to a second type of suit called the Atlet. For this flight, both suits were designed with long-term wearing in mind, while the original penguin suits were designed for relatively short-duration use. Grechko wore the penguin, while Gubarev wore and evaluated the Atlet to see which was more effective.

Reports in the media indicated that everything was proceeding quite well with the two cosmonauts as they were getting a lot of work done. Indeed, their workload was so high that they were starting to eat an extra meal a day to keep up with their nutritional requirements. They were at risk of consuming some of the supplies set aside for the next crew. Eventually, the ground had to order them to ease off their workload and take some previously unscheduled rest days in order to bring the workload more in-line with a reduced diet. On 2 February the crew broke *Soyuz 11*'s record, but it was not publicly acknowledged, probably so as not to bring up any bad omens. Ultimately, the crew would return to Earth on 9 February after spending twenty-nine and a half days in space. The undocking, reentry, and landing took place without any problems.

Physically, the crew was in pretty good shape, although it was determined that while the exercise regime was effective, it wasn't quite up to the task of preparing the cosmonauts for readaptation to Earth gravity. So the exercise regime was tweaked a bit for the next planned mission. For the crew of *Soyuz 17*, doctors also made use of a lower-body positive-pressure device to help keep blood from pooling in the lower legs of the cosmonauts during their first few days back on Earth. Neither cosmonaut suffered any long-term effects from the flight, and each man would fly again on future missions.

The "April 5th Anomaly"

For the second flight to *Salyut 4*, the crew of Vasili Lazarev and Oleg Makarov would ride *Soyuz 18* into orbit. Both cosmonauts were already space veterans and no strangers to Soyuz, as both had earlier flown *Soyuz 12*'s test flight. This flight would be far from uneventful, however. Launch occurred on 5 April 1975. The performance of the strap-on boosters and the core rocket stage proceeded as normal, but a problem occurred during separation and ignition of the third stage at about five minutes into the flight.

Due to a failure of the electrical system needed to fire the locks that secured the core stage of the booster to the third stage, the locks only released on one side. The spent core stage was still attached to the third stage by half of the open-web support structure between them. When the third-stage rocket motor ignited, the two attached stages pivoted around the struts, bending the linked stages into a banana shape, until the third stage finally broke free. The strain of this maneuver and the off-nominal trajectory were detected by the spacecraft's guidance computer, and it triggered an automatic abort. As part of the abort sequence, the Soyuz spacecraft used its propulsion module rockets to separate from the third stage. The Soyuz next jettisoned both its orbital and its propulsion modules from the descent module. The descent module reentered Earth's atmosphere on a ballistic trajectory after only achieving a peak altitude of about ninety miles.

The cosmonauts experienced about 14 to 16 g's of deceleration (with a peak of just over 21 g's) as the craft came in at a steeper angle than it would have experienced during a normal reentry, but at least the parachutes opened without problems. After twenty-one minutes of flight, the Soyuz descent module came to rest in the Altai Mountains near the Soviet town of Aleysk, about five hundred miles north of the Soviet border with China.

The touchdown itself didn't occur without incident either, as the craft landed on a snow-covered slope and began rolling down the hill toward a nearly five-hundred-foot vertical drop before the parachutes got snagged on some scrub trees and halted the slide. The crew knew they were going to land close to China on their descent, but they didn't quite know exactly how close. After landing, Lazarev reportedly burned some important papers connected with an experiment he was going to conduct on orbit, as he didn't want their contents to fall into Chinese hands. It didn't take long for recovery forces to make contact with the cosmonauts, though, and confirm that they were still inside Soviet territory. Even when the capsule was visually located, it still took about a day before the cosmonauts could be airlifted out, due to their remote location.

Vasili Lazarev and Oleg Makarov seemed to weather their encounter in relatively good spirits, although Lazarev had to lobby First Secretary Brezhnev directly to get the spaceflight bonus pay since it wasn't going to be awarded due to the aborted flight. Makarov would fly in space again, but Lazarev apparently suffered a hidden injury to his back that wasn't detected initially. He left the Soviet cosmonaut program in 1981 after failing a regular physical. Officially, the Soviets referred to the incident as the "April 5th anomaly." They also designated the mission *Soyuz 18A* (or *Soyuz 18-1* in some publications) and gave the original *Soyuz 18* title to the next spacecraft.

Second Crew

It was only a little under two months later when the Soviets tried again to launch a second crew to *Salyut 4*. For this mission, the backup crew of Pyotr Klimuk, who had originally flown on *Soyuz 13*, and Vitaly Sevastyanov, a veteran of *Soyuz 9*, would fly the new *Soyuz 18* into space. All went well as the booster sent *Soyuz 18* into orbit on 24 May 1975. A perfect docking was achieved, and the crew entered the Salyut to begin their residency. It was a milestone in Soviet spaceflight; for the first time, a single Salyut station was successful in hosting two crews without any problems.

The first order of business was maintenance, of both the repair and the preventative varieties. The philosophy that prevailed from the early days of the DOS-1 program was that equipment on the station should easily be accessible for on-orbit repairs. The crew performed maintenance such as changing out air filters and elements in the water condenser system and fixing one of the spectrometers with spare parts ferried up from the ground. All things considered, *Salyut 4* was in great shape for a station that had spent nearly five months in orbit to that point.

For this mission, the planners decided to alter the schedule of science gathering. So rather than spending some time on different experiments over the course of a week, the crew would dedicate their time to a single scientific discipline and experiment, gathering data for several days during a specific portion of the mission. For example, they would begin with astronomy, then move on to a different scientific discipline for a period of time, and so on. This was mainly done to help ration the propellant usage, so the station didn't have to be oriented to conduct one type of experiment and then get oriented to another position right after. The rescheduling also helped with the science being conducted. By having an experiment ready to go on consecutive days, a cosmonaut could get right to work on the data collection without spending time on set up or disassembly.

While the *Soyuz 18* mission had its fair share of biological and astronomical science gathering, Earth studies made up one of the largest portions of the mission as the crew collected data on pollution levels in Earth's upper atmosphere. Many photographs of the Soviet Union were taken to help with planning a route for a 3,500-kilometer railway that was intended to run from Lake Baikal in southern Siberia to the Amur River on the Manchurian border. Given the remote location, a survey from space was the best way to locate any hidden tectonic features, which bridge and tunnel designs would have to account for.

Increased solar activity in mid-June had the *Soyuz 18* crew pointing *Salyut 4*'s solar instruments toward the sun to study a solar eruption. On the station's night passes, the cosmonauts also observed the solar eruption's effects on Earth's atmosphere in the form of increased aurora activity, which made for quite a light show. The solar instruments shot over six hundred images of the sun during the *Soyuz 18* mission, thanks to *Soyuz 17*'s successful repair efforts of the primary telescope.

At the end of June, the crew surpassed the *Soyuz 17* crew's thirty-day milestone. The go-ahead was given to press ahead with an ultimate goal of trying to double it, even though this mission extension would result in *Salyut 4* being occupied for a little longer than its projected service life. Sevastyanov celebrated his fortieth birthday on 8 July, and Klimuk celebrated his thirty-third birthday two days later. The Soyuz ferry was used to conduct a reboost of the station's orbit, whereas previous orbital maneuvers had only been conducted with the Salyut's onboard engines.

On 15 July the control centers of Kaliningrad and Yevpatoria were in control of two separate manned missions as *Soyuz 19* lifted off from Baikonur. *Soyuz 19* was the Soviet half of the joint American and Soviet Apollo-Soyuz Test Project docking mission. The day after *Soyuz 19* entered orbit, its commander Alexei Leonov communicated to the *Soyuz 18* crew that if anything broke down, he would be happy to fly over and fix it. This of course was an inside joke, as Leonov and *Soyuz 19* crewmate Valery Kubasov had both trained for flights to three previous DOS stations and never had a chance to visit any of them in orbit. Kaliningrad was traditionally the control center used for civilian missions, but ground control of the *Salyut 4* mission was transferred over to the second control center to allow Kaliningrad's use for *Soyuz 19*. Once *Soyuz 19* returned safely to Earth, control of *Soyuz 18* and *Salyut 4* was transferred back to Kaliningrad.

During one week in July, the orbit of *Salyut 4* meant that it was practically in sunlight almost all the time, and this led to concerns that systems aboard the Soyuz ferry craft might overheat due to the lack of a cooling period in Earth's shadow. So the crew routed additional cooling hoses and ventilation into the Soyuz to help keep the internal systems nice and cool. As a by-product of this orbital period, a little more solar power was available for use with the station's experiments since the storage batteries weren't needed as often.

Toward the end of the mission, *Salyut 4*'s systems were beginning to show their strain as humidity levels were higher than what they were at the beginning of the flight. Late in the mission, the crew were unable to see anything out of the station's windows due to moisture condensation on them, and the walls were also sprouting bright-green mold growth.

Vitaly Sevastyanov's previous spaceflight was the long-duration *Soyuz 9* mission, and he paid the price of a longer postflight recovery when the crew neglected the exercise regime. For this mission, both cosmonauts stuck to their strict exercise regime. They would soon get their chance to find out if the physical work helped them to recover more quickly when they returned home.

On 26 July the *Soyuz 18* crew closed up shop in the Salyut and made a successful return to Earth. There were no plans to send up another crew as the station's onboard consumables had been used up. When the recovery forces arrived at *Soyuz 18*'s landing site, the doctors attached to the recovery team wanted to have the cosmonauts carried to the waiting helicopter, but the crew ignored the advice and decided to climb out of the capsule and on board the recovery helicopter under their own power. However, they still spent a few days recovering from their ordeal. While the new exercise regime helped, it wasn't a cure-all for everything. The pair's prolonged period in space also led to some curious behaviors, as Klimuk reported that he awoke during one of their first nights back and observed Sevastyanov sleeping in bed with his arms outstretched above him, as though he were still weightless.

After this mission, Pyotr Klimuk would fly in space again. But for forty-year-old Valery Sevastyanov, this was the last spaceflight, as the engineer and scientist-cosmonaut was pulled from active flight status in 1976 to serve as a ground controller in support of later Salyut missions. Sevastyanov had been involved in Soyuz spacecraft design prior to his selection as a cosmonaut; he returned to that job in the early 1980s, as he was involved in designing the Soviet space shuttle (eventually known as the Buran).

Outside his involvement with the Soviet space program, Sevastyanov was also the host of a Soviet television program about space exploration in the early 1980s and served as president of the Soviet Union Chess Federation. One of his lasting legacies, though, was in partnership with cosmonauts Leonov and Grechko and astronaut Rusty Schweickart; they formed the Association for Space Explorers in 1984. Membership is open to all people who have flown in space, regardless of nationality. Vitaly Sevastyanov unfortunately passed away on 5 April 2010 after a prolonged illness.

Making Progress

While *Soyuz 18* was the last manned mission to *Salyut 4*, the station would host one more spacecraft. One additional feature that made *Salyut 4* different from its predecessors was that its Igla system made it capable of allowing an unmanned spacecraft with a compatible system to dock with the station. On 17 November 1975 an unmanned Soyuz spacecraft lifted off from Baikonur. Rather than giving the spacecraft a Cosmos designation, the Soviets instead called the spacecraft *Soyuz 20*; over the next two days, it performed a series of maneuvers to rendezvous and dock with *Salyut 4*. The spacecraft took about thirty-four orbits after launch to rendezvous and dock with the station, while previous manned flights had done it in seventeen orbits. This revised rendezvous procedure would eventually become the established norm for future missions to Soviet space stations. Once docked with *Salyut 4*, *Soyuz 20* was powered down, and it would not return to Earth until 20 February 1976.

Soyuz 20 was a test mission for future Soyuz craft. Analysis of the descent module's systems and telemetry from the flight led to the designers imposing a mission life limit of ninety days on the Soyuz craft in a stored state. The ninety-day maximum would potentially allow for future manned missions of longer duration than what had previously flown from either the Soviet Union or the United States.

The testing of the automatic approach and docking would also benefit a new unmanned cargo ferry called Progress. The Progress spacecraft outwardly resembled a Soyuz and also launched on an r-7 carrier rocket (without the launch escape tower found on the manned Soyuz booster). But the new spacecraft was quite a bit different internally. The orbital module section was pressurized and could be outfitted with dry cargo for the crew of a space station. The Progress featured a single module that combined a modified Soyuz propulsion module at the rear with an unpressurized fuel-storage section in the spot normally occupied by a descent module on a Soyuz. The larger fuel supply of the Progress could be used to periodically reboost the station's orbit using the craft's engines, but fuel could also be transferred from the Progress to the onboard fuel tanks of future Salyut stations, thanks to a modified docking collar.

Not only could a Progress fly supplies to a Salyut station, but it would also allow for crews to throw things away. A lot of trash can be generated, and disposal becomes quite a problem as storage space runs out. Trash can't be thrown over the side, either, as it could become a debris hazard. Only pieces that are big enough to be tracked from the ground can be released, provided that their orbits decay quickly. At the end of its mission, the Progress would be carefully loaded with trash. Once full, the Progress spacecraft would undock and execute a deorbit maneuver in order to burn up harmlessly. It would be another two years before a Progress was ready to fly, but when it did, the new supply craft added versatility to the Salyut program.

Salyut 4 itself remained in orbit for over two years before it was finally commanded to reenter and burn up on 2 February 1977. While it was only occupied for ninety-two days, it taught the Soviet engineers a lot about long-term spaceflight. Much of what they had learned would be put into practice on the next generation of DOS stations.

Salyut 5, the Last Almaz

In the meantime, Chelomei's bureau was ready to have another go with the Almaz. The Almaz OPS-3 station was launched into orbit on 22 June 1976. Launch and orbital deployment took place without problems. Once it reached orbit, it became known as *Salyut 5,* but the radio frequencies it used immediately revealed its military role to Western agencies.

The first crew scheduled to fly to *Salyut 5,* on *Soyuz 21,* was made up of veteran Boris Volynov, the survivor of the harrowing *Soyuz 5* reentry, and rookie cosmonaut Vitaliy Zholobov. Zholobov joined the space program as a cosmonaut engineer in 1963 as part of a group of cosmonauts selected from the Soviet Air Force ranks. The pair lifted off on 6 July 1976 and docked with the station a day later.

As with *Salyut 3,* there was no live media coverage of *Salyut 5*'s mission, but the Soviet press did say that the station was going to conduct an extensive series of scientific experiments and observations, with a large portion of those dedicated to photography of regions deep in the heart of Russia and Siberia to look for natural resources. In preparation for this assignment, both cosmonauts reportedly attended classes on geology. Materials-processing experiments were reportedly also carried, including a furnace for smelting metals and studying their behaviors in zero gravity. *Salyut 5* also apparently carried the first sealed water aquarium into orbit to study the embryonic development of primitive species of fish. Fluid dynamics were also studied, and the data was considered important for developing the fuel-transfer system on the Progress.

Western observers considered the scientific experiments carried on *Salyut 5* as little more than set dressing. Most everyone figured out that the primary mission was intelligence gathering. But even with a manned station, not every orbital pass is ideal for photography of intelligence targets of opportunity. Each ninety-minute orbit of Earth shifts farther west while the planet rotates under the station. So after a few days of flight with daylight over possible reconnaissance targets in the United States, eventually it would give way to U.S. darkness, with daylight over the Soviet Union providing ample opportunity for scientific-data gathering.

To this day, the reconnaissance data gathered by *Salyut 5* is still considered classified in Russia, and even the crew doesn't talk about their experiences publicly. Boris Volynov, who was willing to talk about his *Soyuz 5* reentry experience when knowledge of it became public, would simply answer *nyet* ("no" in Russian) and fold his arms when asked questions about *Salyut 5*.

Early Return

All indications were that the crew would remain aboard *Salyut 5* for a sixty-day period and that things seemed to go rather well in the beginning. What happened next isn't entirely clear, though. According to the "Astrospies" television program, sometime during the station's forty-second day in orbit, the electrical system on board failed during a nighttime pass, plunging the crew into complete darkness. The crew were able to restore power and resumed normal operations, but not long after that, Zholobov apparently began to experience audio hallucinations. He would hear things that weren't there, and nothing could be done to help his condition.

Six days later, the mission ended abruptly as *Soyuz 21* undocked and returned to Earth on 24 August at night with a normal, albeit slightly bumpy, return to Kazakhstan. Considering that previous Soyuz returns had taken place in the daytime whenever possible, this raised a few eyebrows in the West. The crew initially was reported to be in not so good health when the recovery forces got to them. Part of the reason was that their abrupt return came before they were scheduled to increase their on-orbit exercise regime in preparation to return home. But after two weeks of recovery time, both cosmonauts seemed to make a complete recovery.

The reasons for the early return are not clear, as references to *Soyuz 21*'s problems are somewhat contradictory, with the "Astrospies" program being the first one to mention an on-orbit power loss. Prior to the announcement of the mission's end, it had been reported on state television that Zholobov was suffering from sensory deprivation. There are indications that Volynov was also starting to show signs of stress as well. After the mission, it had been said that both crewmembers may have been suffering side effects of nitric acid vapors from the station's reaction control system entering the cabin, as both cosmonauts apparently reported to program managers that they could smell an acrid odor days after the power loss. Indications were that the station continued to operate normally while unmanned, and preparations were made to fly another crew to it.

It would be a while before the next crew attempted to fly to *Salyut 5*, while engineers and doctors on the ground tried to determine what had happened. Since the crew left the station so abruptly, they apparently did not have time to load their exposed film aboard the cargo return capsule. So another crew would need to go up to at least finish that particular task. To help with any possible toxic fumes, the crew would carry gas masks.

The Turbulent Flight of *Soyuz 23*

With *Soyuz 22* scheduled to fly a stand-alone mission, *Soyuz 23* would be the next mission scheduled to fly to *Salyut 5*. A pair of rookie cosmonauts were selected to fly this assignment. The commander of the mission was Vyacheslav Zudov, a Soviet Air Force cosmonaut selected specifically for the Almaz program in 1965. Joining him was Valery Rozhdestvensky, a member of the Soviet Navy. During the early days of Almaz, the plan was to select military specialists. As such, a naval cosmonaut might have an easier time identifying specific Western naval ships and other subjects of interest in ports as opposed to an air force specialist. Rozhdestvensky's naval experience likely came in handy for this mission, but in a manner not even he could have predicted.

Soyuz 23 lifted off for orbit on 14 October 1976. It was scheduled to be a relatively short mission of only about two weeks, probably because neither the crew nor the engineers on the ground would know exactly what the living conditions aboard *Salyut 5* were like until they had docked. If nitric acid fumes had contaminated the station's atmosphere, then they could also have contaminated both the station's food and its water supply.

Once the craft acquired a visual sighting of the station, the Igla rendezvous problems reared their ugly head again. It was determined that *Salyut 5*'s transponder was operating properly. But for some reason, the Soyuz was unable to lock on; conditions were not ideal for conducting a manual docking, because the distance wasn't close enough to even make an attempt. The limited battery life of the Soyuz ferry meant that the crew had to shut down all nonessential systems and spend the next few orbits in drifting flight before the next reentry window over the prime recovery zone opened up.

After another day in orbit, the crew executed a proper retroburn and were on their way home. But that wasn't the end of their problems. Due to the aborted flight, the Soyuz came home to darkness and blizzard conditions in the prime recovery area. What was worse is that it splashed down almost dead center in the northern part of Lake Tengiz, a shallow, saltwater lake located in northern Kazakhstan. Where the descent module touched down, the crew were eight kilometers from shore, and the craft broke through the ice layer into the water. The Soyuz craft had been designed for possible water landings, and the crews had trained for them. But no trials had taken place during a blizzard in freezing conditions. It would take extreme effort from the recovery crew to retrieve the cosmonauts.

The ice clogs on the lake meant that boats could not make it to the craft and fog kept recovery helicopters from flying. The crew got out of their Sokol pressure suits and donned their cold weather survival gear to await rescue. With a craft made out of metal floating on the surface of a partially frozen lake, it didn't take long before the interior reached freezing conditions as the outside temperature was about -20 degrees Celsius (-4 degrees Fahrenheit). Plus, the capsule only had very limited battery power for the reentry. So the crew turned their heaters off and used the remaining power to keep an internal light and the ship's location beacon operating. At least the crew wouldn't suffocate as the valves that doomed the *Soyuz 11* crew worked as advertised here, allowing fresh air to come in while keeping the water out.

Many years later, it was revealed just how harrowing the rescue was. Saltwater intrusion under a cover caused the reserve parachute to deploy and it acted like a large anchor that tried to pull the capsule underwater. Only the shallow bottom of the lake kept the capsule from sinking below the waterline. Recovery forces found the capsule purely by chance, but they couldn't get to it. They eventually lost contact as the antennae iced over. Many in the recovery crew fully expected the cosmonauts to be dead from exposure. Finally, at daybreak the weather cleared enough for a rescue attempt to be made, as divers attempted to attach a lifting cable to hoist the capsule out of the lake by helicopter. While the copter was unable to completely lift it, the capsule was dragged to shore, where a cold and exhausted yet very much alive crew was recovered. Neither cosmonaut flew in space again, yet both remained in the program for a few more years.

Soyuz 24, the Last Almaz Mission

On 7 February 1977 another attempt was made to launch a crew to *Salyut 5* for a short-duration visit. Commanding the *Soyuz 24* mission was Viktor Gorbatko, and he was accompanied by engineer Yuri Glazkov. The launch proceeded as normal; a day later, the Soyuz executed a perfect rendezvous with *Salyut 5*. At eighty meters out, the Igla system failed, but Gorbatko was able to take over and executed a perfect manual docking.

The crew donned their gas masks and entered the station. But to their surprise, they found the atmosphere to be clean and quite breathable with no sign of the acrid odor experienced by the *Soyuz 21* crew. This would only be a two week stay, and the crew got right to work on the tasks originally assigned to the previous crew.

Viktor Gorbatko conducted a pretty extensive interview for the "Astrospies" program on his *Salyut 5* mission. Gorbatko said of his mission, "When we flew over the United States, I looked down and immediately recognized New York. We could see human beings on the streets. I would say we could see objects about one meter in size. I had enough time to count planes on the ground when we flew over military bases. We just had to shoot film of any weapons we could spot. That was about all we had to do. Our main assignment with the Agat system was to film ships and planes on the other side. There was some military tension in Israel, so we had to count how many planes they had."

One test that took place late in the mission had *Salyut 5*'s air supply purged and completely replenished with onboard tanks. This was a precursor to EVAs planned for the next DOS station. Both the DOS and OPS station cores had a side hatch built into them. These could allow for space walks to be conducted, but none of the stations that flew up to that point were equipped with EVA pressure suits. The Sokol pressure suits used for launch and recovery could be used in a pure vacuum, but only for short time periods; they lacked the thermal protection required for EVAs.

After a two-week mission, the *Soyuz 24* crew returned to Earth. Since it was still the middle of winter, the weather was still quite cold even though the descent module came down in daylight. The vehicle overshot its recovery area by a little bit, so the crew had to wait about an hour before recovery forces arrived. As a result of these cold-weather landings, officials made sure to equip future Soyuz craft with additional survival provisions. Preparations for intensely cold weather also became part of Soyuz training and continue to this day with current ISS crews.

The Legacy of Almaz

A few weeks after the *Soyuz 24* crew returned home, *Salyut 5* jettisoned its film container, and the capsule was recovered after a nominal reentry. The "secret" capsule was kept in Soviet and Russian hands until it was sold at an auction in New York City in late 1993. *Salyut 5* remained in orbit for another six months after the departure of *Soyuz 24*, when it was deorbited over the Pacific Ocean in August of 1977.

Chelomei's bureau had come a long way since the designer had first proposed Almaz over a decade earlier. At the end of the *Salyut 5* mission, they still had one Almaz station left in storage, and plans were underway to fly it sometime in the early 1980s. The TKS spacecraft would also soon be ready for flight. But flight of another manned Almaz station was not meant to be, due to behind-the-scenes events taking place while *Salyut 5* was in orbit. In late April 1976, Dmitry Ustinov became defense minister for the Soviet Union after the death of Andrei Grechko. The new title made him one of the most powerful men in the Soviet Union with a say in defense and space matters.

While *Salyut 5* was allowed to continue, once its final manned mission was concluded, the decision was made to not fly any more manned Almaz missions. Some of the reasons behind the decision were obvious, such as the capability of unmanned Soviet reconnaissance satellites, which by this point had steadily improved since the program's inception. Secondly, flying the Almaz missions as part of the Salyut program meant that they were more high-profile than what a classified military program should be. Soviet space station efforts had become as prestigious for the Soviets as the moon landings had been for the United States, making it harder to maintain the cover story.

But regardless of any technical arguments for the conclusion of Almaz, the decision likely was also influenced by Ustinov's personal bias. What had kept Chelomei in the manned space business after Khrushchev's removal were his allies in the military, such as Grechko. Now that Grechko was dead, Ustinov was the voice of the military.

The cancellation of Almaz meant that Chelomei's bureau was demoted to acting only in a support capacity only for space activities. There was still a demand for the UR-500 Proton rocket, and portions of Chelomei's bureau were involved with the construction on the next-generation DOS station cores, even if Chelomei had no direct say in their design. But it paled in comparison to the grand plans envisioned a decade earlier. Chelomei would still continue to supply armaments and missiles for the Soviet military, but Ustinov had Chelomei right where he wanted the designer: under tight control. For the foreseeable future, Glushko's NPO Energia design bureau would be in charge of the Soviet space station program.

6. On-Orbit Diplomacy

In 1974 NASA's space efforts were focused on a new mission. But rather than going to the moon or doing another flight to *Skylab*, the plan was for an Apollo CSM to launch and dock with a Soyuz spacecraft. To many observers, the mission didn't seem like much, as it was only demonstrating the rendezvous and docking techniques that had been practiced by both countries since the late 1960s. But the Apollo-Soyuz Test Project (ASTP) would have far-reaching impacts to both nations' space programs, even if they weren't fully understood at the time. It would be the first directly collaborative effort of equipment and government personnel from both the United States and the Soviet Union since the end of World War II.

The Early Path to Cooperation

The building blocks for NASA and Soviet cooperation were first laid in the early 1960s by NASA associate administrator Hugh Dryden, former director of the National Advisory Council on Aeronautics. Dryden was one of the great aeronautical scientists of the twentieth century during the 1930s and '40s. He was well known and highly regarded among Soviet circles as well, since his published papers were considered required reading by aerodynamicists.

While international relations between the two countries cooled in the 1950s, Dryden maintained a dialog with his Soviet counterparts. One Soviet scientist whom Hugh Dryden frequently exchanged letters with was academician Anatoly Blagonravov, a designer of rocket and artillery systems during the Second World War. Blagonravov was an orchestrator of space policy for the Soviet Union during its program's early years.

Informal contact between scientists of the United States and the Soviet Union was also maintained through the foundation of COSPAR, the International Committee for Space Research. COSPAR had been founded to help maintain a level of international cooperation that had begun during the International Geophysical Year (the IGY took place from 1 July 1957 to 31 December 1958), during which scientific studies of Earth's atmosphere and the space environment were made all over the globe. It can be said that the space programs of both the United States and the Soviet Union were spawned by the IGY studies, given that both low-altitude and high-altitude sounding rockets were used to fly payloads in and above the atmosphere. In the Soviets' case, they announced their intentions to fly the first Sputnik as part of their own IGY efforts. Yet few took notice until the first Sputnik flew successfully.

NASA administrator James Webb publicly endorsed more of a hardline stance from NASA and touted American space efforts as a competition against the Soviets with no hint of anything being done jointly. This position resonated just fine with Congress, who continued to fund NASA's efforts throughout the mid-1960s. But back channels were kept open behind the scenes at the encouragement of Webb.

Right after NASA was chartered in 1958, an International Programs Office was opened. Publicly, it was created with the goal of maintaining relationships with other countries that might be able to provide logistical support for space missions (such as land for tracking stations). It also served as a point of contact for other countries who wanted to conduct space science jointly with the United States. It was a point of contact available to the Soviets as well, should they have chosen to utilize it.

When John Kennedy became the U.S. president in 1961, public perception was that the United States was behind the Soviets in the missile-development race, though the Eisenhower administration had previously touted that there was no such problem. Kennedy endorsed increased involvement in space for political purposes, and this was best illustrated when he announced the goal of a moon landing by the end of the decade. But Kennedy also attempted to extend the hand of cooperation to the Soviets in the development of satellites and space probes to other planets, perhaps thinking that his goal of the moon would score some diplomatic points with Soviet leadership. These overtures didn't seem to impress Soviet premier Nikita Khrushchev all that much.

One interesting little footnote to Kennedy's space-cooperation diplomacy occurred in the fall of 1963. There were some hints that Soviets might not take part in a manned lunar program after all and that they were instead hinting at an unmanned lunar program. Kennedy publicly suggested the idea of inviting the Soviets to take part in the Apollo program. This didn't exactly go over well in Congress or with the American public. The announcement also took place less than a year after the Cuban Missile Crisis of October 1962. After hearing the idea, Khrushchev reiterated his country's firm commitment to its own manned program as their half of the space race. The matter was quietly dropped. When Kennedy was assassinated in November 1963, any further intentions he might have had for Soviet space cooperation also died.

Dryden and Blagonravov at least had some success as both men successfully negotiated an agreement on the exchange of scientific and medical findings in each country's respective space programs, provided that they mutually benefited both nations' space efforts and didn't reveal any secrets about their own programs. This was done while Blagonravov served on the United Nations Committee for the Peaceful Uses of Outer Space with Dryden acting as a negotiator on behalf of the U.S. government. This initial step would be a small one, but it was at least a first step. As an additional gesture, the Soviets also were invited by NASA to take part in the Echo II communications satellite program.

The joint Echo II experiments took place in January of 1964. *Echo II* was essentially a giant aluminum-covered Mylar balloon that served as a passive signal reflector between ground-based radio telescopes on Earth. It was launched into a polar orbit, meaning it could fly over Earth's entire surface within a twenty-four-hour time period. The Soviets supplied NASA with some tracking data on the satellite and photographs of its inflation in Earth orbit. They also took part in communications experiments between the Jordell Bank Observatory in England and a Soviet tracking station while *Echo II* passed overhead.

Dryden and Blagonravov continued their exchange of letters and had occasional face-to-face meetings throughout 1964 and 1965 to refine the early agreements, but there wasn't much additional headway reached. Dryden was living on borrowed time, as he had been battling cancer for years, although he maintained a high level of productivity even as his health worsened. With Dryden's death in December 1965, another doorway to potential cooperation closed, and both nations got on with their respective space efforts independently of one another for the next few years.

In late 1968 things began to change with James Webb's retirement from NASA. Incoming NASA administrator Thomas Paine had a little different agenda than Webb. Upon taking office, Paine began to downplay the competitive aspects of NASA's moon program against the Soviets. Behind the scenes, NASA knew that the Soviets were close to launching a moon rocket of their own and had several problems getting it off the ground, information that was not made public until many years later. By the flights of *Apollos 9* and *10*, it was a foregone conclusion that the United States would likely win the race to the moon. The United States could once again extend the arm of cooperation. This time, they could do it from a position of superiority.

As administrator, Paine extended an invitation to Blagonravov in a letter and asked him if a Soviet contingent would like to attend *Apollo 11*'s launch. The invitation was respectfully declined, but the Soviets publicly sent congratulations and acknowledgment to the United States and NASA for their successful efforts when *Apollo 11* landed on the moon. With that official acknowledgment, a crack in the wall between East and West began to form.

Thomas Paine felt that NASA's long-term future in spaceflight might rest in international cooperation, and this mirrored Richard Nixon's détente approach to international politics. With encouragement from the Nixon administration, Paine continued to exchange letters, this time to academician Mstislav Keldysh, president of the Soviet Academy of Sciences. Unlike the National Academy of Sciences (America's participant in COSPAR), which only had an advisory role in U.S. government affairs, the Soviet Academy of Sciences was directly responsible for Soviet policy regarding scientific and engineering matters. Keldysh was a scientist in the fields of mechanics and mathematics. He was one of the architects of early efforts by the Soviet space program and was nicknamed the Great Theoretician to coincide with Sergei Korolev's label of chief designer. As president of the Soviet Academy of Sciences, Keldysh was essentially the top man in matters of Soviet space policy.

Nixon's Space Task Group had also generated a report indicating that the time might be right for international cooperation in future space endeavors, thanks to Apollo's success. So Paine submitted copies of this report, along with additional reports on NASA's plans for its future in space, directly to the Soviet Academy of Sciences. Paine also wrote how he would welcome a face-to-face visit with Keldysh. Keldysh replied that such a meeting on matters of international cooperation was possible, but it would take time to arrange.

Paine got a chance to meet with Dryden's old contact, academician Blagonravov, in early May 1970 when they met at an informal dinner in New York City. While the U.S. State Department briefed Paine prior to the dinner to not to expect much in response, Paine's own advisers at least felt that while Blagonravov wasn't necessarily involved directly in matters of space policy anymore, he could at least report his experience back to Keldysh about Paine's sincerity.

During their brief talk, joint cooperation in a future space venture came up, but Blagonravov replied that he couldn't really speak on such matters. The two men did make headway on allowing astronaut Neil Armstrong, already scheduled to attend the next COSPAR meeting in Leningrad within the month to deliver a paper, the chance to visit Star City afterward.

Neil Armstrong received a very warm welcome in Leningrad when he delivered his paper and had a productive visit to the Star City complex. While the COSPAR meeting took place, NASA deputy administrator George Low met with Soviet officials in Leningrad for behind-the-scenes meetings. On the second day of these meetings, Low met with Keldysh in person. While a plan was not yet in place as to how the two nations would cooperate in a space project, both men agreed to study it further. And the direction that mission would take would be influenced in a small way by a recently released motion picture.

The Saga of *Ironman One*

At this time, another "space program" of sorts was capturing public attention. It was the motion picture *Marooned*, based on the novel written by Martin Caidin. In the movie, three astronauts aboard an Apollo spacecraft called *Ironman One* at the conclusion of a mission to an orbiting space station are stranded in orbit when their service module's main engine fails to perform a deorbit burn. NASA and Air Force technicians rush to prepare a rocket with an experimental spacecraft to rescue the stranded crewmembers before their oxygen runs out. At the climax, a Russian spacecraft appears, and its cosmonaut renders aid to the stranded astronauts until the rescue craft arrives. While the movie was not a big financial success in the fall of 1969, it did strike a chord with many people, and it is considered a cult classic to this day.

One of the men it influenced was Dr. Philip Handler, then president of the National Academy of Sciences. Handler took part in an official visit to the Soviet Union to meet directly with Keldysh a few weeks before the COSPAR meeting in Leningrad. Like his colleagues at NASA, Dr. Handler also welcomed increased cooperation with the Soviets in scientific endeavors. Dr. Handler had been encouraged to discuss space cooperation with the Soviets in informal talks he had with both Webb and Paine but to not necessarily get his hopes up that anything would come of it.

While NASA's official handling of such matters with the Soviets tended to be done with kid gloves at high levels of diplomacy and protocol, Handler was rather blunt in his approach when he brought up the movie's portrayal of a fictional cosmonaut to the Soviets. As Handler described the incident in a NASA history interview, the fact that "an American film should portray a Soviet cosmonaut as the hero who saves an American's life came to them as a visible and distinct shock."

The topic of pursuing a common docking system that could be used in the rescue of crewmembers from each country's spacecraft was discussed. A group of young scientists who attended the meeting with Handler and Keldysh expressed enthusiasm for such a project, while Keldysh himself was more guarded, as he was not in a position to give an answer on the matter just yet. But Keldysh asked Handler if he could perhaps wait for a response in a few days. Handler agreed and kept quiet on what was discussed when he returned to the United States. A few weeks later at his office at the National Academy of Sciences, Handler took delivery of a note from the Soviet Embassy. The note specified that after consulting with the appropriate groups, the Presidium of the Soviet Science Academy was prepared to discuss common docking mechanisms for space stations. This was the potential opportunity that NASA was looking for.

Paine decided to go a step further and actually proposed that a Soyuz-compatible docking device could be installed on *Skylab* to potentially allow Soviet spacecraft to dock with it. While Paine didn't necessarily believe the Soviets would support that idea, he submitted it to show that NASA was sincere in its efforts and willing to go further than just simple discussions where nothing would actually happen in the end. Many letters were exchanged between Keldysh and Paine as the issue changed from a personal matter between two officials to an official dialog between two countries. Paine retired from NASA soon after this, but he reiterated in letters to Keldysh that he was retiring for personal reasons only. Officially, NASA still wanted to take part in a joint mission with the Soviets.

Proposal of a Docking Mission

As part of an agreement with the Soviets, five men from NASA journeyed to Moscow. From Houston, the group included center director Bob Gilruth; Caldwell Johnson, a mechanical and electrical systems engineer of exceptional skill from the original Langley NASA Space Task Group; and flight director Glynn Lunney, who was knowledgeable in orbital mechanics and rendezvous techniques. Joining them would be George Hardy from the Skylab program at Marshall and Arnold Frutkin, NASA assistant administrator for international affairs. Frutkin had the most knowledge of previous NASA efforts with the Soviets, as he had been involved in running NASA's Office of International Relations since the early 1960s. Accompanying the group was William Krimer, a Russian-language interpreter from the State Department. The intent for this first group was to provide the Soviets with information about NASA's capabilities, without necessarily giving too many technical details, lest accusations be made that NASA was giving away secrets.

This first contingent flew to Moscow on 24 October 1970. They wondered what would greet them as they were briefed by intelligence and State Department officials not to get their hopes up or to expect much. The concerns of a cold reception abated once they arrived. On that first night in Moscow, the Soviets gave their guests a welcome dinner and took them to see several sites, including a Soviet space museum.

On the next day at their first visit to Star City, the delegation met high-ranking members of the Soviet manned space program. General Kunetsov was there, as were Soviet cosmonauts Beregovoy and Shatalov. The American contingent was allowed to tour the simulators used by the cosmonauts in their training. This was the first chance anyone from NASA, outside a couple of astronauts, had to inspect this equipment up close, and the American engineers considered it a highlight of their trip.

NASA officials met scientist-cosmonaut and Soyuz spacecraft designer Konstantin Feoktistov as both sides began two days of hardware and capability briefings. Lunney gave the first presentation into NASA's capabilities, while Feoktistov discussed the Soviet ones. Caldwell Johnson's presentation about American docking hardware came next. He also showed images of a proposed androgynous docking system in addition to the Apollo ones.

The term "androgynous" means possession of both male and female characteristics in life science matters and has engineering and electrical applications. The docking elements of both the Apollo and the Soyuz vehicles used a probe and drogue mechanism, with the probe being the "male" element and the drogue housing being the "female" element. The androgynous system replaces both elements with a universal mechanism found on both docking spacecraft. Guide petals on the outside of the docking tunnel align the mechanism, and capture latches secure the system into place for a hard dock. At the time Johnson made his proposal to the Soviets, this concept was being considered for the shuttle program.

After Johnson's talk, Soviet engineer Vladimir Syromyatnikov briefed the Americans on the previously used Soyuz docking system and revealed the one they intended for use in their DOS station program with the internal tunnel. This was a pretty bold revelation, given that the DOS program was still in development. But the Soviets only said the design would be used in the near future, and they didn't make mention of the DOS program to the Americans. After a lunch break, George Hardy gave a talk on the Skylab program, which the Soviets watched with much interest.

During a party at the U.S. Embassy, the two engineering teams got to know each other in a more relaxed environment. Indications were it was less like two countries' teams meeting each other and more like two groups of colleagues in the same field who hadn't met in a long time getting together. Bob Gilruth and Feoktistov hit it off well, as both had been involved in their respective space programs since the earliest days.

The next day included further discussions about *Skylab* and guidance equipment. At the conclusion, further agreements were hashed out on additional topics needing discussion and what information should be exchanged on future technical matters. Both sides decided to create three integrated working groups of engineers and specialists from both countries to handle the technical matters. Each group would have a different set of technical challenges to overcome if a docking mission were going to happen.

An agreement was drawn up by Feoktistov and, with minor revisions, was signed the next day. Even with the concerns of bureaucratic red tape getting in the way, things seemed to move pretty smoothly. While there was a language barrier, both teams found that they shared more similarities than differences. If all went well, the engineering teams would meet again in six months, this time in Houston.

This didn't mean that the project was a go, as it was still only the first meeting. NASA would still have to justify the mission in budget proposals to Congress. NASA engineers also had to determine if a proposed hardware development schedule drawn up by Feoktistov could be met in the appropriate time. Given NASA's full plate in late 1970, with work taking place on *Skylab*, on redesigning hardware after the near disaster of *Apollo 13*, on space shuttle design proposals, and on plans for the next lunar flights, it is amazing that a proposal for such a joint mission was even being considered at the time.

George Low journeyed with a NASA negotiating team to Moscow in January of 1971 to meet with Keldysh. Among their topics of discussion was a possible docking mission. Feoktistov was also present at these meetings, to discuss the Soviet hardware's technical aspects if such a mission were flown. While both Keldysh and Feoktistov said they didn't have the authority to commit the Soviet Union to such a mission, both expressed that they were in favor of doing it.

In preparation for this meeting, NASA drafted some docking-mission studies using the new data acquired from the Soviets. The ideas considered were either for directly docking an Apollo craft with an internally modified Soyuz or for using a separate module between the two spacecraft. Whichever method was selected, some sort of an airlock would be needed since Apollo spacecraft still operated with a pure-oxygen environment at an internal pressure of 5 psi, while the Soyuz used oxygen and nitrogen at about 14.7 psi (atmospheric pressure at sea level). An airlock is required because if a person goes directly from high pressure in an oxygen-nitrogen environment to lower pressure, nitrogen gas bubbles form in the blood and tissues, causing decompression sickness. To prevent this, a cosmonaut would need to prebreathe pure oxygen for a couple of hours inside the airlock to flush nitrogen from the body before reducing the pressure.

The Apollo spacecraft could not handle anything over 8 psi, or the internal-pressure vessel would rupture. The Soyuz spacecraft could theoretically operate at a lower pressure, although the potential fire danger brought about by a pure-oxygen environment concerned the Soviets greatly. Lowering the pressure in a Soyuz too far would require some major rework, and the mission's intention was to try to minimize changes as much as possible.

Ultimately, it was decided that a new airlock and docking module would be developed. That way any atmospheric changes during crew transfers could be localized to this module. Inside it, crewmembers could prebreathe as needed, and the environment to one spacecraft could be sealed off while open to the other one. The module could also act as an additional space for mounting communications and guidance equipment unique to the mission, meaning that the Apollo and Soyuz spacecraft themselves wouldn't need as many modifications. As a side benefit, some additional experiments could be flown in the new module as well. So after the docking mission had concluded, the Apollo could stay aloft to conduct a few more days of dedicated science gathering before returning home.

A nineteen-person Soviet delegation led by academician Boris Petrov arrived in Houston for their next scheduled meeting less than a week after the *Soyuz 11* crew successfully docked with the first Salyut space station. It was during this second meeting that the Soviet delegation revealed that going with a universal docking system of a type similar to what Caldwell Johnson had proposed was preferable to using an off-the-shelf one. Their decision surprised the Americans, but Johnson was encouraged by what he heard.

Over the next few days of joint design and discussion, it looked like a docking mission was possible. The language barrier between the two groups was eased somewhat by the fact that some members of the Soviet delegation both spoke and wrote English. But none of the Americans spoke Russian, and they had to make their points through interpreters to the Soviet team members who did not speak English. One final formality before the Houston sessions were concluded was the appointment of project directors from each country. Representing NASA would be Glynn Lunney, and representing the Soviet side would be Konstantin Bushuyev. It would be up to these two men to jointly approve the various design and hardware aspects of the docking mission, should the project officially be approved for flight by their respective governments.

There was also the question of what measurements to use. American aerospace engineering had settled on using the English-based inch-and-foot system for measurements, while the Soviets used metrics. Early on in the discussions, it was decided that metrics would be used by both sides, so it would take a bit of conversion work on the American side during equipment design. Metric measurements would also be used on orbit for callouts of speed and distance during joint-docking operations.

After the meeting, the NASA officials held a press conference to discuss what had taken place. There were concerns by some in the press that NASA might be giving away knowledge to the Soviets. That stance was downplayed, though, as these early meetings were merely discussions about how such a joint mission might take place if it were approved rather than about an approved program with a mandated technology exchange. One reporter also questioned what these discussions with the Soviets might mean for the world. Bob Gilruth replied, "Well, I think you'd have to decide that for yourselves. None of us here are politicians or politically inclined people. I think we are all impressed with the fact, however, that we have been able to meet with the delegation from the Soviet Union in an area of great technical difficulty, work together, and with a friendly atmosphere come to a number of important general agreements, and I think that it's always good when people can meet and work together in harmony."

When the Soviet delegation returned to the Soviet Union, the cosmonauts of *Soyuz 11* had spent twenty days in space. But a few days later, the crew was dead after their craft depressurized during reentry. There was much speculation in the West as to what had happened. While the Soviets had conducted a thorough investigation into the cause, they weren't as completely open with the results as many members of NASA and Congress would have liked at that time. Still, the Soviets gave NASA a lot of details on what had happened. Assurances were made that the tragedy would not affect continued plans for a possible docking mission. Ultimately, *Soyuz 11* was a reminder of the unforgiving nature of spaceflight, where a simple oversight can have lasting repercussions and can cause death to people, regardless of country of origin.

Docking System and Airlock Module

The working group tasked with development of a universal docking system settled on an arrangement similar to Caldwell Johnson's androgynous concept. Either side of the docking collar could be used in active or passive modes during docking operations. When Johnson visited Moscow a few months later with the American members of his working group, he was pleasantly surprised to find that Vladimir Syromyatnikov and the Soviet members had been working hard on the design, using elements discussed since the first meeting in October 1970. Like the American system, it made use of a petal arrangement to align the docking collars, but it utilized three petals instead of Johnson's original four. While the new design incorporated some additional changes, Johnson felt that the configuration was sound and that the detail differences could be worked out easily enough. The new docking device would be unlike anything that had been used before and would be a true meshing of American and Soviet engineering principles, both literally and figuratively.

The docking module itself was also taking shape. It would be a cylindrical-shaped module capable of holding two crewmembers. One side would dock with the Apollo craft via Apollo's normal probe-and-drogue assembly, while the other end would use the new docking system. The docking module would be carried aloft by a Saturn 1B launch vehicle, in the same spot normally occupied by a lunar module. Once orbit was achieved, the Apollo CSM would dock and then extract the module from the S-IVB rocket stage before beginning rendezvous maneuvers with the Soyuz.

Naturally, a couple of big concerns on the U.S. side questioned what this program would cost and what off-the-shelf equipment it would use. The Skylab program had been assigned a Saturn V and three Saturn 1B boosters. If all went well with Skylab plans, it would leave two surplus Saturn 1B boosters available for the docking mission (one primary and one backup). Top NASA management decided that the cost of this project should not exceed $250 million.

There were four Apollo CSMs that could potentially be used for this mission. One was the rescue CSM assigned to Skylab (CSM 119); one was a leftover Apollo spacecraft similar to what was used on the early lunar missions (CSM 111); and the other two were unfinished Apollo CSMs, each with a SIM experiment bay in their service modules (CSM 115 and 115A). While it would have been nice to use the SIM bay capabilities for additional science experiments, the projected cost to finish these spacecraft would have exceeded the available budget. So CSM 111 became the primary craft used in the joint mission; CSM 119 would act as the backup, provided it wasn't called on for a *Skylab* rescue. At least three docking modules would be built for testing, flight, and backup purposes.

Top NASA officials felt that it was a good idea to at least push for one more mission utilizing Apollo hardware. NASA was ramping up its efforts to get Congress to approve the development of the space shuttle. The most optimistic projections showed that the shuttle would not be ready for its first flights until five years after the last planned *Skylab* crew's launch. If NASA endorsed a docking mission with the Soviets, the hardware could be ready to fly by 1975. From a support standpoint, one more mission would help keep the engineering, support, and design teams at NASA and the contractors together for a little longer during the gap between the last Apollo and the first shuttle missions.

Apollo Salyut?

During the early discussions between NASA and the Soviets after the Salyut station's early success, the working groups had proposed an Apollo spacecraft docking with a Salyut. Given the fact that Salyut had evolved with the military Almaz space station project and its relatively short gestation period as a civilian station, the proposal to use it in a joint project is somewhat surprising. The idea put forth was that a Salyut could be equipped with two docking ports. One would be the normal type to allow for docking with a Soyuz craft, and a second port would be androgynous and able to accommodate an Apollo spacecraft.

Salyut was preferable to Soyuz thanks to its mission-duration capabilities. An original-configuration Soyuz vehicle was only capable of independent flight of up to a week at most. Granted, *Soyuz 9* had flown an eighteen-day mission, but the rendezvous and docking equipment were removed to handle the increased consumables and air-purification equipment. A Soyuz equipped with a new docking device could not be stripped down that way, as the R-7 launch vehicle still had a weight limit.

In April 1972, after much negotiating, things were in place for the formal mission proposal to be drawn up, which the leaders of the United States of America and the Union of Soviet Socialist Republics would sign at a summit meeting the following May. NASA officials visited Moscow to help negotiate the final elements of the agreement, but before discussions got underway, a statement was made to the Americans. The Soviets had concluded that it was technically and economically unfeasible to modify a Salyut for use in the docking mission by a planned 1975 mission date. The current Salyut design only had one docking port, and it would cost both time and money to modify it for a second one. Therefore, it was recommended that they proceed with using just the Soyuz if all other aspects of the mission could be met with no further difficulties. The Americans were surprised by this change but accepted it. An Apollo craft would dock with a Soyuz, not a Salyut, and it would be a relatively short-duration mission.

ASTP Is Born

On 24 May 1972 President Nixon and Premier Kosygin signed the Nixon-Kosygin accord on space. With this agreement, the proposal between NASA and the Soviets became an official space mission with a target launch date of mid-1975. There was still a lot of work to be done, but now the engineering teams could get on with the tasks of turning designs into hardware, refining procedures, and planning for actual spaceflight. In June 1972 the mission officially became known as the Apollo-Soyuz Test Project, or ASTP.

The plan on paper was simple. An Apollo spacecraft and a Soyuz spacecraft would both liftoff from their respective launch sites and then rendezvous and dock in orbit. They would carry out various ceremonial activities to commemorate the event and conduct joint scientific experiments. All of this would all be carried out on live television internationally.

For this mission, not only would Soviet citizens get to watch live coverage of the mission on the American side, international press would be covering the mission live from the Soviet Union as well. This caused some problems between the Soviets and NASA's public affairs people, and many of these problems weren't fully ironed out until only a couple of weeks before launch day.

One of the dictates of the agreement was that the flight crews and the support crews for each side of the mission would have to be chosen two years before the planned launch date. The primary reason was that these astronauts, cosmonauts, and controllers would have to begin training together as a team. This is not something that could be thrown together at the last minute. The language barrier would also be a problem until each team learned the other team's language well enough to communicate critical commands at a moment's notice.

During the next meeting of the working groups, held in July of 1972, the Soviets revealed that they would have two Soyuz spacecraft ready for launch on the first day of the mission. This seemed a little odd since the Americans knew the limitations of the Soyuz included not being able to stay aloft for as long as an Apollo CSM. Logically, the Americans considered it a good idea to launch Apollo first and have the Soyuz launch second. The Soviets didn't agree, and they felt it was better to launch a Soyuz first, be it the primary spacecraft or the backup if something went wrong. The backup spacecraft and crew would be kept operational and ready to go. That way, if a launch delay occurred on the American side with the Apollo unable to be launched before the Soyuz had to come home, the second craft would fly. The Americans didn't fully understand the reasoning, but they accepted it. Prestige issues may have been at work here. The Soviets were first with Sputnik and with Gagarin, so why shouldn't they launch first?

The Soviet members also questioned why NASA wanted to fly the mission above 230 miles in orbit. Since no specific reason was given, the Soviets wondered why it couldn't be lowered a bit. Finally, after a direct question was asked, the real reason came out. The Soyuz could not orbit that high. So the orbital altitude for the mission was lowered. The Soviets weren't necessarily trying to hide their capabilities intentionally. Pride likely got in the way, as the Soviets seemed to be reluctant to reveal that the Soyuz craft had certain limitations compared to the Apollo one.

For the rendezvous, it became apparent early on that the Apollo craft would be the primary vehicle involved with the task of locating the Soyuz and docking with it. After the first docking, the pair would undock, and the Soyuz would maneuver to dock with a passive Apollo craft. But the pair wouldn't get far enough away to require a new rendezvous procedure. Because of his experience in rendezvous with the *Gemini 6, Gemini 9*, and *Apollo 10* missions, astronaut Thomas Stafford was assigned to the first working group to help hash out the rendezvous details. Soviet cosmonauts Nikolayev and Yeliseyev were assigned to the same working group thanks to their combined experience. The first training sessions for the assigned flight crews for ASTP were scheduled for mid-1973.

Work on the joint-docking hardware went ahead with only minor problems. The NASA people didn't consider the problems to really be any different than what might happen at the level of NASA and its contractors. But for ASTP, both countries' teams were acting as equal partners and learning to work together. With a contractor, NASA can unilaterally dictate what has to be done, and the contractor does its best to accommodate the decision. For ASTP, there were elements of give-and-take from both sides, but each team worked the problems and found innovative solutions along the way.

In October 1972, work continued at the next meeting in Moscow. The teams did preliminary reviews of the docking hardware and made some decisions about the atmospheres flown on each spacecraft and the docking module. During docked operations with the Apollo CSM, the pressure in the Soyuz would be lowered, and the gas mixture would be 40 percent oxygen and 60 percent nitrogen. Any higher oxygen content would increase the fire danger. The reduced pressure aboard the Soyuz would mean that crewmembers in the airlock transferring to Apollo would only need to prebreathe oxygen for about an hour instead of the nearly three hours required when transferring from sea level pressure to an environment with pure oxygen at 5 psi.

The Crew of ASTP

In January of 1973 the flight crews for ASTP were announced. On the American side, the mission commander would be Gen. Thomas Stafford. This would be his fourth spaceflight. Stafford had no plans to fly in space after *Apollo 10* and instead moved into management. But his participation in the state funeral for the *Soyuz 11* crew and being one of the heads of the astronaut department meant that Stafford was known and well respected by members of both countries. His involvement in development work on ASTP as part of the working groups also meant that he had direct experience in developing the rendezvous and docking procedures intended for the flight.

The command module pilot for this flight would be Vance Brand. Brand was a former U.S. Marine Corps fighter pilot and a test pilot with lots of experience. At the time of his selection to ASTP, Brand was performing duties as a backup commander for one of the Skylab missions, so he wouldn't be available for ASTP training for about a year.

The third member of the Apollo crew waited the longest to get his flight. He was Mercury astronaut Donald K. "Deke" Slayton. An issue with a minor heart fibrillation had grounded Slayton from his assignment to fly the second Mercury orbital spaceflight, as the flight surgeons acted ultra-conservative when the problem cropped up. His fellow Mercury astronauts nominated Slayton for assignment as chief of the astronaut office, a duty that Slayton performed with skill and professionalism. Slayton, of course, maintained a high fitness level and got as much stick time as he could in NASA aircraft to try to give himself a chance to fly in space, should his medical grounding get lifted. Finally, in the early 1970s, after the latest physicals revealed that his heart hadn't skipped a beat in quite a long time, Slayton was successful in getting his flight status restored, with help from NASA administrators and sympathetic doctors.

Even with his flight status restored, the remaining Apollo and Skylab assignments had been spoken for. That left ASTP or early shuttle flights as possible spaceflight opportunities. Slayton wanted to command ASTP, but NASA felt Stafford would be a better choice with his experience. Slayton was assigned to the crew as docking module pilot (DMP). For launch and landing, he would occupy the flight engineer seat originally assigned to lunar module pilots in Apollo, but he would also get stick time flying the Apollo during the second of two planned dockings with the Soyuz and during one of the formation flying experiments.

For the Apollo backup crew, Al Bean would be the mission commander, *Apollo 17* veteran Ronald Evans would assume command module pilot duties, and the docking module pilot assignment would go to Jack Lousma. Like Brand, Bean and Lousma wouldn't really be able to participate in joint training exercises until after their mission to *Skylab* had returned home, but they would be available for training before Brand.

On the Soviet side, Alexei Leonov and Valery Kubasov were assigned as crew of the Soyuz. Like Slayton, Kubasov had some not-so-good experiences with doctors, thanks to the botched tuberculosis diagnosis that scrubbed both him and Leonov from *Soyuz 11*. Both cosmonauts also had trained for Salyut flights that ended in launch and deployment failures. The Soviet half of ASTP seemed like the perfect mission for two cosmonauts who had already been in training for quite a while.

In an interview he gave for NASA's oral history, Stafford mentioned that he was expecting Shatalov or Nikolayev to be assigned due to their rendezvous-and-docking experience. But ultimately he understood why Leonov was selected for command, since assigning to this mission the first person to conduct a space walk provided some excellent public relations opportunities for the Soviets internationally. For the ASTP cosmonaut backup crew, the Soviets selected Anatoly Filipchenko as commander and Nikolai Rukavishnikov as flight engineer. Rukavishnikov was a veteran of *Soyuz 10* and, like

26. The official ASTP crew portrait. Seated (*left to right*) are Slayton, Brand, and Kubasov, while standing behind (*left to right*) are Stafford and Leonov. The model in front represents the Apollo and Soyuz craft with the specially designed docking module between them. Courtesy NASA.

Leonov and Kubasov, had spent a long time in perpetual training thanks to the back-to-back failures of two DOS stations.

The Language Barrier

The astronauts and cosmonauts knew from the day they were selected that they were going to have to learn each others' languages. The Soviet crewmembers had already begun classroom sessions to improve their English-language skills. Stafford grew concerned when he learned of this. If the Americans didn't get up to speed quickly, they would quickly fall behind. The astronauts began private tutoring sessions on their own until NASA finally made time in the training schedule for dedicated language classes. Russian is not an easy language to learn, and it takes a fair amount of effort to speak properly, let alone to read and write. Plus, with any language, there are little subtle nuances and phrases that crop up. A concept that might have a one-word description in one language might not necessarily have exactly the same meaning, let alone a single-word translation, in the other language. Ultimately, Stafford estimated that during the last sixteen months leading up to the mission, he conducted about a thousand hours of Russian-language study alone. It was a similar situation for both Slayton and Brand. According to Stafford, the crew needed every bit of language training they could get.

There was also the question of what language would be used once the flight had begun. Tom Stafford was apparently the person who came up with the idea that the Apollo and Soyuz flight crews would communicate with one another in the other country's language. So during communications to the Soyuz, Apollo crewmembers would talk in Russian, and the Soyuz crewmembers would talk in English. The benefit here is that for a crewmember speaking in his nonnative language, he speaks slower and more clearly to get the point across and to be understood more easily. Speaking in a native tongue—with astronauts speaking in English and cosmonauts, in Russian—might cause a critical detail to be missed if things are spoken too quickly. This became the accepted standard by both sides during joint training sessions as well as in phone conversations between the astronaut and cosmonaut crews. All crewmembers practiced as much as they could.

Tom Stafford's own efforts to learn Russian became a source of humor. Stafford was born and raised in Oklahoma, and his distinctive Oklahoma twang carried over into the Russian language, causing a few curious grins and chuckles among the cosmonauts. In one conversation with the press, Leonov is reported to have said that with Stafford, they don't speak "American-ski" or "Soviet-ski," but instead they speak "Oklahoma-ski."

During visits to the United States, Leonov had become something of a media darling, as his joyous smile and willingness to converse with members of the American media were infectious, as was his famous quote at one press conference, "I want to be [a] movie star." Plus, his artistic skills had already made him something of a minor celebrity in European circles. For most Americans growing up at the time, Leonov and Kubasov were the first cosmonauts they had ever gotten to meet or see up close, and Leonov had a commanding presence wherever he went.

Snowball Fights and Fireworks

All members of the primary, backup, and support groups seemed to bond together quite well with their colleagues. On a visit to the Soviet Union in

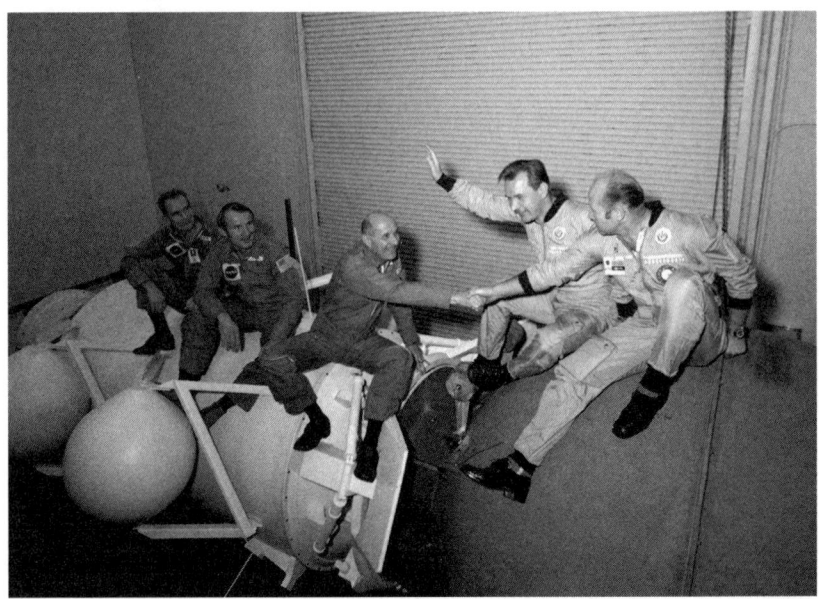

27. The lighter side of training! Both crews practice the handshake on top of the docking module and Soyuz mockups. Courtesy NASA.

the winter, both sides took part enthusiastically in an impromptu snowball fight during one stop on a bus trip. On a summer visit by the Soviets to Houston a few months later, both sides participated in an outdoor cookout as if they had done it before. Because this mission was as high profile as it was, a lot of eyes were watching to see how international relations were proceeding. But very few problems cropped up, and things seemed to run smoothly.

The Soviets also seemed to spare no expense in showing good hospitality to their American comrades. For earlier visits, NASA representatives stayed at a hotel near central Moscow and had to endure a long drive to and from Star City each day. The distance between the two locations was short by Western standards, but Soviet roads meant that a trip of only fifty-four kilometers usually would take over an hour one way. And today it is still a long drive. So the Soviets built a new hotel near Star City just for the Americans.

The Americans had a little fun at the hotel, as they figured every room was likely bugged and watched around the clock by the KGB. So when one American suggested that it might be nice to have a pitcher of water and some drinking glasses to no one in particular, a pitcher of water and drinking glasses were on the table the next day when they returned from work. Other "requests" made to "the walls" were granted in a similar prompt fashion.

Some awkward moments did occur, though. For example, during one visit to Moscow in July, backup Apollo command module pilot Ron Evans managed to sneak half a case of fireworks into the country. So on 4 July, the NASA contingent began setting off firecrackers and bottle rockets. Soviet neighbors became alarmed, and the police were called. Soviet security assigned to protect the Americans was on high alert, while people were wondering what in the world was happening. But all was understood when the Americans explained that they were celebrating "the day of our revolution," as Stafford described it.

A Science Directive

Although ASTP had endorsements by the governments of two nations and seemed to be on track for a successful outcome, there were concerns in Congress that something on the Soviet side might topple the plans to fly the mission at the last minute. To guard against this possibility, several members of both the House of Representatives and the Senate mandated that NASA should add a secondary goal that could be conducted independently, should the docking mission itself be called off. NASA briefly entertained thoughts of conducting a short-duration mission to *Skylab*, but the idea was dropped, primarily because the docking module they had spent time and money developing wouldn't be used on a visit to *Skylab*.

It was decided that both the CSM and the docking module would carry scientific experiments. Ultimately, the Apollo portion of ASTP carried equipment to conduct twenty-eight different experiments on orbit. Five of them would be joint experiments with the Soviets, twenty-one would be American experiments to be conducted unilaterally, and two experiments would come from West Germany.

The Apollo service module contained experiment packages to study soft X-ray sources from celestial bodies, a telescope to study extreme ultraviolet light from sources other than Earth's sun, and an instrument to study the sun's helium glow. As part of a joint experiment, the Apollo would be used to produce an artificial solar eclipse that could be observed by the Soyuz. The Soviets would conduct observations of the sun's corona while the Apollo observed the eclipse's shadow on the Soyuz craft. Another joint experiment called for the Apollo spacecraft to beam light to a reflector on the Soyuz in order to study the concentrations and properties of atomic oxygen and nitrogen at the orbital altitude of both spacecraft. It required precision flying to do this. The results of this study would be used to help with the design of similar equipment for Earth-resource satellites.

In the command module, gamma-ray detectors were carried to study the effects of cosmic rays. This was similar to an experiment carried on *Apollo 17*. Cosmic rays, especially heavy high-energy ones (HZE particles), are filtered out by Earth's atmosphere. But in space, they can be detected easily, as there is nothing to stop them until they hit something. NASA scientists in life science fields had also taken an interest in the possibly destructive effects of cosmic rays on living cells. Starting with *Apollo 8*'s flight, astronauts had mentioned that when their eyes were closed, they could occasionally see a bright flash, likely associated with a cosmic ray penetrating the vicinity of their optic nerves. The ASTP astronauts were asked to write down any experiences they had when noticing a similar flashing-light phenomenon. This data would be compared with readings from the radiation detectors to see if there was indeed a correlation.

From West Germany, the Apollo CSM carried an experiment called Biostack III. It was a container that carried various types of dormant living cells, from microbe spores to insect and shrimp eggs to plant seeds. Radiation detectors were located adjacent to the cell samples. The cells would be analyzed when the mission was over by allowing them to grow and develop in order to determine the effects of cosmic rays on living tissue. Earlier Biostack experiments had been flown on Apollo missions to the moon.

A joint Soviet-American experiment would obtain similar data on active fungus cultures in orbit to see how cosmic rays and zero gravity would affect them. Some fertilized killifish eggs would also be flown and hatched in sealed water bags to see how the fish developed in zero-g conditions. The combined ASTP crews would also be experiment subjects as well. Given that both groups of space travelers lived in different parts of the world, blood and tissue samples were taken before and after the flights to see what type of microbes (if any) they exchanged with one another in Earth orbit. Studies were made of the ASTP crews' immune systems as well.

ASTP also flew some biological and materials-processing experiments in the docking module. Studies were carried out using a process called electrophoresis with special equipment developed by West Germany. In electrophoresis, an electric charge is used to separate either different biological cells or different molecular compounds from one another in a gel medium thanks to differences in their electrical properties. The process can be used to extract pure compounds or biological cells of a certain type or for removing impurities from a sample. It was hoped that performing this experiment in zero gravity would produce better results than on the ground, where gravity can alter the samples, potentially making them not as pure. ASTP was the first time an electrophoresis experiment would fly in orbit, but it wouldn't be the last, as several shuttle missions conducted similar experiments over the next decade to study practical applications of this technology.

An electric furnace was also carried in the docking module, and this was used to conduct several experiments in materials science. The principle behind using a furnace is that materials that don't normally combine in space under heat might be able to do so in a zero-g environment without gravity causing the heavier material to separate during cooling. The problem can be illustrated easily with a common oil-and-vinegar salad dressing. Shake the bottle, and the two materials mix together. But let the bottle sit, and eventually the heavier oil separates to the bottom while the lighter-weight vinegar settles on top.

Taking Shape

On both sides, there were some concerns about crew safety. For the most part, these concerns stemmed from the unfamiliarity each team had with equipment from the other country. The Americans and Soviets had differences in how they certified hardware and spacecraft for flight. For the most part, each side received assurances that the proven hardware flown before would do the job. But engineers made sure that there were no potential issues, just in case. For ASTP's new equipment, the Soviets adopted a strategy similar to what NASA expected from a contractor in order to certify flight hardware for use. As designs were finalized and hardware developed, a safety review panel would check to see if any potential problems might crop up with the equipment. This kept things progressing along a positive path.

Soyuz 16 lifted off in December 1974 with ASTP backup cosmonauts Filipchenko and Rukavishnikov on board. It was a full dress rehearsal and test of the ASTP-configured Soyuz craft, with the new docking system attached to a docking collar acting as a stand-in for the American docking module. *Soyuz 16* also included the internal-communications and radar equipment intended for use in ASTP. Deployment and retraction of the docking collar was tested in orbit, and tests were made with the communications systems on the frequencies that would be used for the ASTP mission itself. The crew also successfully tested lowering the internal cabin pressure with the atmosphere at 40 percent oxygen and 60 percent nitrogen for an extended period. Finally, the attached docking collar was jettisoned to test that explosive bolts could be used to separate the two spacecraft in an emergency. *Soyuz 16* successfully returned to Earth after almost six days in orbit, which was the planned duration for the Soyuz half of ASTP.

With the return of *Soyuz 16*, that left two Soyuz missions to fly before *Soyuz 19* flew the Soviet half of ASTP. Preparations were going well until *Soyuz 18A* experienced a launch abort that resulted in cosmonauts Vasili Lazarev and Oleg Makarov landing in the mountains of central Russia after a very short flight.

Under normal circumstances, the Soviets likely would have kept the mishap secret. But with the launch of *Soyuz 19* scheduled for only three months later, the Soviets took an unprecedented step and told the Americans what had happened during the "April 5th anomaly." The R-7 booster being used for this mission was part of an older series than the one planned for use in ASTP. While the fault was still being investigated, the newer booster already had one modification made to the electrical circuit controlling the locks between the core and third stages. If a similar failure occurred with only half the locks firing, they would be staggered around the ring instead of only on one side. So if the thrust of the third stage was needed to separate the two stages, it should have done so cleanly. NASA was satisfied with the explanation, and preparations continued with no further delays.

In Congress the launch abort plus lingering concerns about Soviet safety didn't sit well with some. Or rather, it could be said that the situation gave some senators an excuse to give the ASTP program further scrutiny for political purposes. Among them was Senator William Proxmire from Wisconsin. The year that ASTP flew was the first year of Proxmire's Golden Fleece Awards, which he would use as opportunities to showcase what he felt was pork barrel spending for dubious scientific studies. Yet even before that, Proxmire was a very vocal critic of NASA on Capitol Hill, and he received the ire of science and space program advocates everywhere. To many, it appeared he only criticized both NASA and the military to further his own career rather than to discourage improper spending practices. It seemed a bit strange to many people that Senator Proxmire was speaking in favor of canceling ASTP due to concerns for astronaut safety after the recent launch abort. NASA representatives spent portions of their regular briefings to Congress after *Soyuz 18A* to allay these safety concerns.

Tom Stafford explained the Senator Proxmire situation from his perspective in a NASA oral-history interview he conducted in 1976: "Proxmire tried everything he could to shoot the mission down. Not that he was against the Russians so much as that he was always against the space program and technology. Anything in technology, whether it was Air Force, NASA, he was against it. And I'll tell that to his face. He likes to grab any issue, just so he can get a headline. That was the first time he was concerned about safety. Well, why wasn't he concerned about all the other times?" Concerning the question of safety in relation to the April 5th anomaly, Stafford continues, "Well, that wouldn't have had any effect on [Apollo] if they had aborted. That wouldn't have affected our safety a damn bit."

Launch Day

Ultimately, preparations came down to the final launch day for both the Americans and the Soviets. Public relations and television were a big aspect of ASTP. Most everything was negotiated in advance to prevent any last-minute substitutions or violations of protocol. To an observer, ASTP's joint operations between the Americans and the Soviets looked like a meeting between two heads of state in space.

In public, the behind-the-scenes technical aspects of the flight would be largely ignored, and in their place would be a carefully orchestrated set of procedures for what would be done and when it would be done. Such procedures covered when the handshake would take place in orbit, who would do it, what materials would be exchanged, and which crewmembers would visit which spacecraft and when they would do so. A lot of these procedures had to be practiced in the joint training sessions, alongside the equipment familiarization. But the crews got through it, the same as any other training before a mission.

The Soviets would launch their half of the mission first, with *Soyuz 19* as the primary vehicle. The backup spacecraft with the backup crew would be sitting on another launchpad, ready to go at a moment's notice, should there be an unforeseen problem or an abort. About eight hours later, the Apollo spacecraft would lift off to begin its chase of the Soyuz for eventual rendezvous and docking. To help with communications in orbit when the craft might be out of range of ground-based tracking stations, the mission would use Applications Technology Satellite 6 (ATS-6) to relay transmissions using the Apollo's high-gain antenna array fitted for lunar flights. This would allow for continuous coverage of the mission during nearly half of each ninety-minute orbit.

For Leonov and Kubasov, the big day came on 15 July 1975. The launch coverage was extensive as the cosmonauts were seen suiting up, riding their bus to the pad, and climbing the first set of stairs on the gantry before heading up the elevator that would take them to the spacecraft level to be strapped in. The countdown proceeded normally, and launch occurred on schedule at 12:20 GMT (08:20 EDT). The backup spacecraft did not have to be used, and *Soyuz 19* entered orbit without any problems. All that was needed now was an Apollo spacecraft.

Stafford, Slayton, and Brand were asleep when the Soyuz lifted off, but when they awoke two hours later, they were informed that everything was proceeding well. It turned out to be a beautiful day with clear blue skies on 15 July. Since this would be the last flight of an Apollo spacecraft, a crowd of thousands flocked to the Space Coast to witness history in the making. KSC also got its share of international visitors, as Soviet ambassador to the United States, Anatoly Dobrynin, was on hand to watch from the Launch Control Center. Earlier that day, he watched the launch of *Soyuz 19* in Washington DC at an auditorium with U.S. president Gerald Ford before flying to Florida.

For this mission, NASA set up a television camera inside the command module to transmit live pictures of the astronauts during their ascent. Countdown proceeded as normal, with no hang-ups. At 19:50 GMT (15:50 EDT) the eight engines of the Saturn IB's first stage ignited, and the Apollo lifted off the milk stool of pad 39B for a smooth climb into Earth orbit with a smoke trail that was visible for quite a distance. About ten minutes later, the spacecraft was in orbit and working perfectly.

And so began the two-day chase as Apollo gradually closed the distance

28. Photos of the ASTP spacecraft taken from each other are combined to show how the rendezvous and docking may have looked from outside. Courtesy NASA, composite by the author.

with the Soyuz. Visual contact was made on 17 July 1975. As the spacecraft grew closer, they spent some time station keeping with one another. The rendezvous was successful, so that just left the docking. Finally, mission control had a message for both crews: "Houston is go for docking. Moscow is go for docking." The two craft achieved a good dock, with Leonov saying, "We have capture," and Stafford replying in Russian, "We also have capture." Finally, Leonov said in English, "Apollo and Soyuz are shaking hands now."

A few hours later, Stafford and Slayton opened the docking module's hatch to the Soyuz. After a little delay, the handshake was made, and so began the joint activities between two spacecraft over a two-day period. After the pomp and ceremony of certificate signings and phone calls from state leaders, a meal between the cosmonauts and the astronauts took place. Somebody on the ground with a sense of humor had decided to decorate the Soviet toothpaste-style food tubes containing borscht with brand-name vodka labels. Later, during a press conference on orbit, Leonov was asked about the food. His reply was, "The best part of [a] good dinner is not what you eat, but with whom you eat."

On the second day, Valery Kubasov hosted Vance Brand in the Soyuz for an introduction of the spacecraft to American audiences while Leonov, Stafford, and Slayton gave a tour of the Apollo for the Soviet viewers. Additional joint activities and experiments took place before the crewmembers retired to their original spacecraft for the final time. Remaining joint operations would take place only by radio. The spacecraft undocked and performed a second docking on 19 July to verify the system with the Soyuz active. Total docked time on orbit was a little over one day and twenty-three hours. The two craft flew in formation conducting the solar eclipse experiment and the ultraviolet atmospheric studies, with Slayton in the pilot seat while the service module's detectors did their job.

Finally, the two spacecraft parted company for the last time. Soyuz successfully reentered and soft landed in Kazakhstan on 21 July with the event being carried on live television around the world. The Apollo spacecraft would stay in orbit for three more days, continuing its experiments. The docking module was jettisoned to gather data for a Doppler gravitation experiment a day before the Apollo spacecraft returned home.

Breathing Exhaust

To the casual observer watching Apollo's reentry and ocean recovery on television, things seemed to go quite well. However, some serious problems cropped up, which people on the ground didn't realize at the time. As part of normal Apollo procedures, the command module pilot flies the craft on its reentry profile. This meant that Vance Brand was in the commander's seat for this phase of the flight, with Stafford in the center seat and Slayton occupying the far-right seat. They had some excess noise in their headsets, in addition to the noises produced by the reentry and the corrective thruster firings from the command module's thrusters controlling the descent trajectory. As a result, the crew got slightly off sequence from the checklist.

One item in the checklist called for activation of the ELS (Earth landing system), which would deactivate the command module thrusters, jettison the cover over the command module parachutes, and fire the drogue chutes. For whatever reason, Brand didn't arm the ELS switches, probably because he was waiting for a callout from Stafford or Slayton. When Brand noticed that the capsule had descended through thirty thousand feet without the drogue chutes deploying, he flipped the switches to jettison the cover and fire the drogue parachutes manually. But the thrusters were still active, and they began to fire when the drogue chutes deployed, in an attempt to stabilize the craft. Finally, the ELS switches were flipped on, which turned off the thrusters.

But prior to deployment of the drogue parachutes, the cabin-pressure relief valve opened to equalize the command module's cabin with outside air pressure. So when the thrusters fired, the rush of outside air into the cabin also sucked in some thruster exhaust. The crew began to cough and feel the effects as their cabin filled with nitrogen tetroxide fumes. Nitrogen tetroxide is highly toxic, and it doesn't take much to kill a person. When Stafford saw the fumes enter the cabin, he flipped closed the thruster isolation valve switches, which helped a little, but that still left the fumes that had already entered the cabin.

The spacecraft continued a normal descent under the main chutes and experienced a hard splashdown that immediately flipped the capsule upside down in the water before the parachutes were jettisoned. The capsule settled inverted in the water in the stable 2 position. This means that it would not flip upright into the stable 1 position until the three flotation bags on top had been inflated. So the crew was left hanging in their seat straps, still coughing from the fumes. Brand called for Stafford to grab the oxygen masks normally worn in case of fire. Stafford unstrapped, fell into the Apollo docking tunnel, and had to contort in an odd position to grab the masks, get the masks to the others, and flip the switches that would inflate the flotation bags. Stafford picks up the story from there:

> *I knew I had a toxic hypoxia because I was having a hard time breathing. I was starting to grunt-breathe, to make sure I had enough pressure in my lungs . . . to keep my head clear. I looked over at Vance, and he was just hanging in his straps. He was unconscious, his arms were [limp] . . . and I said "Vance? Vance?!" He was out cold. . . . His mask had slipped off his face. So I got the mask back on his face [and] hit the high-flow valve. He came to, and he just clobbered me like a mule. People do this when they are unconscious. He just clobbered the hell out of me and knocked me back like that . . . knocked the mask back off his face and passed out again. At this time, I got a hammer lock on his head and put it [back] on him. He struggled some, but not near as bad. Fortunately with the hammer lock I got him back to [consciousness]. And then we got the [capsule] right side up real fast.*

Once the capsule was upright, the crew was able to expel the rest of the fumes before the onboard emergency oxygen ran out. Nobody else knew what had happened until after the Apollo command module was on the deck of the USS *New Orleans* and the crew got out. The welcoming ceremonies were underway when the captain of the *New Orleans* learned about what transpired and ended the ceremonies early. The only visible sign that something might be wrong was that Vance Brand looked a little green, as he was still a bit ill. He apparently got a higher exposure of fumes due to his proximity to the air vent. Once the crew got to the ship's sick bay, Brand passed out again. None of the crew seemed to have difficulty breathing. But there was some damage done to their lungs, and they weren't taking in as much oxygen as they should with each breath. As a result, the Apollo crew had to spend two weeks in the hospital in Honolulu, Hawaii, before being released. Slayton stayed in the hospital a little longer when a lesion was discovered on his lung. Surgery to remove the lesion revealed that it was noncancerous. The crew recovered normally after that.

ASTP's Legacy

Of the ASTP crewmembers, Leonov, Stafford, and Slayton would not fly in space again. In 1976 Leonov accepted the position of chief cosmonaut, giving him the responsibility of crew training at the Gagarin Training Center in Star City; he held that position until 1982, before being promoted. He finally retired from the space program in 1992 to continue painting, to write articles for books, and to pursue business opportunities.

Tom Stafford stayed active in the U.S. Air Force, being promoted to major general after the ASTP mission, and took command of the U.S. Air Force Flight Test Center at Edwards AFB during the early testing of the space shuttle *Enterprise*. After he retired from NASA and the U.S. Air Force, Stafford took up a few jobs in the private sector, including heading up a consulting firm that maintained contact with leaders in both the U.S. and Russian space programs to advise them on respective spaceflight matters.

Deke Slayton stayed on at NASA to manage the approach and landing test phase of the shuttle program, but he retired in 1982 not long after the first successful flights of the space shuttle *Columbia*. He pursued some ventures in the private sector, including development of a commercial space payload launcher, before being diagnosed with a malignant brain tumor. Deke Slayton died of cancer in 1993.

Both Valery Kubasov and Vance Brand would return to space, with Kubasov flying one more Soyuz mission before retiring in 1993 and Brand flying three space shuttle missions before leaving the astronaut office in

1992. After that, Vance Brand held various administration positions at NASA until he finally retired from the agency in 2008. Kubasov recently passed away on 19 February 2014. He was seventy-nine years old, and his death was reported to be from natural causes.

The end of the flight didn't mean the end of ASTP, as there were further meetings between the engineering teams after the mission to analyze the data collected. But future joint manned spaceflights would have to wait, as both countries' leaders took on stances that became progressively more hard-line in the years that followed. Many observers felt that while ASTP was an excellent joint effort and fostered a spirit of cooperation, it was unlikely to be repeated anytime soon. However, the words of U.S. president Gerald Ford, from a speech he read to the ASTP crews during their first day of docked operations, would prove prophetic in the decades that followed: "It has taken us many years to open this door to useful cooperation in space between our two countries, and I am confident that the day is not far off when space missions made possible by this first joint effort will be more or less commonplace."

7. Salyut Endurance!

Once the Apollo command module representing the American half of ASTP returned successfully to Earth on 24 July 1975, it would be almost six years before NASA would fly astronauts in space again. For half a decade, the Soviets would have manned space exploration all to themselves. They would devote that time almost completely to development of the DOS Salyut stations and rack up many successes along the way.

Soyuz 22, the Last Independent Soyuz Mission

During the flight of the *Salyut 5* OPS station, the Soviets launched *Soyuz 22*. They made use of the backup spacecraft built for ASTP. The APAS docking collar was removed from the orbital module, and in its place was mounted the MKF-6 multispectral camera built by the East German firm Carl Zeiss Jena. The system could take photographs in both visible-light and infrared wavelengths and could take up to six photographs at once of a point of interest in a staggered fashion. Once the film was developed, the pictures could be put together like a mosaic. It was capable of photographing an area up to 165 kilometers wide on Earth from orbit, and it could image up to 500,000 square kilometers in ten minutes. The photos were similar to those from a Landsat satellite, with false color imagery showcasing details that might not be seen normally. *Soyuz 22* was launched on 15 September 1976 and lasted until 23 September. On board were *Vostok 5* veteran Valery Bykovsky and rookie flight engineer Vladimir Aksyonov. Bykovsky flew and oriented the Soyuz to achieve the proper attitudes for collecting imagery, while Aksyonov operated the camera system.

In addition to the camera experiment, *Soyuz 22* also carried an aquarium to study the behavior of primitive fish in zero gravity and a small centrifuge to see if artificial gravity would have a positive effect on plant development and growth. The cosmonauts also kept a log to see if they encountered any cosmic-ray light flashes, as the Apollo astronauts had. The camera system imaged over thirty geographic areas in Soviet and Eastern European territories and took over 2,400 photographs. The images collected were of very good quality and were used in Earth science and resource investigations for several years after. *Soyuz 22* would be the last scheduled independent flight of a crewed Soyuz. If any craft ever flew free after that, it would be either an unmanned test flight or the result of a docking failure.

Salyut 6

The next Soviet DOS station to fly didn't look all that much different from its *Salyut 4* predecessor, but internally *Salyut 6* was a much more mature vehicle, making use of every bit of design knowledge that was learned up to that point. The hardware was beefed up and provisions were made for equipment change outs. Like *Salyut 4*, *Salyut 6* had three main solar arrays that could track the sun independently.

One major change made on *Salyut 6* was an extension ring built onto the back of the core module replacing the Soyuz-based propulsion module. The new extension housed the station's main engines; within the ring, designers fitted a second docking port. A second Igla antenna was also fitted facing aft. Soyuz craft could dock at either port, but the aft port would mainly be used by the supply craft since it also contained fuel lines in the docking collar to allow a Progress to refill the Salyut's onboard fuel tanks.

To accommodate refueling, the fuel and oxidizer supply of the station was changed from nitric acid and hydrazine to unsymmetrical dimethylhydrazine and nitrogen tetroxide. The internal plumbing and fuel-pressurization system of *Salyut 6* featured a set of internal bladders pressurized by nitrogen gas. During a refueling procedure, the bladders would provide negative pressure in the Salyut's fuel and oxidizer reservoirs while the fuel system of the Progress would remain at a positive pressure. As a result, the Salyut would suck fuel and oxidizer out of the Progress. The process would take hours to perform, but it would guarantee an orderly transfer of fuel and oxidizer between spacecraft. With this capability in place, *Salyut 6* could stay operational as long as its systems were functioning and its fuel supply could be topped off. The system also meant that the station didn't have to carry as much of a fuel reserve, so *Salyut 6* could be reboosted more frequently into higher orbits.

Salyut 6 (DOS-5)

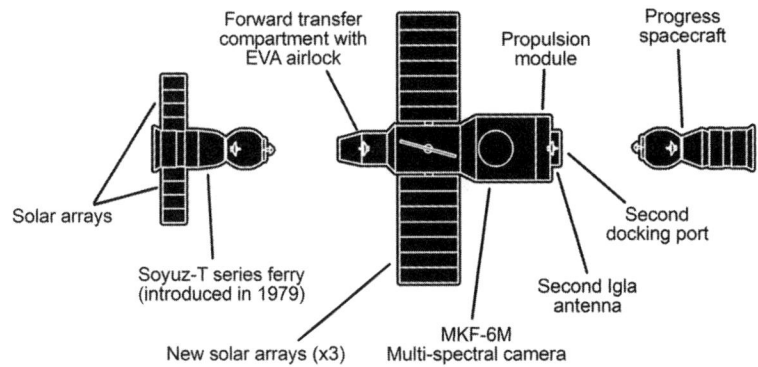

29. With two docking ports, *Salyut 6* could be resupplied frequently.
Courtesy of the author.

Internally, *Salyut 6* shared about the same layout with its DOS predecessors, but with some minor equipment changes. *Salyut 6* was the first Soviet station to make use of color television cameras, while previous stations used only black-and-white cameras to transmit video to Earth. Scientific equipment was also developed for *Salyut 6*, but not all of it was launched with the station, in order to keep the flight weight down. Instead, Progress craft would bring up the equipment on regular supply runs.

Enter the Orlan

Two space suits were also at *Salyut 6*'s disposal. While the DOS and OPS core always had a side hatch intended for EVA use, the previous stations were not equipped with space suits. *Salyut 6* contained two Orlan (a Russian word meaning "Sea Eagle") suits. The Orlan was originally designed for spacewalk use in the Soviet lunar-landing program. When the lunar program was canceled, work still continued on refining the Orlan. By the time *Salyut 6* launched, the Zvezda design bureau had ten years of development and testing invested in the Orlan, making it a very mature system.

The Orlan is a semirigid space suit featuring a solid torso and integral helmet. Cosmonauts climb inside through a hatch resembling a refrigerator door in the suit's backpack. The backpack also houses the suit's life-support system. Different gloves can be fitted, while the length of the arms and legs can also be adjusted for the occupant by using a set of internal straps. A cosmonaut can don the suit in as little as five minutes, after putting on a liquid cooling garment to both regulate body temperature and minimize perspiration.

Internal atmosphere in the suit is maintained using a chemical oxygen-generation system that also removes exhaled carbon dioxide and moisture. The Orlan also includes a backup, bottled oxygen supply for use in emergencies. For the Salyut missions, the suit was given up to a five-hour single-EVA capability, and it could perform as many as four EVAs for up to ten hours total use with proper maintenance and filter changing between space walks. While it doesn't have the dexterity or finesse of NASA's EMU (extravehicular mobility unit) designed for the shuttle program, the Orlan suit provides an important capability. The Orlan design has gradually been upgraded over the years, improving its capabilities. The latest versions of the Orlan suit are still used to this day aboard the ISS.

Salyut 6's exterior was fitted with EVA handholds. Each cosmonaut would remain connected to the station through a safety tether and an electrical and communications line that powered the suit and allowed for a voice link with the ground. Two cosmonauts would perform each space walk and keep track of one another using the buddy system. For EVA training, the Soviets built an NBT at Star City known as the Hydrolab and outfitted it with an exterior mockup of the station. Versions of the Orlan for underwater use were developed; thanks to regular training sessions, cosmonauts had plenty of experience by the time they reached orbit. From that point on, while not all Salyut missions had a scheduled EVA, cosmonauts would train for them in case the need for one arose.

Salyut 6 was launched on 29 September 1977. Launch and orbital insertion proceeded as planned. Ten days later, *Soyuz 25* lifted off with Vladimir Kovalyonok in command and Valery Ryumin as the flight engineer. While both cosmonauts were rookies to space, they had both been on the cosmonaut program for several years. Kovalyonok joined the program in 1967. Ryumin was a stocky individual and looked very much like a stereotypical Russian, even at his relatively young age. Prior to joining Korolev's bureau in 1966 as an engineer, Ryumin had served in the Soviet Army as

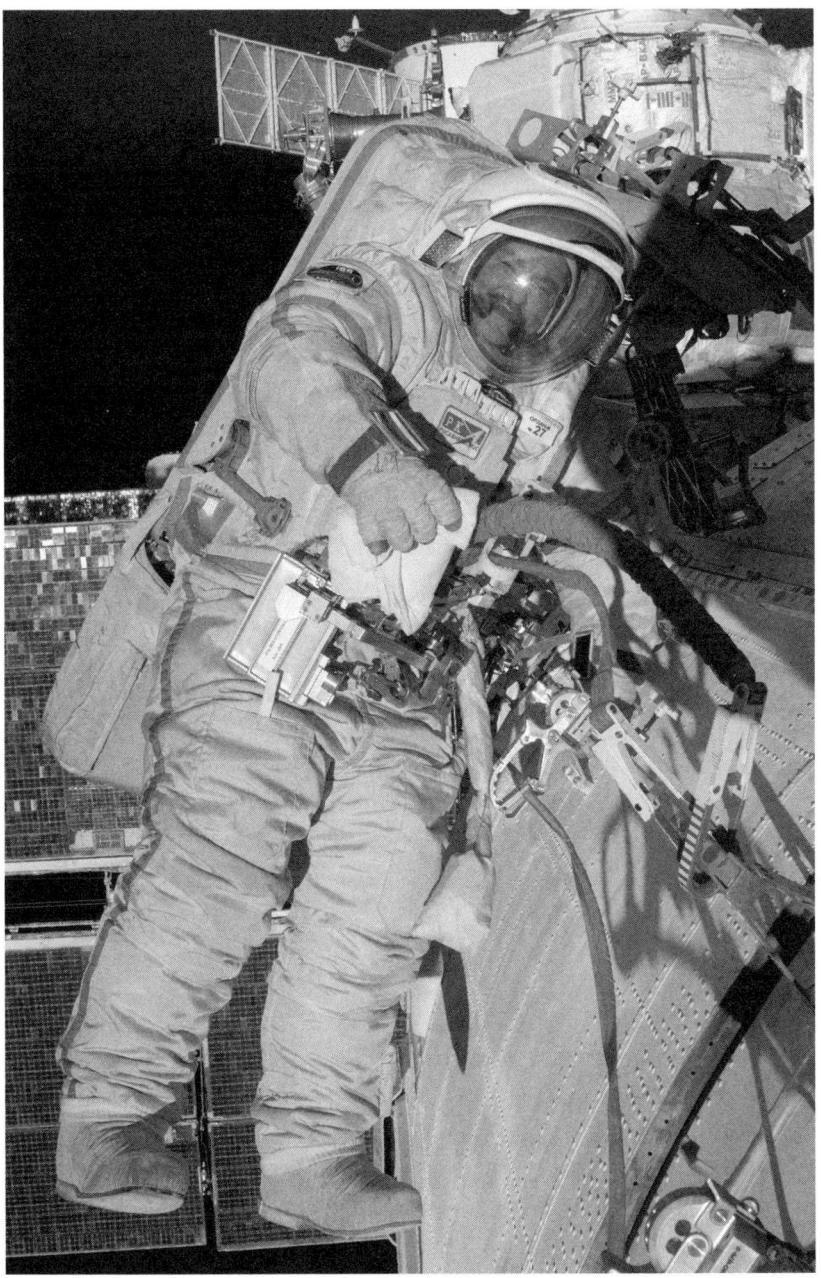

30. The Orlan suit was first used on *Salyut 6*. Pictured here is an Orlan used on the ISS. Courtesy NASA.

a tank commander before going back to school to get his electronics and computing degree in spacecraft control systems. Ryumin became a cosmonaut in 1973.

There were high hopes for the *Soyuz 25* mission, as it was to take place during the sixtieth anniversary of the October Revolution. Unfortunately, hardware problems tend to have little concern for dates and anniversaries. Rendezvous took place with no problems. But although the craft achieved a successful soft dock on the front docking port, they were unable to secure a hard dock. The next few tries were no more successful than the first one, and the mission was called off after five docking attempts. The crew asked controllers if they could attempt an aft-port docking, but permission was denied, to avoid damaging the aft port if the Soyuz probe was the source of the problem.

After yet another docking failure, it was decided that at least one crewmember with previous spaceflight experience would fly successive missions. This meant that the next crew, also made up of spaceflight rookies, was split up and that its two cosmonauts were divided into new crews. *Soyuz 26* launched next on 19 December 1977 with Yuri Romanenko and flight engineer Georgy Grechko on board. The veteran Grechko was no stranger to fixing balky hardware, after coming up with the work-around for *Salyut 4*'s solar telescope problem. In addition to a normal long-duration mission, the crew was also tasked with inspecting *Salyut 6*'s front docking hatch and conducting repairs as needed. Their Soyuz craft eased in for a successful manual docking with *Salyut 6*'s aft port.

Upon entry, the crew got right to work in making sure their home for the next few months was in good shape. Medical experiments would be the primary task of the mission. The crew would try to beat all previous space endurance records set by both the Americans and the Soviets. For scientific studies, *Salyut 6* included an expanded and revised MKF-6M camera system, similar to the one flown on *Soyuz 22*. Additional telescopes were fitted for celestial and solar observations. Additional equipment would arrive later.

First Salyut EVA and "A Good Joke"

About a week after docking, Romanenko and Grechko conducted an EVA to inspect *Salyut 6*'s front docking port. This required them to don the Orlan suits for the first time and to depressurize the station; Grechko could then conduct the EVA while Romanenko remained inside, acting in a support role. Only the front docking port hatch would be opened, not the side door. Grechko's legs would remain inside the hatch while Romanenko held on to him. The EVA would serve as an excellent first operational test of the Orlan suits.

Both cosmonauts sealed themselves inside the forward transfer compartment's airlock and opened the front docking port's drogue assembly to space. Grechko spent the next twenty minutes inspecting the drogue assembly, using special tools to actuate the docking latches to make sure there was no damage. He also took the time to inspect the station's Igla radar system and found everything to be in good working order, confirming that *Soyuz 25*'s docking problem was due to its probe assembly and not *Salyut 6*'s port. As a final EVA task, Grechko also mounted a Medusa experiment cassette outside the station to expose some biopolymer material directly to space. It was retrieved on a future EVA.

Near the end of the EVA and after Grechko reentered the station, Romanenko decided to stick his head outside and have a look around. This was not planned, and the crew kept it quiet. Apparently, Romanenko was not properly anchored and began floating out the hatch until Grechko grabbed his legs. For years after their mission, Grechko rather coyly joked that he had saved his comrade from a certain death of floating free in space with no way to get home. Many members of the international press took Grechko's joke seriously, and the story of Romanenko's "near-death" experience gained traction in published accounts for years to come. In reality, the early Orlan suits were firmly linked to the station at all times since the Orlan needed the tether to supply power. So even if Romanenko had drifted completely free of *Salyut 6*'s docking hatch, he still could have climbed back in using his tether, or Grechko could have pulled him back in.

With a successful EVA, the crew settled in for the long haul. One big change in the scheduling from previous Salyut missions had the crew operating strictly on Moscow time, even on orbits when they would not be in range of Soviet tracking stations during daylight hours. This was made possible by improvements in data tape gathering and storage on board. The findings from scientific experiments could be recorded and transmitted to the ground when the Salyut was in range. The crew's wake period lasted from 08:00 to 23:00 and was logically scheduled to include time for meals, exercise, mission tasks, and station maintenance. No attempts were made to have one crewmember sleeping while the other worked, and only an emergency would require crewmembers to be awakened during normal rest periods. The crew was in good spirits and got along well with one another, even though they had been put together at almost the last minute.

Company's Coming!

On 10 January 1978 history was in the making as *Soyuz 27* lifted off from Baikonur with cosmonauts Vladimir Dzhanibekov and Oleg Makarov on board. This was Dzhanibekov's first spaceflight, but Makarov had been in space before on *Soyuz 12*'s test flight. Makarov was hoping to spend more time in space than two days, though, as his previous mission was the aborted *Soyuz 18A* mission. This time, the Soyuz spacecraft made a successful dock with the Salyut station.

When the hatches were opened, it marked the first time three spacecraft had successfully docked with one another. The visitors were greeted with warm wishes and a bit more, as they were initially a little overwhelmed by the smell of the station. While the *Soyuz 26* crew had only occupied the station for a month and had maintained good hygiene, they could do nothing about their flatulence. Excess gas from their intestinal tracts due to normal digestion produced a bit of an odor. When two people live in that for a while, their olfactory senses get used to it. But for newcomers, the smell was a little overpowering until their noses tuned it out. Since that time, many cosmonauts and astronauts have commented that the first "greeting" they get from the opening of the hatches is the assault to their olfactory senses as the environments from two spacecraft intermingle for the first time.

The *Soyuz 26* crew welcomed the distraction from the day-to-day routine, and it made for a very happy ship for the next few days. Unlike the crew that launched in December, Dzhanibekov and Makarov would only be on board for about a week, as their primary job was to ferry up a new Soyuz and take the old one home, freeing Romanenko and Grechko to conduct a long-duration mission without having to return home before their original spacecraft expired its flight design life. That isn't to say the new crew would be just taking in the sights, though, as the pair were put through a battery of medical tests to see how well they adapted to space. Future short-duration Salyut crews would conduct similar experiments to add data. The two newcomers also acted as on-orbit repairmen, fixing a couple of minor systems that had failed soon after *Salyut 6*'s activation and performing some preventative maintenance on other systems.

The docked configuration of two Soyuz spacecraft and the Salyut core also had engineers on the ground evaluating whether *Salyut 6*'s control system was compensating properly for the revised center of mass. These tests were fairly simple, as they involved the crewmembers making jostling movements inside the station while engineers on the ground monitored how well the forces would dampen out and see if any resonance vibrations remained. Resonance vibrations at the right frequencies, when left unchecked, can sometimes lead to metal fatigue and equipment failure, two things engineers do not want to see.

All too soon, it was time to end the joint mission. The crews transferred their Sokol pressure suits and seat liners between the spacecraft, and *Soyuz 26* was powered up for a systems check on 15 January. The next day, Dzhanibekov and Makarov bid their friends farewell and undocked for an uneventful return journey to Earth. The main reason for the mission's short length was due to conditions in the prime recovery area. While technically a rocket can launch everyday when the station's orbit crosses over the launch and recovery areas, there was only a ten-day period where the spacecraft can land in daylight conditions. This daylight period occurred only once during each fifty-six-day orbital cycle of *Salyut 6*. For most future missions, the spacecraft would launch at the beginning of the daylight window and land at the end. *Soyuz 26* with its ferry crew landed safely with no problems. With their comrades safely home, Romanenko and Grechko continued their long-duration mission, but they would not have long to wait for their next "visitor."

Progress Docks

On 20 January the first Progress cargo craft lifted off from Baikonur. Inside the pressurized cargo section of the Progress were food and water supplies, in addition to the experimental Splav materials-processing furnace. The Progress also included fuel to top off *Salyut 6*'s tanks. Two days after launch, the Progress was visible in the station's window as it made an automatic rendezvous with the station. For safety reasons, the crew retreated to the Soyuz ferry and sealed the hatch, in case a problem during docking caused a hull breach. Thankfully, the Progress glided in for a proper docking with no mishaps. A few hours later, after the seal integrity was checked, the cosmonauts opened the hatch to their new care package.

Many crewmembers on the Salyut stations and on *Mir* and the ISS have described the arrival of a Progress as the equivalent to a mail call on an oceangoing ship or at a remote base. It is also a little like opening a present on Christmas. In addition to vital equipment, consumables, food, and scientific supplies, a Progress can also be packed with letters and reading materials, along with fresh fruits and vegetables, for the crew. The fresh produce is prized by crewmembers, not just because of its nutritional value, but also because the aromas help to give them a nice reminder of home. To a person assigned to spend months inside a sealed environment the size of a bus where every smell, inside view, and taste is familiar, the introduction of fresh food can improve a crewmember's morale in many intangible ways. The crew unloaded the supplies, and the first fuel transfer in orbit between a Progress and a Salyut also took place with no mishaps. Everything was shaping up to be a textbook mission.

Over the entire course of *Salyut 6*'s life in orbit, it would host a total of twelve Progress spacecraft dockings on the rear port. Some would come in very quick succession for critical mission supply needs, while others were spaced out in regular intervals between manned dockings of Soyuz ferries. The Progress has become a familiar sight since its first use, and the design has given decades of sterling performance with very few problems.

The Splav furnace was up and running within a few days. It would eventually be joined by two Kristall furnaces on successive Progress flights. Spare parts were used to fix some balky equipment, and the cosmonauts continued their daily routine of exercise, experiments, and photography of regions in Siberia with the MKF-6M system. *Progress 1* was eventually loaded up with trash from the station and undocked soon after. It was then commanded to reenter the atmosphere, where it burned up on 8 February.

"Guess Who's Coming to Dinner?": The Intercosmos Program

The next spacecraft to visit *Salyut 6* was *Soyuz 28*. Since the current Salyut crew's mission was drawing to a close, no ferry swap would be needed. The task of flying *Soyuz 28* would go to *Salyut 4* veteran Alexei Gubarev. The Soyuz ferry could be flown by one person, meaning there would be

31. Remek from Czechoslovakia (*left*) and Gubarev (*right*) would fly the first Intercosmos mission to *Salyut 6*. Courtesy of the author.

an empty seat available for a passenger acting in a simple support role. The second seat would be occupied by a Czechoslovakian cosmonaut named Vladimir Remek, flying as part of the Soviet program Intercosmos.

The Intercosmos program was started in the late 1960s and created to allow nations allied with the Soviet Union to actively take part in space research. In the 1970s it was expanded to include manned spaceflights. This was likely in response to NASA's intentions to fly noncareer astronauts from Western countries on the space shuttle as payload specialists. The Soviets were able to start their Intercosmos program first, while NASA's space shuttle was experiencing development delays.

Vladimir Remek came from a military family; his father was a general in the Czech armed forces and a deputy minister of defense. Remek became a military pilot in 1970 and held the rank of colonel in the Czech Air Force. For this flight, he was given the title of research cosmonaut. This title has been used since to describe Soyuz crewmembers on short-duration missions who are not career cosmonauts. Research cosmonauts typically undergo training for about two years before being assigned to a flight.

Soyuz 28 successfully launched on 2 March 1978 and successfully docked with *Salyut 6* a couple of days later. Romanenko and Grechko welcomed their new visitors with open arms; in Grechko's case, it made for a nice reunion with his old crewmate from *Salyut 4*. Like the previous short-duration crew, the newcomers took part in a battery of medical tests to see how well they adapted to space. Remek also took part in a brief set of experiments for Czech researchers, including taking images of his country with the MKF-6M system, but he wasn't allowed to integrate too much with the Salyut crewmembers, as his instructions were to not interfere with their research.

Indeed, Remek remarked rather wryly after the mission that his hand got red soon after he arrived on board. If it got too close to a piece of equipment, one of the Soviet cosmonauts would slap it and say, "Don't touch that!" It isn't entirely known if he was joking or serious. Even with the apparent personality disconnect, the Soviets got a lot of good public relations from the mission and announced plans to fly research cosmonauts from other countries on future missions. *Soyuz 28* undocked and successfully returned to Earth with Gubarev and Remek on 10 March.

Six days later, Romanenko and Grechko turned over *Salyut 6* to automated control and boarded *Soyuz 27* to return home. Most of their experiment results had been sent home with the two previous short-duration crews, so there wasn't much left for them to take home. They had set an orbit record of ninety-six days and were none the worse for wear when they landed, except for a severe toothache that Romanenko had been suffering from for almost a month, which required a dental visit after their landing.

Physically the two men were in excellent shape with no apparent lasting effects from their space mission, aside from the challenges of having to readapt to Earth's gravity. As with the many cosmonauts and astronauts who came before them and after them on long-duration spaceflights, waking up during the first few days back on Earth had Romanenko and Grechko trying to swim out of bed as if they were still weightless and finding that objects on Earth don't float like they do in zero gravity when they are let go of in midair. But other than having to relearn how to live in one gravity again, doctors felt it should be possible with proper preparations and exercise on orbit to have cosmonauts live up to a year in space and perhaps longer.

Further *Salyut 6* Successes

While Salyut was designed for long-duration missions with multiple crews, the next crew would not be launched until three months after Romanenko and Grechko returned and all their test results had been evaluated. Some tweaks were made to the training program based on the findings and the experiences of this record-setting crew.

The next long-duration crew of Vladimir Kovalyonok and Aleksandr Ivanchenkov launched into orbit aboard *Soyuz 29* on 15 June 1978 and successfully docked the next day. For Kovalyonok, it was his first mission to *Salyut 6* since the failed *Soyuz 25* docking, making him the first two-time visitor to the station, even though he was unable to enter Salyut on his last visit. The crew got down to business and continued the work started by their predecessors, finding that *Salyut 6* had weathered its three months of hibernation just fine. Periodic reboosts of the station's orbit took place, experiments were conducted, and the crew continued their rigorous physical exercise regime.

Kovalyonok and Ivanchenkov hosted the crew of *Soyuz 30*, who launched on 27 June. Aboard the Soyuz was veteran Pyotr Klimuk and Polish Intercosmos participant Mirosław Hermaszewski. Hermaszewski was born in 1941 in the town of Lipniki, which is part of present day Ukraine. He witnessed the horrors of World War II firsthand as a young child and became a military pilot in 1965. He was selected for the Intercosmos program from five hundred pilot candidates; to this day, he is the only citizen from Poland to fly in space. This short-duration mission went well, although cloud cover prevented the Pole from shooting good pictures of his native country with the MKF-6M system. He shot images of the Ukraine and Russia instead. Images of Poland would later be taken by the resident cosmonauts after the *Soyuz 30* crew returned to Earth on 5 July.

The crew conducted a scheduled EVA on 29 July. For this task, they opened up the side airlock door of the forward transfer compartment for the very first time. The mission tasks were simple: retrieval of the Medusa cassette left outside by the first crew and retrieval of additional exposure cassettes that had been mounted outside the station since the day it launched. They also evaluated the EVA handholds built on the station's exterior. With all their tasks finished sooner than planned, the cosmonauts spent the final portion of their allotted EVA time outside enjoying a view in their helmets that was more spectacular than what *Salyut 6*'s windows could produce.

The next Progress module to arrive was loaded to the gills with supplies. In addition to the normal food and consumables, a guitar had been flown up at Ivanchenkov's request, along with two sets of fur boots to help with their cold feet. The chill was a side effect of reduced blood flow to the lower extremities of the body in weightlessness. The guitar was a nice morale booster for the crew, as both it and the boots were put to use immediately.

The crew next hosted *Soyuz 31*, flown on a regularly scheduled ferry-swapping mission starting on 27 August. In command of that mission was *Vostok 5* veteran Valery Bykovsky flying his third and final space mission, and accompanying him was East German Intercosmos research cosmonaut Sigmund Jähn. As with the other Intercosmos flights, Jähn conducted much of the same research as his predecessors; like them, he was a military pilot before being selected for cosmonaut training. Upon return to Earth, the German Democratic Republic proclaimed him the first *German* space traveler, rather than saying he was simply the first *East German* citizen to venture into orbit. The newcomers returned home uneventfully with the older spacecraft on 3 September.

On 7 September, Kovalyonok and Ivanchenkov boarded *Soyuz 31* to undock from the station's aft port and redock with the front port. This was necessary, as the fuel transfer lines used by Progress were only present in the aft docking port. The crew prepared for undocking as if they were coming home from a mission, in case a situation developed that would prevent them from redocking. This involved equipment shutdowns, loading the Soyuz with their experiment results, and putting the station in hibernation mode. But the undocking, the rotation of the station 180 degrees on command from the ground, and the manual flying by Kovalyonok to realign *Soyuz 31* with the front docking port took place smoothly. The crew redocked and resumed their habitation of the station with no problems. Using the station to realign its orientation with its onboard control-moment gyros meant that the Soyuz ferry only used its thrusters to back up, station keep, and then close in to redock with the front, while saving precious fuel in the process. It was the first time such a maneuver had taken place with a space station, but it would not be the last.

After a regular Progress visit in October, the crew spent the final few days before their 2 November 1978 return putting the station in hibernation mode once again. The crew had maintained a steady exercise program to stay in peak physical condition. Upon their return to Earth, they had spent 139 days in orbit, which shattered the previous mark by over a month. Physically the crew was in very good shape, and the scientists put the cosmonauts through a battery of tests when they returned. Except for a finding that red blood cells produced entirely in weightless conditions are smaller than normal ones on Earth, the crew suffered no long-term effects. The door was open for future long-duration occupancy.

Salyut 6 Save

While the next crew got ready to fly to *Salyut 6*, the fuel-pressurization system was being evaluated by the ground, as it had apparently developed a leak, causing nitrogen gas to enter places it shouldn't. If the reboost engines were fired, they would likely run rough due to nitrogen gas contamination. There was an additional concern that the corrosive fuel could seep into the nitrogen-pressurization bladders through the same leak and cause damage. While a previous station might have been abandoned after developing a problem like this, it was decided to attempt repairs. *Salyut 6* had both a primary and a backup fuel system. If the leak could be isolated and the suspect system closed off, the station's propulsion system would still operate. The repair procedure would require a Progress craft and cosmonauts to perform the needed tasks. Other pieces of equipment would soon need replacement as well, since they were nearing the end of their service lives.

The *Soyuz 32* crew of space rookie Vladimir Lyakhov and *Soyuz 25* veteran Valery Ryumin would be tasked with doing the checkout and repair work. They would try to set a new space endurance record if they could, but for the first few days of the mission, they would make sure that *Salyut 6* was ready to host additional crews by conducting methodical inspections of every piece of critical equipment. They would replace what they could and make a list of additional spare parts to be flown up on the next Progress. *Soyuz 32* launched on 15 February 1979 and achieved a successful docking two days later.

Progress 5 arrived at the station on 15 March with supplies and the requested spare parts. A week later, the crew got to work isolating the damaged fuel system. The procedure was an involved one, as *Salyut 6* was commanded to begin an end-over-end spin, producing a small amount of gravity from centrifugal force. Nitrogen tetroxide fuel is heavier than nitrogen gas, so the gravity in combination with pressurizing the lines caused the fuel to empty out of the damaged tank and into the Progress. Once the damaged tank was purged, the spin was stopped by thrusters on the Progress, and the damaged lines were opened to space for about a week to boil off any remaining fuel vapors. Once that was completed, the lines were repressurized with nitrogen and then completely sealed off. *Salyut 6*'s engines worked just fine after the repairs were completed.

Experiments conducted for the first time on this mission included the hatching of fertilized quail eggs launched with the crew to monitor the development of complex embryos in weightlessness. The crew also conducted the first repair in orbit using a soldering iron to fix a malfunctioning video recorder, in addition to their other tasks of furnace processing, plant growth, and imagery of Earth. It was shaping up to be a normal long-duration mission once again. As an added bonus, *Progress 5*'s supply load also included a television monitor, so the crew could conduct two-way television transmissions for the first time and see their colleagues and family on the ground, as opposed to just hearing them over a speaker. This would be a great boost to morale for future crews. *Progress 5* was undocked on 3 April and burned up, its mission completed.

Soyuz 33 Rendezvous Failure

On 10 April 1979, *Soyuz 33*, with cosmonaut Nikolai Rukavishnikov and Bulgarian research cosmonaut Georgi Ivanov, launched into orbit to conduct a short-duration Intercosmos mission to *Salyut 6* and a Soyuz ferry transfer. About one thousand meters out, the Soyuz was supposed to conduct a six-second firing of its engines for its final rendezvous, when a problem developed. The craft began to shake violently, and the engine shut down after only a three-second burn. The Salyut crew reported seeing a strange plume from the Soyuz as the rocket exhaust seemed to fire sideways instead of from behind the craft. It was impossible to tell if the plume was coming from the Soyuz propulsion system itself or from an ignited cloud of propellant behind it.

Rukavishnikov asked if he could continue the approach with thrusters and achieve a docking, but he was overruled by controllers. If the Soyuz had hidden damage, it would be a crippled craft and potentially useless in an emergency. There would be no docking for *Soyuz 33*. Coincidentally, Rukavishnikov was on the *Soyuz 10* crew that failed to achieve a hard dock with the first Salyut. His second mission, on *Soyuz 16*, was an ASTP test flight. Rukavishnikov was again denied a chance to enter a Salyut in orbit.

For the return to Earth, the Soyuz had a backup engine designed to fire only once, continuously for the required time needed to deorbit the craft. When the deorbit time came, the engine fired but had to be turned off manually when the automatic timer didn't shut it down. *Soyuz 33* landed safely, its crew none the worse for wear after their spaceflight experience. Rukavishnikov reportedly commented that he felt as though he had been in space for a month thanks to the problems that cropped up. Neither Rukavishnikov nor his Bulgarian colleague Ivanov would fly in space again, but both were branded space heroes and awarded the standard Soviet medals upon their return.

Because the investigation into the *Soyuz 33* engine failure would take a month and parts modifications would take a little longer, *Soyuz 32* would be past its ninety-day service life by the time another Soyuz was ready to go. There was also a slight risk that *Soyuz 32* might have the same design fault as *Soyuz 33*, although it operated with no problems during its launch and rendezvous. When the next Soyuz was ready to go, the decision was made to launch it unmanned, since a ferry crew would be unable to return in the first Soyuz due to mission rules. Provided that no emergency took place that would require the crew to abandon the station, the *Salyut 6* mission could continue, albeit with a suspect craft. *Progress 6* was flown next in May to resupply the station, and it performed this task with no problems. Since the Progress used a different engine design, they weren't required to go through the same engine modifications as the Soyuz.

Once the engineers were satisfied with the final tests of *Soyuz 34*'s engine, it was launched unmanned on 6 June and spent the next two days flying a Progress-style rendezvous-and-docking trajectory toward the station. It docked without incident, with the modified engine working normally. A week after *Soyuz 34*'s arrival, the crew loaded up *Soyuz 32* with 180 kilograms of material, over three times the amount they would have normally sent home during a ferry exchange with a crew on board.

In addition to experiment results, exposed film, and materials-processing samples, the crew used the added cargo space to return some expired equipment to Earth so that engineers could evaluate how well it had performed in orbit. Without this unique opportunity, the equipment would have been dumped along with the rest of the trash in a Progress vehicle. *Soyuz 32* returned to Earth on 13 June with no difficulty. After its return, the cosmonauts boarded *Soyuz 34* and transferred it from the station's aft docking port to the front docking port, using the same procedure as the *Soyuz 31* mission.

Radio Telescope Experiment

On 30 June *Progress 7* docked with the station. In addition to the normal fuel and supplies, the Progress was packed with a stowed radio telescope experiment. The folded-up parabolic dish of the KRT-10 radio telescope would be deployed in a unique fashion that would not require a space walk. After the Progress was loaded up with stowed trash, the dish was fixed to the drogue assembly of the aft docking port. The probe assembly of the Progress was not fitted and its orbital module was left pressurized. When the command to undock the Progress was given, the cabin venting pushed it away from the station without thrusters and exposed the folded dish. The Progress was then commanded to station keep with the Salyut at a safe distance in order for its cameras to monitor the dish's unfolding.

The KRT-10 antenna dish was ten meters in diameter when fully deployed. It was used for the next few weeks to see how well it could detect radio waves from celestial bodies. Its results were compared with a telescope bank on the ground in Crimea. For a first experiment, it provided valuable data, albeit with limited results. Power requirements on the station meant that it could not be used continuously.

When the time came to jettison the antenna, things didn't go as planned, and the assembly got stuck on some antennae in the back of the station like a kite stuck in a tree. Attempts to jostle it free by shaking the station with thruster firings didn't help, either. It would have to be removed before the next visitor could dock at the aft port. A brief space walk had been scheduled for the crew to retrieve experiment packages from the outside of the station, but the crew gave an enthusiastic yes when the ground asked if they also wanted to take care of the stuck antenna. It would be the first time an EVA had been conducted to the rear of a Salyut station; while EVA handholds were setup there, nobody knew what the results would be. Repair procedures were tested in Star City's NBT to see what methods of antenna removal could be tried.

On 15 August Lyakhov and Ryumin got to work. After Lyakhov retrieved the experiment cassettes, Ryumin studied the problem and communicated his intent to cut the assembly away using a set of sheers. Since it was the first time a Soviet cosmonaut had attempted to do such a thing, controllers were understandably cautious. They eventually gave Ryumin a go, and he made the required snips before pushing the freed radio telescope assembly away from the station.

Four days after conducting the space walk, Lyakhov and Ryumin returned home in *Soyuz 34* on 19 August after setting a new endurance record of 175 days. They had not only shattered previous endurance records, but they also did it with no visitors to help relieve the boredom. Five-month-long station missions would soon become commonplace.

Final Long-Duration Crew

At the end of the *Soyuz 32* and *34* mission, *Salyut 6* had pretty much exhausted its original service life, but most of its onboard systems were still performing well. Just before the end of the year, a new variant of the Soyuz spacecraft was launched unmanned and made two approaches to *Salyut 6* before finally docking with it. This new craft was called *Soyuz T-1* (*T* standing for "Transport"). It was a third-generation Soyuz craft reintroducing solar arrays and the ability to fly three cosmonauts in pressure suits. The most important feature of this improved Soyuz, though, was the digital Argon computer, derived from the Argon system intended for the Almaz station a decade earlier. This new system would allow totally automated rendezvous and docking maneuvers to be performed by a Soyuz, with no input from the crew or the ground, except in an emergency. After spending one hundred days docked with *Salyut 6* in a hibernation state, *Soyuz T-1* was used to reboost the station's orbit before undocking.

For the final long-duration crew, *Soyuz 35* would be commanded by Leonid Popov on his first space mission. *Soyuz 13* veteran Valentin Lebedev was supposed to join him, but he suffered a bad knee injury in a trampoline accident and was scrubbed from the flight. The backup cosmonaut was considered not experienced enough, so Valery Ryumin agreed to do another long-duration mission. After a successful launch, *Soyuz 35* docked with *Salyut 6* on 10 April 1980. The crew got right to work doing what they set out to do, and Ryumin reportedly readapted to weightlessness like a fish to water. For this mission, the crew would do the usual experiment work. But repair and replacement of aged equipment was the priority, and systems were monitored closely for signs of breakdown. *Salyut 6* was originally designed for two years of service. At the end of this mission, it would be over twice that old.

Over the course of their stay, Lebedev and Ryumin hosted several visitors. The first, *Soyuz 36*, was a scheduled ferry swap in late May. Joining ASTP veteran Valery Kubasov was Hungarian Intercosmos cosmonaut Bertalan Farkas. After a week of occupancy and experiments, they returned home on *Soyuz 35*. The next mission was the first manned flight of the Soyuz T spacecraft. *Soyuz T-2*, with the crew of Yuri Malyschev and *Soyuz 22* veteran Vladimir Aksyonov, launched only two days after *Soyuz 35* departed from the station. The new Soyuz performed well, although Malyschev overrode the Argon computer and conducted a manual docking since he didn't like the approach path the automatic system was taking. Postflight analysis revealed that the computer likely would have docked just fine if allowed to continue. *Soyuz T-2*'s crew only spent three days on board before departing. These visits in rapid succession earned the station the nickname Hotel Salyut.

Soyuz T-2 was also the first mission use of a new version of the Sokol pressure suit. This version of the Sokol streamlined the design and cut the flight weight of each suit to only eight kilograms. It could be donned more quickly in emergencies, was easier for crewmembers to operate in, and its redesigned helmet and torso allowed it to be tailor fit to a larger range of body types. The revised Sokol suit continues to be used in the ISS program to this day, with over three decades of successful service under its belt.

After the *T-2* crew left, Lebedev and Ryumin moved the *Soyuz 36* up to the front docking port. A Progress delivered yet more supplies to the station, including a color television monitor to replace the black-and-white one used previously for two-way visual air-to-ground transmissions. On this monitor, the crew got to watch and take part in the opening ceremonies of the 1980 Olympic Games held in Moscow, but it didn't receive much coverage internationally since many Western nations boycotted the games after the Soviets invaded Afghanistan in 1979.

On 25 July *Soyuz 37*, with Viktor Gorbatko in command and Vietnamese

Intercosmos cosmonaut Phạm Tuân in accompaniment, docked with the station to conduct a ferry swap, replacing *Soyuz 36* with a fresh spacecraft. Tuân was the first space traveler of Asian descent and the first research cosmonaut to not come from a member nation of the Warsaw Pact. Tuân was a MiG-21 pilot who had flown in the Vietnam War. The primary experiments Tuân conducted were photographic surveys of Vietnam's water features for mapping purposes using the MKF-6M camera. The visitors returned home a few days later.

On 19 September Lebedev and Ryumin hosted their final visitors as *Soyuz 38* docked at the aft port with *Salyut 6* veteran Yuri Romanenko in command and the first Cuban cosmonaut, Arnaldo Tamayo Méndez, as his passenger. Méndez was the first space traveler to Salyut from a country in the Western Hemisphere and the first of African descent. Due to the orbit of *Salyut 6* and because of mission-recovery constraints that required daylight in Kazakhstan for a Soyuz return, there were no daylight opportunities for Méndez to use the MKF-6M system to conduct a photographic survey of Cuba, so those images were taken after he had left. *Soyuz 38* departed with its launching crew since the resident Salyut crew would soon return home themselves.

The *Progress 11* spacecraft docked with the station just before the crew left. After transferring its fuel, it was left docked with Salyut to conduct station reboosts while engineers evaluated sending additional crews to visit. Things were in remarkably good shape on the station when Lebedev and Ryumin returned home on 11 October 1980, and both men amazed the recovery forces by climbing out of the capsule and walking to their reclined recovery seats without assistance after 185 days in orbit. Part of this was due to the physical labor required during their repair work. After two back-to-back long-duration flights, Ryumin was in excellent physical shape. Doctors were encouraged by the findings.

Final *Salyut 6* Missions

For the final two missions to *Salyut 6*, the goal was to see if intensive maintenance could be undertaken to perhaps extend its service life even further. *Soyuz T-3* launched on 27 November 1980 with a crew made up of rookies Leonid Kizim and Gennadi Strekalov plus *Soyuz 18A* and *Salyut 4* veteran Oleg Makarov. This was the first three-person Soyuz crew since *Soyuz 11*.

Konstantin Feoktistov himself had hoped to fly, as he had been involved heavily in the mission planning. But he was disqualified from the flight on medical grounds, so Strekalov took his place. The Soyuz ferry was allowed to conduct an automatic docking on its own with the new Argon computer system and did so without any mishap.

The crew's prime task was to service the station's glycol coolant loops. The old system was showing signs of failure and was not designed for repair in orbit. Accessing it involved pulling up panels and insulation blankets to get at the pipes, which the crew did with some difficulty since the lines were located in places not meant for servicing. The plan was to completely replace the station's original coolant circulation pump with a new one, and the crew did that by cutting into the original coolant lines to bypass the old unit. This was risky work since glycol coolant can cause kidney damage if ingested, and glycol contamination in the cabin could force the station to be abandoned. However, the crew was able to finish their tasks safely. With that out of the way, the crew performed some scientific experiments and tested a new holographic imaging camera before returning home on 11 December. *Progress 12* was sent up shortly after with more supplies and fuel to keep the station's orbit stable until the next mission.

Before the next crew could depart for *Salyut 6*, one of the station's solar arrays jammed and was no longer able to track the sun. Combined with the degradation of the remaining arrays, it meant that *Salyut 6* did not have enough electrical power to keep its internal environmental systems running properly, and the station's internal temperature dropped to only ten degrees Celsius. The next crew would try to restore the station to operational use if they could. *Soyuz T-4* launched into orbit on 12 March 1981 with veteran cosmonaut Vladimir Kovalyonok and rookie Viktor Savinykh on board. The spot normally used by a third cosmonaut was occupied by equipment needed for the repair attempt.

After docking, the crew got down to work. Thankfully, once the jammed solar array's control system was replaced, it began tracking the sun again, and the station's internal temperature was restored to a comfortable twenty-two degrees Celsius. After unloading and jettisoning *Progress 12*, the crew hosted their first visitors, the *Soyuz 39* crew of Vladimir Dzhanibekov and Mongolian Intercosmos cosmonaut Jügderdemidiin Gürragchaa, for the standard short-duration visit. No Soyuz ferry swap took place.

Two months later on 14 May 1981, *Soyuz 40* lifted off and docked with *Salyut 6* a couple of days later. Leonid Popov, who had flown the most recent six-month tour of *Salyut 6*, was in command, and joining him was Romanian cosmonaut Dumitru Prunariu. Prunariu was a scientist and conducted many space physics experiments while on board. His visit was one of the more extensive research missions of the Intercosmos program.

Two days after *Soyuz 40* returned home, Kovalyonok and Savinykh climbed back aboard *Soyuz T-4* to begin their return journey. They would be the last human visitors to the station when they returned home on 24 May 1981. *Salyut 6* had been occupied for 685 days. It had taught many lessons, and its achievements were a high point of the Salyut program. But its work was not yet over.

While the *Soyuz T-4* crew occupied *Salyut 6*, Cosmos 1267, a Chelomei-designed TKS spacecraft, lifted off from Baikonur on a Proton booster. It was the third unmanned flight test of the TKS design and flew autonomously for about a month before detaching its VA reentry capsule, which returned to Earth successfully. On 19 June the spacecraft's FGB node successfully rendezvoused and docked with *Salyut 6*'s aft docking port. The pair spent the next twelve months docked, with the FGB periodically reboosting the station and showing how robust a system it was. Ultimately, *Salyut 6* was commanded to reenter and burned up safely on 29 July 1982 after nearly five years in Earth orbit.

Salyut 7 Flies

On 19 April 1982 *Salyut 7* was launched into orbit on a Proton rocket booster. The spacecraft, internally known as DOS-6, was the engineering backup to *Salyut 6*. The decision was made to fly it after delays were encountered with the DOS-7 station, a revised design fitted with additional docking ports. Thanks to experience gained on *Salyut 6*, many refinements were made to *Salyut 7*'s systems. Internal logistics were improved, and the life-support system was beefed up for larger crews. Progress ferry transports were also now capable of transferring water to a Salyut station, in addition to fuel, through transfer lines in the docking collar. This would free up the pressurized section in the Progress for other cargo.

Salyut 7 was equipped with refined versions of scientific equipment that had flown previously, including an MKF-6M camera system and a holo-

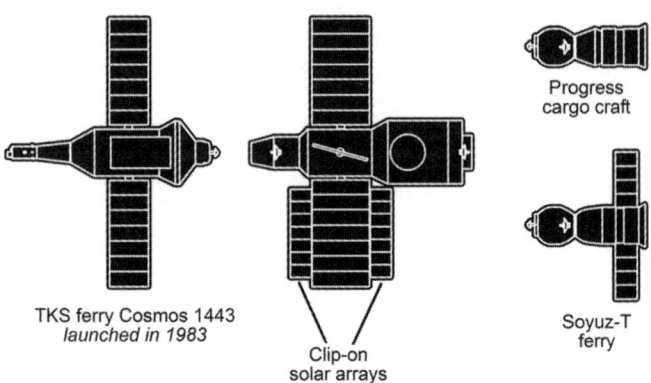

Salyut 7 (DOS-6)
with support spacecraft

Progress cargo craft

TKS ferry Cosmos 1443
launched in 1983

Clip-on solar arrays

Soyuz-T ferry

32. Both *Salyut 6* and *7* docked with Chelomei's TKS spacecraft. Courtesy of the author.

graphic imaging system tested on the *Soyuz T-3* mission. Plant cultivator experiments and materials-processing furnaces would be sent up later on Progress missions. As before, the station had a treadmill and a bicycle veloergometer, except the bike was bolted to the ceiling to save floor space. Improvements were made to the food packs. On prior missions, complete meals were packaged with everything in them. So if an entrée was opened, then so were the side dishes, regardless of whether or not crewmembers wanted them. This time, the food was all packaged separately, so the cosmonauts could choose to open only what they wanted. Crews had access to a warming oven, and a small refrigerator was provided to store fresh foods.

To help keep the portholes on the station free from strike damage by orbital debris, they were equipped with protective shutters that could be closed when not in use. *Salyut 7* was also equipped with additional EVA handholds and a rudimentary winch system that could be used to clip additional solar panels on to the larger arrays when they began to lose their efficiency from age. Several experiment packages were also placed externally on the station before launch for materials testing after long-term space exposure.

For missions involving two-man crews, the Salyut could orbit higher than three hundred kilometers. But if a three-person crew visited, the station's orbit would have to drop below three hundred kilometers, since the new ferry couldn't orbit any higher with a heavy load. Fuel resupply from Progress ferries would be very critical. *Salyut 7* was designed to remain operational for a minimum of four years. It was hoped that regular exchanges of crewmembers would allow for constant occupation of the station, but things didn't exactly happen that way.

Early Success and Visitors

Anatoly Berezovoy and Valentin Lebedev made up the first crew of *Salyut 7* when they docked *Soyuz T-5* with the station on 15 May 1982. They got right to work setting up and launched a microsatellite a couple of days later. They didn't have to wait too long for the first visitors, as *Soyuz T-6* docked on 25 June with Soviet cosmonauts Vladimir Dzhanibekov and Aleksandr Ivanchenkov on board, along with French CNES (Centre national d'etudes spatiales) astronaut Jean-Loup Chrétien. While the French CNES space program was connected with the European Space Agency, France's political contacts with the Soviet Union were more open than European countries that were part of the NATO alliance. The Soviets extended an invitation to France for Chrétien and his backup, Patrick Baudry, to train for a Salyut mission. This program was similar to Intercosmos but not connected with it.

For the week that the *T-6* crew spent on *Salyut 7*, it was a very busy time for Chrétien. He conducted echocardiograph measurements of the crewmembers to monitor how the heart migrates in the chest cavity in zero gravity and motion detection studies of how body posture alters in orbit. Chrétien also studied the effects of zero gravity on microorganisms and the effects of antibiotics on them. Due to his workload, he ended up sleeping very few hours each night and was exhausted by the time the *T-6* crew returned to Earth on 2 July. No ferry exchange took place since the T-series Soyuz could stay docked for longer periods.

The resident crew performed a space walk in late July, mounting experiment cassettes, retrieving others, and inspecting the external condition of the vehicle. A Progress delivered a new materials furnace along with other heavy equipment, which the crew activated not long after that. The crew found that by not hard-bolting the furnace to a rack and letting it float while only attached by its power cords, it could generate purer materials samples since vibration was kept to a minimum.

On 20 August *Soyuz T-7* lifted off to make history. Joining veteran Leonid Popov and rookie Aleksandr Serebrov was only the second woman to fly in space, Svetlana Savitskaya, flying nineteen years after Valentina Tereshkova's *Vostok 6* mission. The flight was likely in response to NASA selecting women as full-time mission-specialist astronauts for the shuttle program and took place a year before Sally Ride's shuttle flight on STS-7. Even today, female Russian cosmonauts are a rarity. Svetlana's father, Yevgeniy Savitzky, was a World War II fighter ace and a high-ranking general in the Soviet military, so his political connections likely led to his daughter's selection. However, Savitskaya was also an accomplished civilian and military pilot who had won aerobatic championships and set many aviation records before her cosmonaut selection.

During her time on *Salyut 7*, Savitskaya was more than just a passenger, as she conducted electrophoresis experiments on biological samples of blood protein and urine. She also processed interferon. To accommodate their female crewmate, the Soyuz was set up as a private living quarters where Savitskaya could sleep and use the toilet facilities, but she opted to move her sleeping bag into the main cabin and slept with her male counterparts. It was an exhausting week in orbit packed with experiments before the newcomers departed in the *Soyuz T-5*, leaving *Soyuz T-7* for use by the resident crew. The new craft was moved to the front port a few days later to make room for the next Progress.

Ultimately, Berezovoy and Lebedev would spend 211 days in Earth orbit. Things did not go smoothly, though. *Salyut 7* encountered failures of its Delta attitude-control computer, and the water-recycling system failed due to a seized bearing. The bearing was replaced easily, but the Delta system would need total replacement since it was required for precise aiming of the station for astronomical observations.

After the crew left, Cosmos 1443 was launched on 2 March 1983. It was a TKS spacecraft, and it docked with *Salyut 7* a few days later. It used its control system to maneuver the complex, and its solar arrays added to the station's power-generating capabilities. The TKS "tug" was also outfitted with 3,600 kilograms of supplies, and the VA capsule would allow for return of 500 kilograms of cargo to Earth.

Docking Failure

On 20 April 1983 *Soyuz T-8* launched into orbit with crewmembers Vladimir Titov, Gennadi Strekalov, and Aleksandr Serebrov. Titov was making his first spaceflight, while Strekalov and Serebrov were veterans. The crew's mission goals were to replace the Delta control system and install replacement solar arrays on *Salyut 7*, along with other repair tasks. But they wouldn't make it to the station. During *Soyuz T-8*'s launch, the Igla rendezvous radar antenna was torn off during separation of the payload shroud. The crew asked for and received permission to attempt a rendezvous and docking using just the onboard optics and radar observations from the ground. The crew was able to close within three hundred meters before the station drifted into Earth's shadow during a night pass. Illumination of *Salyut 7* from Soyuz spotlights made Titov think his closure rate was too high, so he aborted the approach. By the next daylight pass, *Salyut 7* was four kilometers away, and Kaliningrad aborted the mission due to fuel constraints. Although the mission was a failure, *Soyuz T-8* provided experience on how to conduct a manual approach, something that would prove invaluable in a few short years.

Second Crew

The TKS spacecraft reboosted *Salyut 7*'s orbit to over three hundred kilometers in preparation for the next two-person crew. *Salyut 6* veteran Vladimir Lyakhov and rookie Aleksandr Aleksandrov were successfully launched into orbit on 27 June and docked with the station soon after. Their goal was to conduct at least a four-month mission, but they would also carry out the repairs originally intended for the *Soyuz T-8* crew, a task made more challenging by having one fewer crewmember.

After two months in orbit, the crew loaded the VA capsule on the TKS spacecraft with about 350 kilograms of experiment results and filled the rest of its cargo hold with failed equipment being sent back to Earth for analysis. This particular TKS was a full-production version of Chelomei's design, and it even had three seats in the recovery capsule. But no crewmember would be riding it home. The entire spacecraft, including its FGB node, was undocked on 14 August, and the VA capsule was deorbited on the twenty-third, making a normal reentry and landing. The TKS flew free for a few more weeks until deorbiting on 19 September. While it would have been nice to keep TKS docked, the aft port had to be freed up for the next Progress to replenish *Salyut 7*'s fuel and water tanks.

Fuel Leak and Launch Abort

Progress 17 came next to deliver supplies in early September. During the fuel-transfer procedure, one of the nitrogen tetroxide lines ruptured and began leaking fuel into the unpressurized portion of the station's engine bay. Engineers and managers on the ground called it "a slight leak," but it was more serious than that, as it meant that the station's reboosting engines were crippled until the line could be repaired. Thankfully, the Progress could perform reboosting operations with its own engines to keep the station's orbit stable. Engineering teams worked on a solution on the ground while the resident crew prepared for its next visitors.

The next mission, originally designated *Soyuz T-10*, was scheduled to launch on the night of 26 September with the two-man crew of Vladimir Titov and Gennadi Strekalov. They were rescheduled to fly this mission because of their extensive preparations to conduct EVA repairs as part of the *Soyuz T-8* mission. Seconds before scheduled liftoff, a valve on the pad failed to close, and leaking fuel caught fire, engulfing the bottom of the R-7 rocket. The fire steadily grew for about twenty seconds into a raging inferno before a launch controller finally triggered the escape tower on the Soyuz, pulling the two cosmonauts safely away on a high-g ride that would ultimately land them about four kilometers from the pad. The rocket exploded soon after. The crew was a little shaken from their ordeal, but at least they were alive. Reportedly, Titov said in subsequent interviews that they turned off the ship's cockpit voice recorder after the escape rockets fired because "We were swearing," which was probably putting it mildly. In keeping with the Soviet tradition of not assigning a full number to a mission that failed to reach orbit, this mission was called *Soyuz T-10A* or *T-10-1* when details of the launch abort leaked out to Western agencies.

Thus, it would fall to the resident crew to carry out the solar array repair. The new clip-on solar arrays had been sent up on Progress ferries and were already on hand. On 1 November Lyakhov and Aleksandrov opened up the airlock to conduct their first EVA. The first clip-on array was added just fine, while the second one took some effort due to difficulties with the winch system, but finally it was put into place. Originally, three people were desired for this repair, because once one set of panels was added, the solar array being upgraded could be rotated 180 degrees by a crewmember inside to allow the second pair of panels to be fitted on one EVA. With only two crewmembers, the EVA tasks took two days to complete since no one inside could rotate the array. The added power from the new arrays added capabilities to the station's onboard equipment. On 23 November Lyakhov and Aleksandrov returned home. While they didn't set a space endurance record, the crew and their Soyuz spent 149 days in space, validating the T series Soyuz module's design life. It would be up to the next resident crew to fix the fuel leak.

New Residents

A new *Soyuz T-10* mission took place on 8 February 1984 with the three-person crew of Leonid Kizim and rookies Vladimir Solovyov and Oleg Atkov. Atkov was a medical doctor who was added to the team so that he could monitor their health on orbit. Once the crew docked with *Salyut 7*, their first tasks were to check over the station to see if it was still suitable for long-duration habitation and to reequip it with supplies from the next Progress ferry. Once those tasks were done and a short-duration crew visited, they would get down to the task of repairing the station's damaged fuel line.

Soyuz T-11 docked next on 4 April with the crew of Yuri Malyschev, Gennadi Strekalov, and Indian research cosmonaut Rakesh Sharma. Third time was the charm for Strekalov, as there was no docking failure or launch abort to deny him this Salyut visit. Rakesh Sharma was the second international crewmember not from a Soviet-aligned nation to fly in space. A pilot in the Indian Air Force, Sharma conducted several medical experiments during his time on orbit with the help of Dr. Atkov. Sharma photographed India with the MKF-6M, as well, in preparation for an Indian hydroelectric dam project. The week of activity left both him and his crewmates exhausted, but they got some good data on the mission before departing, swapping ferries for the *T-10* module. The resident crew redocked *Soyuz T-11* on the front soon after to make room for the next Progress.

Progress 20, which had docked only six days after *Soyuz T-10* left, contained the usual load of supplies and spare parts, but it was also fitted with a small platform to help with the repair of the fuel system. Kizim and

Solovyov suited up for an EVA on 23 April. Their first task was to set up the work site for the engine module with a special platform and pre-stage tools for the repair. The task took longer than expected, so the next phase didn't begin until three days later, after a rest and servicing of the Orlan suits. Performing the simple task of repairing a leaky pipe was anything but simple. After the suspect pipe was exposed, nitrogen was pumped through it from inside to verify the leak. The plan was to isolate the damaged pipe and its valve while replacing both with new ones. They only just got the new valve on before EVA time expired, so the crew rested for another three days before venturing out one more time to complete the repair task by fitting the new pipe.

After *Progress 21* docked with the final set of tools they needed, the pair conducted their fourth space walk in less than a month and finished their tasks to bring the fuel system for *Salyut 7*'s engine back online. They next fitted a second pair of clip-on solar panels on to one of *Salyut 7*'s arrays and cut a sample from the original panel for analysis on the ground.

The next visiting crew of Vladimir Dzhanibekov, Svetlana Savitskaya, and Igor Volk docked *Soyuz T-12* with the station on 19 July 1984, bringing the crew tally up to six. For this mission, Savitskaya was a flight engineer, while Volk occupied the third seat. Volk was a test pilot in the Soviet space shuttle program (later known as Buran), and he was flying a space mission to see how a week in zero gravity would affect his piloting abilities during simulated return of a shuttle from orbit.

Savitskaya's primary task was to conduct a vacuum welding experiment during an EVA, beating American Kathy Sullivan to the milestone of first EVA conducted by a woman by about three months. Dzhanibekov and Savitskaya performed their task on 25 July, with Savitskaya taking up position at a special workstation built by the Institute of Electrical Welding in Kiev. She used the specially designed electron-beam welder to both cut metal plates and weld them together. When her task was completed, she exchanged places with Dzhanibekov, and he conducted the same tasks. Funny enough, after the jobs were finished, academician Boris Paton, director of the Institute, commented that the test work would lead to robotic welding processes that could be carried out automatically during future space missions. To date, no other welding tasks have been conducted in orbit.

After the short-duration crew departed following ten days of joint activity, Kizim, Solovyov, and Atkov continued their long-duration assignment, eventually setting a new endurance record of 237 days in orbit. They made numerous photography passes of Earth with the MKF-6M, conducted celestial observations with a new X-ray telescope, and processed numerous material samples in the furnaces. Dr. Atkov conducted a thorough on-orbit study of the cardiovascular systems of his two crewmates. This, in combination with the crew conducting six of the station's seven EVAs during the mission, meant that it was a very productive flight on all fronts. When *Soyuz T-12* returned to Earth on 2 October, the stage seemed to be set for future long-duration crews. There was even talk of having an all-women cosmonaut crew occupy *Salyut 7* on a future mission.

Saving a Dead Space Station

In the beginning of 1985, *Salyut 7* was in a stable, high orbit. While the station was not quite as reliable as its immediate predecessor, all indications were that it could support additional crews. However, all those plans were put on hold when contact was lost on 11 February. *Salyut 7*'s orbit was still stable, and observations from the ground indicated that the station was intact. But it was dead in space. In early March the Soviet news agency TASS reported that *Salyut 7* had completed its mission. So it came as a surprise to many when *Soyuz T-13* lifted off on 6 June 1985 with Vladimir Dzhanibekov and Viktor Savinykh on board. In place of a third cosmonaut, their Soyuz was crammed with equipment for an attempt to bring the crippled station back to life. The Soyuz flew a two-day rendezvous trajectory to save propellant. If docking succeeded, the T series Soyuz could stay docked for up to ten days, acting as the controlling spacecraft, before a lack of consumables would force the crew to come home.

With the experience gained from *Soyuz T-8*'s manual rendezvous, *Soyuz T-13* closed within visual range of the station, using tracking data from the ground. *Salyut 7* was doing a very slow roll of about .3 degrees a second, enough for the Soyuz to match; the structure looked intact otherwise. The Soyuz periscope system had been equipped with a laser rangefinder for this mission, to give rate-of-closure data for a manual docking. Dzhanibekov carefully guided the Soyuz in for a soft dock and then achieved a successful hard dock when the latches were retracted. After opening a valve to determine that the Salyut was still pressurized and finding no toxic fumes, the crew opened the hatch. There was still atmosphere in the station, but it was completely dark inside with all the window covers shut. The two cosmonauts were greeted to a surreal sight of frost on almost every surface and zero-g icicles. It was well below freezing; even with cold-weather gear on, the crew had to retreat to the Soyuz for warmth after each hour of work. For the first few days, the two men would eat and sleep inside the Soyuz. They ran a ventilation hose into *Salyut 7* from the Soyuz since the stagnant air contained floating pockets of carbon dioxide.

The fault was traced to an electrical switching system that controlled whether the station was powered by batteries or solar energy. After a night pass, it had failed to switch back to solar power; the batteries ran down, ironically, while trying to keep the arrays tracking the sun. Four of the six batteries were still useable; once the station was in a stable attitude, they began taking a charge from the solar arrays.

As the hours went on, the crew slowly began to restore *Salyut 7*'s attitude control and environment systems. The temperature inside was brought up slowly so as not to overwhelm the moisture-collection system with too much melting ice. *Salyut 7* went through a slow thaw for the next several days. Once the humidity levels began to drop, the heaters were finally activated to warm the station up fully. The ice had damaged some of the onboard equipment, as split water pipes needed replacing and the Orlan suits were written off as being too damaged for EVA use. A Progress ferry brought up spare parts on the next supply run and docked with no difficulties. *Salyut 7* was saved.

Final Resident Crew

On 17 September *Soyuz T-14* launched with a three-man crew of Vladimir Vasyutin, Georgi Grechko, and Aleksandr Volkov. They docked with the station the next day. For this mission, Vladimir Dzhanibekov would return home with Grechko, leaving Viktor Savinykh to attempt setting a new space endurance record with two new crewmates who hoped to be in orbit for six months themselves. Grechko and Dzhanibekov departed on 26 September; after swapping the ferry's docking port, the crew settled down to receive their next visitor, Cosmos 1686, which docked on 2 October.

The Cosmos craft was a TKS ferry, but the VA capsule section was heavily modified with additional instrumentation, making it incapable of independent flight. There has been speculation that this new module was an attempt to conduct a military mission similar to Almaz. Vasyutin was a military-research cosmonaut and specially trained to operate the systems on this particular TKS ferry. With new equipment and supplies delivered by the TKS, all seemed to be going rather well.

It wouldn't stay that way, however; by mid-October, Vasyutin was experiencing symptoms from an unknown ailment resulting in anxiety problems, little sleep, and a loss of appetite. The crew kept the news quiet at first, but by the end of October, they consulted medical specialists. Preparations were made to depart, but Vasyutin seemed stable enough to wait it out until mid-November, when conditions at the recovery site were ideal for a daylight return. By 17 November, Vasyutin was experiencing sharp abdominal pain, and the crew was ordered to close down *Salyut 7* and return home.

The crew departed the station on 21 November, far short of the planned mission duration. Soviet news sources downplayed Vasyutin's ailment, saying that he was in good health and that he was admitted to a hospital for precautionary reasons. Since then, it has been reported that Vasyutin was suffering from a prostate infection and that he may have started showing signs of it before launch. But he kept it quiet from the doctors during physicals. Whatever the cause, Vasyutin was forced to retire from the cosmonaut program on medical grounds. *Salyut 7* would ultimately host one more visit from a crew in 1986, but other than that, no additional crews would take up residence.

While *Salyut 7*'s operational history was a bit more checkered than *Salyut 6*'s, its contributions to Soviet long-duration spaceflight were no less important. All that experience, both good and bad, would come into play when the next station flew.

8. European Participation

After the Skylab program, NASA spent the next few years designing and building their next space vehicle. This would be something very different from Apollo, a reusable system known as the space shuttle. Reusable space vehicles had been a dream of many for decades. Wernher von Braun considered it a vital element to any long-term space-faring effort in order to ferry both men and material into Earth orbit. Film and literature of the time (such as the appearance of a commercially flown passenger shuttle in *2001: A Space Odyssey*) helped to make the concept popular with the general public. But concepts are one thing, while reality can end up quite different.

The shuttle's promise was that with the spacecraft being reusable, the months of preparation needed to produce, stack, and launch an expendable rocket could theoretically be reduced down to a few days' time to refurbish the craft and its boosters between missions. In theory this would cut costs and give NASA and the air force (who were brought in as partners on the project to use shuttle as a launcher for DOD payloads) expanded capabilities over what the Soviets could do with their fleet of space launchers.

When the 1972 space appropriations bill was passed by Congress, it included funding for development of the shuttle. But the amount of funding was not as much as NASA had hoped, and some changes had to be made to the design. NASA's compromise had the shuttle whittled down from a winged orbiter with a fully reusable winged booster to a winged orbiter, a large external tank (ET) serving as the fuel supply for the shuttle's LOX-and-LHX-burning rocket engines, and two solid-rocket boosters (SRBS) strapped to the sides of the ET to give the vehicle additional thrust for roughly the first two minutes of flight. Once the fuel was used up in the ET, it would be jettisoned to burn up on reentry, while the orbiter and SRBS were recovered and reused.

A unique shuttle feature was its capability to carry payloads both into orbit and back via a sixty-foot-long by fifteen-foot-diameter payload bay encased by a pair of giant doors. This design was in keeping with NASA's modular approach for future programs. The shuttle would be able to deploy satellites, launch probes to other planets, take specialists up to repair satellites, build structures in space, and return cargo from orbit. It was a progressive, logical approach to spaceflight in comparison to a one-destination program, such as Apollo.

There was a downside to this approach, though, as the shuttle would only be as productive as the payloads it would fly. While NASA anticipated a rich future launching satellites for commercial firms and flying payloads for the DOD, frequent missions would be needed to help reduce mission costs to the point where the shuttle could be considered a viable alternative to expendable rockets.

Another concern was that NASA would be spending almost all its manned spaceflight resources to develop the shuttle and practically nothing else. The shuttle would be an orbit-going pickup truck with no specific destination it could fly to repeatedly. The logical destination for a space shuttle was a space station, but NASA had no funding left to build a station. It was hoped that *Skylab* would stay in orbit long enough to allow a shuttle to dock with it, but its reentry in 1979 before the shuttle was ready killed that idea.

During the space shuttle's initial development, ideas were drawn up to have the shuttle fly an experiment module to conduct research in orbit similar to what a space station could do. This module could be flown on missions lasting from about a week to ten days. The early concept was referred to as the "Sortie Lab." It featured a combination of a pressurized module and an unpressurized instrument section fitted in the cargo bay of the vehicle, essentially turning the shuttle into an orbiting space laboratory. The lab could be configured with modular instrumentation and fly different experiment packages in many scientific fields.

From a budget and practicality standpoint, the Sortie Lab made a lot of sense. Developing the concept into hardware would potentially be less costly to build and operate than a separate space station. Frequent missions could be flown on an as-needed basis, so astronauts would not have to spend weeks or months in space. Early missions could gain valuable data, meaning that follow-up missions could fly revised experiments more quickly, using

33. Concept artwork of what would become known as Spacelab. Courtesy NASA.

the lessons learned. Sortie Lab missions would also increase the frequency of shuttle flights, helping to lower operating costs. The idea made sense in many ways. The only question left to answer was who would build it.

Early European Space Efforts

Many Western countries in Europe had contracted the "space bug" after seeing the successes achieved by the United States and the Soviet Union during the early years. But while ambitions were high, the budgets of many of these nations in the 1960s were not enough to generate space programs of their own. Many European economies were still recovering from the lingering effects of World War II. The Cold War also made defense expenditures a higher priority. Space rockets also require an ocean expanse or a large, sparsely populated territory to fly safely over in order to reduce the risk of spent rocket stages landing on someone's head.

Only two European countries had proper launch facilities. Of the two, the United Kingdom made an attempt to develop its own satellite launcher in the form of Black Arrow. But the program's one and only success in launching a satellite came after the program was canceled in favor of launching payloads on less expensive American rockets. France, on the other hand, orbited small research satellites aboard the Diamont series of rockets, but they lacked a heavy booster to fly large payloads.

For Europe's situation, cooperation between multiple countries seemed to be the best answer for space operations. Two collaborative space organizations were formed. The first was the European Launcher Development Organization (ELDO). ELDO was a consortium of six European nations: Belgium, the United Kingdom, France, the Netherlands, West Germany, and Italy. It was formed with the purpose of creating a space launch capability for Europe. Australia was brought in as an associate member, because it provided a test range that could be used for space launches. The second group was the European Space Research Organization (ESRO). It was formed initially along the same lines as the European nuclear research organization CERN and consisted of ten member nations. ESRO's members included the six ELDO partners in addition to Denmark, Sweden, Spain, and Switzerland. Its stated goal was to develop hardware strictly for space research.

Things got off to a slow start with both organizations. The biggest hurdle to forming consortiums is that the member countries don't all necessarily work toward the same goals. The early years of each group were spent figuring out how to work together for a common good and altering the organizational structure to take care of bureaucratic stumbling blocks while minimizing political infighting behind the scenes. A second problem was that each country had its own unique culture, working habits, and in many cases language. So it took a while for the countries' representatives to figure out the idiosyncrasies of one another while also seeing what each nation could bring to the table in terms of scientific contribution, hardware development, and most importantly financial support.

The ELDO organization managed to create the Europa rocket booster. It featured a UK-developed first stage, a second stage built by France, and a third stage built by West Germany. While the Europa showed promise in early testing of its separate stages, all eleven tries to launch the complete rocket ended in failure. In 1971 the United Kingdom pulled out of the project, and the rest of the member countries tried to salvage what was left.

ESRO didn't have much early success either, as politics and economics of the member nations produced many stumbling blocks to steady financing. Over time, though, they got their act together. ESRO was able to create its own spacecraft-tracking network and initiated a successful program in launching scientific payloads aboard sounding rockets. ESRO also developed research payloads that were flown into orbit on American rockets and began cooperative programs with NASA to develop research satellites. However, ESRO still craved more involvement.

By the early 1970s, ESRO's mission policy had been altered to include development in practical applications of space technology with potential profit in mind, specifically telecommunications satellites. This produced an increase in funding and more involvement from European corporations in the aircraft and defense industries. ESRO was also considering manned spaceflight. The problem was that it didn't have a manned spacecraft or launch vehicle, and it also lacked funding to start an astronaut program. So it looked to NASA to see what might be possible.

Representatives from ESRO, ELDO, and NASA had been meeting one another off and on for a few years in the late 1960s. NASA's International Programs Office set up by Arnold Frutkin provided opportunities for European countries to have direct relations with the agency. One benefit included releasing research data collected by NASA space projects to scientific organizations in Europe. By 1969, while NASA administrator Thomas Paine was trying to drum up support for the agency's grand plans after Apollo, he gained enthusiastic interest from representatives of both ELDO and ESRO to potentially become full partners in a future space project.

ELDO was hoping to develop hardware for a proposed space tug, a vehicle designed to take payloads from low Earth orbit, where the shuttle flies, to higher orbits and back down as necessary for repair or return to Earth. Given the difficulties with the Europa, many ELDO officials felt that a space tug project would give the organization a future and keep the consortium intact. While Europa itself was a failure, there were many positive aspects of ELDO's experience that could be applied toward a similar endeavor. ESRO, on the other hand, was primarily interested in contributing scientific-research payloads for a future space station. Building operational hardware didn't interest them as much, unless it involved scientific research or some sort of practical application that could directly benefit its European members.

While the money from Congress was not enough to build a space station and the space tug in addition to the shuttle, there was at least support from both Capitol Hill and the White House to allow international cooperation in the shuttle program to help lower the development costs. But there was a reluctance to have European industry take direct part in construction of the shuttle itself due to concerns about technology transfer. Any spin-off technologies developed might be used in direct competition with American interests.

NASA's Sortie Lab proposal seemed like a perfect alternative. If Europe was to take part in building the lab module, they could handle most of the development costs with guidance from NASA. In exchange, the European consortiums would get a bigger foot in the door on future manned space projects. They would also have the opportunity to fly their own astronauts on NASA spacecraft to conduct dedicated research in orbit.

In 1973, NASA and the European agencies entered into an agreement to develop the Sortie Lab concept into hardware. Having two space consortiums involved in the mix might seem like too much bureaucracy for the program's own good, but the decision had already been made in Europe to merge both ELDO and ESRO into one agency. The result would be known in the coming years as the European Space Agency, or ESA.

For all intents and purposes, NASA would be dealing with the ESA. Though the final merger of the two consortiums would not take place until 1975, steps had already taken place to streamline the ESA's management structure, before the formal name change, in order to begin hardware development. In addition to participation in the Sortie Lab project, the ESA was also tasked with the creation of a new family of space launchers for Europe, the Ariane series. The Ariane 1 and 4 rockets would go on to become some of the most successful commercial satellite launch vehicles in history (followed by the Ariane 5 in 1996).

The Birth of Spacelab

The Marshall Spaceflight Center became the primary NASA facility responsible for coordinating the lab program with the ESA, thanks to their success with the Skylab program. Marshall would also be in charge of the finished modules. JSC would be responsible for managing the development of the shuttle's integration with the lab and flight-crew training, while KSC would handle ground preparation of the modules for flight. All three centers already were coordinating with each other on the space shuttle program since the shuttle was a combined system featuring the integrated hardware of a rocket booster, an aircraft, and a spacecraft in one package. Similar coordination was just as important for the lab's development, since it would be an integral part of the shuttle when mounted in the payload bay. Several of the shuttle's systems would have to be designed for laboratory support early on since, unlike a space station, the lab could not fly as an independent spacecraft.

At the start of the project, the ESA had elected to change the name of the system from Sortie Lab to Spacelab. Initially NASA was reluctant to use the new name since it was close to Skylab, but the name stuck and described the purpose of the system perfectly. The Spacelab system would not just be a one-size-fits-all, single-module concept. While the pressurized laboratory module is the item most people identify with the Spacelab program, it is only one part of an entire family of hardware. In addition to the pressurized module, the ESA also developed a system of pallet racks that could be fitted with experiments in the shuttle's payload bay. These pallets could be used in support of a pressurized lab module or fly independently. For a pallet-only mission, a pressurized "igloo" was developed to help house computer subsystems for the experiment racks.

The reason for the igloo's use is due to heat transfer characteristics in the pure vacuum of space. Air circulated by fans can be used to cool electronics, while waste heat in a vacuum has to be removed by other means. There are the temperature extremes of space to consider as well. If the computers are sealed in their own little air pocket, they can be cooled conventionally without the need for more costly engineering. Waste heat from the igloo could then be removed through the shuttle's own cooling systems. For research projects being conducted on limited budgets, this approach made sense. In the end, though, the igloo had its fair share of engineering problems to solve during development.

Whether the Spacelab systems required a pressurized laboratory or the smaller igloo, they were dependent on the shuttle for power, environment (in the case of the lab module), and the ability to expel a percentage of waste heat into space with the shuttle's cooling loops and the payload bay's radiators. The Spacelab would supply additional heat radiation capability. Communications and telemetry with control centers on the ground would be routed through the shuttle's radio and television antennae. The shuttle would also have to be flown in the attitudes required for specific missions by the flight crew. Essentially, when a Spacelab was flying in the payload bay, the shuttle itself would become the laboratory.

The initial budget for the Spacelab project was approximately $250 million. Given the nature of several European partner nations having their own monetary currencies in the 1970s (this was about two decades before the creation of the Euro), the ESA and NASA came up with a standardized accounting unit as a stand-in for a single European currency. This would help streamline the account books somewhat. The Spacelab project still ran into cost overruns as drawing board concepts became actual hardware, resulting in a total cost of about $1 billion when the first Spacelab finally flew. Compared to the shuttle program's cost overruns, the Spacelab's budget problems were relatively small.

While NASA would provide invaluable assistance to the ESA for creation of Spacelab, it would not pay for the first pieces of hardware, as these were given to NASA in exchange for the training and flying of Europe's first astronauts. The contract called for one pressurized module, five pallet racks, and an engineering model for use in integration and testing of Spacelab systems on the ground. The engineering model would essentially be a functioning Spacelab, although it would never fly in orbit. Priority was also given in the shuttle program to fly three joint NASA and ESA Spacelab missions at the earliest opportunities. NASA would have to fund creation and building of ground support facilities to service the Spacelab systems between missions. If NASA liked the equipment, the Americans would have to pay the ESA for additional hardware.

The fact that the ESA would be building the hardware and "giving" it to NASA did not sit well with many in the European press and general public, though. It seemed like the ESA was spending a lot of money for a Spacelab in exchange for training payload specialists and only three joint missions, with money having to be spent if the ESA wanted to fly "their" Spacelab on future missions. Even with this public discord, the managers in charge at the ESA knew that if they were going to have a long-term future in manned space projects, they had to make some commitments to get their foot in the door and gain experience at building hardware.

Many at the ESA also hoped that NASA would give a price break to them for future shuttle flights after Spacelab's delivery. But NASA had to explain that the price being charged to the Europeans for flights was unfortunately going to be the same as the price charged to American customers, even though many of the same potential customers had also spent money and resources developing the space shuttle. This was standard practice in U.S. government contracts, and they couldn't waive the process for Europe. This resulted in some bruised feelings between NASA and the ESA during the early years.

Working Together

On the surface, the relationship between NASA and the ESA was similar to that of a customer and a contractor, but what made Spacelab different is that European industry prior to this point had no experience in designing equipment for manned space missions. They had to get up to speed quickly since the initial timeline for Spacelab called for its first flight to take place sometime in 1979.

NASA temporarily assigned several engineers from all the NASA centers to work at locations in Europe during Spacelab's initial development, giving valuable guidance to the ESA on how things should be done. Several ESA engineers also made frequent visits to the United States to sit in on shuttle design meetings, inspect relevant shuttle hardware, and analyze the impact of shuttle design changes on their own efforts. Gradually, the Europe-based NASA workforce was reduced as the ESA and their contractors got up to speed. Improvements in long-distance communications also streamlined the coordination. But for important meetings, such as design reviews, many high-ranking NASA managers and engineers had to travel to Europe to attend those sessions in person.

There are universal constants in the development of hardware for spaceflight, and the primary one is that weight will go up when a design is turned from a paper drawing to a piece of hardware. The Spacelab had to undergo several phases of refinement during its development in order to shed the pounds it gained as the first pieces of hardware were built. This was no different than what happened during the Apollo program.

Another constant is that the command and control computers selected for a spacecraft during its design are going to be considered obsolete by the time they fly. Computers needed for spaceflight operations don't need to be too advanced, just very well understood so that no unforeseen bugs crop up during use. But older computer hardware can limit processing capabilities as experiments often have to be adapted to use the older equipment.

While NASA initially wanted an off-the-shelf computer system based on the one flown on *Skylab* for use in the Spacelab, the Europeans wanted a homegrown computer instead. Ultimately the unit selected was a French system of similar capability to the U.S. system but still not as advanced. The computers of both the shuttle and the Spacelab were developed just prior to the PC revolution in the United States, so they did not benefit from the constantly evolving microprocessor architecture of the next decade.

The prime industry contractor for Spacelab, with its major responsibility being the pressurized laboratory module, was a consortium of several German aerospace companies that were part of VFW Fokker and ERNO (Entwicklungsring Nord, a German phrase meaning "Northern Development Circle"). In the mid-1980s, this company came to be known as MBB-ERNO. Co-contractors for the project included Aeritalia of Italy, responsible for the module structure and thermal control; Engins Matra of France, responsible for the command and data computers; AEG Telekfunken of Germany, responsible for the electrical systems; Dornier Systems of Germany, responsible for environmental control; Hawker Siddeley Dynamics of England (which today is part of British Aerospace, also known as BAE), responsible for the pallet structures; Bell Telephone of Belgium, responsible for ground support electrical equipment; Inta of Spain, responsible for mechanical ground support hardware; Fokker VFW of the Netherlands, responsible for airlocks; Sabca of Belgium, responsible for additional structures; and KAMPSAX of Denmark, responsible for computer software.

There would be U.S. contractor work as well. TRW would offer design and liaison support to their European colleagues, since it had vast experience in aerospace projects going back to the early days of ICBM development. Making sure the ESA hardware would be fully compatible with the shuttle would be the task of the McDonnell Douglas Technical Services Corporation (MDTSCO), acting as NASA's integration contractor. The combined experience from MDTSCO's parent companies, the recently merged McDonnell and Douglas corporations, made the firm a valuable resource. MDTSCO would also develop the transfer tunnel that would allow astronauts to enter the Spacelab's pressurized module from the middeck of the shuttle. Their design experience on the MOL transfer tunnel for the Gemini-B came in handy for this project.

NASA and the ESA selected joint Spacelab program directors. On the

NASA side, Douglas R. Lord would represent the program until his retirement in 1980. Lord would primarily direct the program from the Office of Manned Spaceflight in Washington DC, with the program's management being run out of Huntsville, Alabama. ESA management became a revolving door, though. The first head of the ESA side (still ESRO at that point) was Heinz Stoewer. Later, the job fell to Bernard Deloffre, who served in that capacity before abruptly resigning in 1976. The job finally was given to Michel Bignier. Prior to joining the ESA, Bignier was the director general of CNES in France from 1972 to 1976 and was considered to be a very approachable individual who also provided a firm hand of guidance for the ESA at the time. He also spoke very fluent English, which NASA program director Lord found to be very comforting. Bignier served in this capacity until 1980 when he became director of the ESA's Space Transportation Systems, from which he retired in 1986.

Early on in the program, personalities between NASA and European industry clashed somewhat, partly due to the cultural differences and the management structure of some of the contractors. The biggest problems in the early days were with ERNO, as the Germans resented the presence of NASA representatives and considered them a distraction to the work being done. NASA tended to favor a sideways management style where different departments could work with one another independent of high management, while ERNO favored a top-down structure instead, with problems being taken up the chain of command and relayed to the other departments by the management teams. It took a while to smooth things over. But ultimately the problems were worked out, and mutual respect was earned as the Spacelab program matured.

European industry also provided a bit of a culture shock to some NASA representatives. At some contractor factories, the work seemed to be a throwback to the early days of aviation, with equipment being hand built by skilled artisans in open-air factory buildings. Other factories featured high-tech manufacturing techniques more familiar to the Americans. Many of the firms working on Spacelab were also involved in building hardware for the European aircraft and defense industries. ERNO was involved in construction of the Ariane launch vehicle for the ESA and had clean rooms that rivaled any U.S. space firm. For the most part, these companies were able to deliver hardware to the required specifications with very few problems cropping up during manufacturing.

Throughout the Spacelab design process, astronauts were brought in to offer their input. Skylab astronauts Ed Gibson, Paul Weitz, and Joe Kerwin, along with Skylab backup astronaut Bill Lenoir, were on hand to inspect one of the Spacelab engineering mock-ups during a preliminary design review in 1976. Bill Thornton and Owen Garriott also provided their input on experiment matters. Astronaut input was critical to the design of the Spacelab systems, since ultimately it was the astronauts who would be operating the experiments in orbit and dealing with any problems that developed.

Europe's First Astronauts

Once Spacelab began flying, there was the question of who would conduct the research. By this time, NASA's astronaut corps had been organized along the ranks of pilot and mission specialist. The pilot group included shuttle commanders, and their job would be operating the shuttle as needed for mission support. The mission specialists would be responsible for the EVAS, running the robot arm and tasks related to shuttle systems. While some mission specialists would be assigned to conduct scientific research on missions, some payloads had a need for a crewmember that had more in-depth knowledge into the hardware being flown and less need to operate the shuttle systems. This required a new classification, the payload specialist.

A payload specialist is essentially a part-time astronaut. They would get a minimal amount of spaceflight training, usually only six months to a year, while the pilots and mission specialists were career astronauts with at least two years of training under their belts. Some payload specialists would be little more than VIPs, on hand to conduct satellite launches for their countries. Others would be integral to operating the payloads that were flown, as they knew the experiments better than anyone else.

In 1978, the ESA selected three astronauts to fly as payload specialists, and the trio journeyed to Houston for training. They were Ulf Merbold, Wubbo Ockels, and Claude Nicollier. Ulf Merbold was born in what became East Germany after World War II, but escaped to the West before the Berlin Wall was erected. He became a physicist and holds a PhD in that field. Prior to joining the ESA, Merbold worked for the Max Planck Institute for Metals Research in Stuttgart. Wubbo Ockels was born in the Dutch town of Almelo and has PhDs in both math and physics. His primary field of research before the ESA was the creation of radiation particle detectors. Before being accepted for a job with the ESA, Claude Nicollier was a pilot with the Swiss Air Force and studied physics while holding down a job as a commercial airline pilot.

The first two men would get to fly as payload specialists, with Merbold taking part in several missions and Ockels only one. Claude Nicollier, on the other hand, wouldn't fly for many years, but he added a diploma from the Empire Test Pilots School at Boscombe Down, England, to his resume in 1988; he became one of the ESA's first mission specialist astronauts. He took part in several high-profile shuttle missions, including the first repair mission of the Hubble Telescope on STS-61 in 1993. A fourth astronaut candidate, Franco Malerba of Italy, was also selected by the ESA, but he didn't journey to NASA for training and remained in Europe. Malerba eventually flew as a payload specialist on STS-46. He holds a PhD in physics, specializing in biophysics.

Mission Simulations

While the hardware was being designed and built, parallel work was underway on the science. New procedures were needed for selecting experiments, operating them, and determining the best methods for data collection. All this work had to take place long before the first operational Spacelab mission, in order to iron out as many bugs as possible. NASA instituted a Concept Verification Test (CVT) program starting in 1974 utilizing a mock-up at Huntsville that mimicked the Spacelab. It had a cylindrical laboratory section and a pallet module. Five simulations were made in the mock-up lab over the next couple of years before the project was canceled due to lack of funding. Even with its brief existence, the CVT program provided invaluable data.

Additional simulation studies were carried out using a Convair 990 passenger airliner converted for use as an airborne science laboratory by the NASA Ames Research Center. Known as the Airborne Science/Spacelab Experiments System Simulation (ASSESS), the aircraft would fly simulated Spacelab mission profiles and test out experiments that were under consideration for spaceflight. The experiment evaluations would be conducted over about a week's time period, simulating a whole Spacelab mission from beginning to end. The aircraft would fly for several hours each day with investigators on board conducting their experiments. At the end of the day, the plane would park next to a housing facility, so the mission participants would remain isolated from the ground support crews.

The Convair 990 aircraft, nicknamed the *Galileo II*, provided invaluable data on many of the celestial-observation systems and spectrograph experiments being considered for Spacelab. NASA and the ESA both got valuable data on how the hardware performed in a sealed aircraft environment, drawing power from the onboard electrical equipment. The ESA also got an excellent run-through on their procedures. After one round of flights in 1975, the ESA asked for and was granted permission to conduct a second set of flights in 1977 to give their experiment managers more experience. Several ESA candidates hoping to become payload specialists took part in these missions, as did some of NASA's astronauts who would be flying as mission specialists on the early Spacelab missions.

Spacelab Hardware Anatomy

The Spacelab system was designed to conduct research missions with either instrumented pallet racks or a pressurized laboratory module, usually as a combination of both. If the lab module was being flown, once the equipment was powered up, astronauts would gain access to it by floating down a long transfer tunnel. Early plans called for the tunnel to be equipped with an airlock on top for EVA activities, but budget cuts meant that this capability was never developed. The shuttles were each equipped with an internal airlock; if an emergency space walk had to take place to service either the shuttle or a Spacelab system outside, the lab would have to be deactivated and a flexible part of the tunnel would have to be partially retracted.

The reason why the tunnel was so long had to do with center of gravity issues of the shuttle. If the pressurized module was mounted too far forward, the shuttle might not be able to glide or land properly. So on every Spacelab flight where the pressurized module was flown, typically the entire complex would only occupy the rear half or two thirds of the payload bay. This left empty space up front except for the tunnel. If heavily loaded pallets were carried with the lab module, it could be located farther forward in the bay. Lab module flights were typically flown with the lab only or a lab module with a single pallet.

The lab module itself was 13.8 feet (4.17 meters) in diameter and made up of two segments just under 9 feet long (2.7 meters) each. The original Spacelab design allowed for use of a single segment, but all sixteen shuttle flights that used the pressurized laboratory module only used the two-segment configuration. Inside the module, both walls were lined with standardized equipment racks, and the roof of the lab was typically equipped with either a view port or a scientific airlock for exposing samples to space. The floor of the lab could be kept empty or have biomedical and exercise equipment mounted to it.

The internal equipment racks could be fitted with experiments of many different types, from life science experiments with microbe samples, insects, mice, and monkeys in self-contained cages to materials-processing and celestial-observation equipment. Any project that required human presence could be flown in the module. After taking delivery of the first pressurized lab module, NASA was impressed with what they saw and purchased a second one from the ESA.

The external pallet racks were 3 meters (9.8 feet) long by about 4 meters (13.12 feet) wide. They were U shaped and could handle many sizes and types of experiment hardware. In addition to the ESA-delivered pallets, NASA also had a couple of engineering-development pallets that were not originally intended for spaceflight. But since the production pallets were not available yet, the development pallets were converted into operational hardware for use on two early shuttle test flights (STS-2 and STS-3). The development pallets could not be used for many missions and had a reduced load limit, but they did the job well. As for the production pallets themselves, they remained in use even after the lab modules had been retired in the late 1990s.

To help with celestial-observation missions, the ESA developed the Instrument Pointing System (IPS). It was a telescope-shaped housing with a multiaxis mounting bracket and star-tracker ports built into it. Its job was to house and point scientific instruments at targets of study on Earth and in space with more precision than what the shuttle alone could provide.

The IPS proved to be the item that caused the most headaches, as problems with its specifications and management led to delivery of the unit falling far behind schedule. Ultimately the unit required a major redesign effort, delaying things further. Eventually the IPS was delivered and utilized with great success, but the slip in the schedule resulted in the delay of its debut on the Spacelab 2 mission (the first pallet-only mission to fly), causing it to be postponed until the pressurized lab module flew for a second time as Spacelab 3.

Like any other program, the Spacelab experienced delays in its produc-

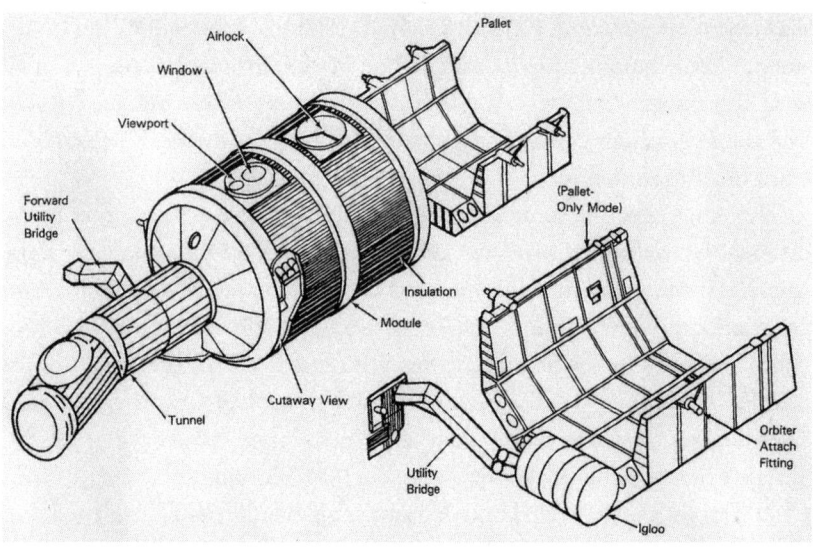

34. A diagram of the hardware that formed the Spacelab system. Courtesy NASA.

tion. The delivery dates for the first pieces slipped by a couple of years. Not all the problems were on the ESA contractor side, though, as NASA's almost-constant revision of shuttle design specifications meant that Spacelab had to change its load parameters. All these changes affected the design and testing of the first pieces of hardware in many small ways, and each of these problems had to be dealt with one at a time.

The ESA wasn't just supplying the Spacelab, either; it was also supplying an engineering model for integration with the testing hardware in the United States. This engineering model was essentially a Spacelab that would never fly, and its sole purpose was to make sure that new experiments selected for flight would work before incorporating them into the flight hardware. As expensive as it is to fly payloads, a spaceflight is not the best time to find out that there is a major experiment problem, or a lot of time and money can be wasted in the process.

Development of the TDRS System

For Spacelab to maximize its potential, it required development of an improved communication and relay system. Previously, spacecraft in Earth orbit had relied on ground-based tracking stations and ships to communicate with control centers. This system had been adequate, but there were several periods when a spacecraft's orbit would take it out of range of most tracking stations. The Skylab and Salyut programs both made use of computer tapes to record and store data collected from the experiments, which could be "data dumped" to the ground when the orbiting spacecraft got within range of a tracking station.

While Spacelab would have a similar capability for data storage, almost-continuous two-way communications would help to maximize coordination between astronauts in orbit and investigators on the ground during relatively short-duration shuttle missions. That way if a problem cropped up, a payload specialist was not sitting around idle waiting for the next communications pass to find out how to correct it. Modifications to experiments could be coordinated with the ground much more quickly as well.

With the success of the ATS-6 satellite's use during the ASTP mission, NASA set out to create the Tracking and Data Relay Satellite (TDRS) network to replace the ground stations. The TDRSes are stationed in geosynchronous orbits around Earth at about 22,300 miles up, meaning they orbit around Earth at the same speed that the planet rotates. One satellite can provide up to forty-five minutes of voice and data coverage between a shuttle and the ground. Two satellites with one each stationed over Earth's Eastern and Western hemispheres can provide 80 percent total coverage. A third satellite accommodates the final 20 percent and acts as a backup to the other two.

The first TDRS was launched on 4 April 1983 on mission STS-6 by the space shuttle *Challenger*. Deployment of the satellite occurred on schedule, but its inertial upper stage (IUS) rocket motor malfunctioned, stranding the satellite in a useless, high elliptical orbit. All was not lost, though. The TDRS was loaded with a large supply of thruster propellant, and a plan was developed to use the thrusters to carefully nudge it into the proper orbit over the next several months. The contingency plan worked, and TDRS-A was open for business, ready to support the first Spacelab mission. The second TDRS wouldn't be ready to fly until after Spacelab had flown its first four missions.

Spacelab Flies

The science crew for Spacelab was originally selected in 1978 in anticipation of a mission sometime in 1980 or 1981, but delays with development of the shuttle meant that it would not fly until 1983. Owen Garriott and Bob

Parker would act as mission specialists, while Ulf Merbold would be the ESA's payload specialist. Joining them on the flight would be payload specialist Byron Lichtenberg, a researcher from the Massachusetts Institute of Technology (MIT) with several doctorate degrees in science. In addition to his MIT role, he was also an air force reserve fighter pilot and Vietnam veteran with over two hundred combat missions to his credit. Given his background, Lichtenberg probably could have joined NASA as a mission specialist, if that had been his desire. He would be the first U.S. payload specialist to fly during the shuttle program.

The flight crew was selected in 1982. The commander for Spacelab 1's mission was John Young, an Apollo mission veteran fresh off his command of STS-1, the first flight of the shuttle program. STS-9 would be Young's sixth space mission. Joining Young would be shuttle pilot Brewster Shaw, an air force veteran combat pilot and one of NASA's first class of shuttle astronauts selected in 1978. Even though the science work would take place in the payload bay, it would be up to the flight crew to make sure *Columbia* was flying in the proper attitudes for the required tests and data collection.

Spacelab 1's mission would conduct a large amount of science in many fields. There were over seventy experiments in five major scientific disciplines from astronomy to life sciences. Nearly half of them would be materials-processing experiments. Garriott's and Parker's Skylab experiences, both in orbit and on the ground, allowed them to have a hand in selection of the first experiments to fly. The mission specialists would also take on the role of on-orbit repairmen. If Spacelab's equipment had problems and repair was possible, it would be up to the mission specialists to do the work.

The principal investigators on the ground and the support teams would occupy a room at the Johnson Space Center known as the Payload Operations Control Center (POCC). This way, they could get real-time data from the lab and maintain two-way voice contact for experiment matters that didn't require a mission control capsule communicator, or CAPCOM. During Spacelab's early days, there was back-and-forth discussion between the NASA centers as to who would host the scientists, since both Houston and Huntsville wanted a stake in that area. For the first missions, only JSC would have a POCC, as the program had a limited budget. But with the sheer number of people crammed into the facility to support the first missions (which required twenty tons of air-conditioning to keep both the ground comput-

35. The Spacelab 1 crew in orbit. Pictured (*clockwise from the top*) are Parker, Young, Merbold, Garriott, Shaw, and Lichtenberg. Courtesy NASA.

ers and the people from overheating), it was finally concluded that the Marshall Spaceflight Center in Huntsville should have its own POCC as well. Known later as the Payload Operations Center (POC), Marshall's facility still operates today in support of the International Space Station program.

After a few minor delays, launch day finally came on 11 November 1983. The VIP crowds for this launch were a little larger than others, given that many representatives from the ESA and NASA, both active and retired, were on hand to witness it in person in addition to the thousands of spectators that lined the roads and beaches around KSC. This would be *Columbia's* fifth trip into space and its first after a minor refit at the conclusion of its first four test missions.

At precisely 11:00 EST, *Columbia* rose from Launch Complex 39A; after an eight-and-a-half-minute ride into orbit, it injected itself into a fifty-seven-degree orbital inclination 187 miles high. At launch, the Spacelab module and its pallet loaded with experiments weighed a little over sixteen tons. It was the heaviest shuttle payload to date. Data collection began immediately as Lichtenberg and Merbold were both wired up with bio-medical headgear to monitor their eye movements during liftoff and ascent. The lab itself was open for business within three hours of reaching orbit.

For missions like this, where around-the-clock observations would be conducted, the crew was divided up into two teams. There would be some overlap, but mostly it would be one team doing experiments in the module with a pilot-astronaut flying the shuttle while the other crew would rest on the shuttle's middeck. The red team consisted of astronauts Young, Merbold, and Parker, while the blue team was made up of astronauts Shaw, Garriott, and Lichtenberg. To do this, the teams had their sleep cycles adjusted on the ground a few days before the mission. Dividing the crew up into two teams and adjusting sleep schedules would become standard practice for future shuttle missions involving around-the-clock science.

During the mission, the science gathering went well, although it was marred from time to time by problems primarily with the Spacelab's computer when it ran too hot. Bob Parker also had to play orbital mechanic when the high-rate data recorder jammed. Without the recorder, data collection for most of the scientific experiments would have been severely hampered, other than the results transmitted to the ground during communications passes (which were limited by the lack of total TDRS coverage). Parker managed to carefully disassemble the recorder, remove the jam, and return it to normal functioning. Experiments early in the mission primarily focused on Earth observation, with the shuttle's payload bay pointed downward, and medical studies into space adaptation syndrome.

The second phase of the flight had the Spacelab oriented toward space and cold soaked to see if any problems would develop during the astronomy experiments. All systems continued to work well, although the lab did make some popping noises as it underwent oil-canning expansion and contraction in the changing temperatures. This startled its occupants somewhat. In his downtime, Owen Garriott performed an experiment of his own, using a ham radio to contact amateur radio operators on Earth, including King Hussein of Jordan.

The mission was originally scheduled to last nine days, with eight of them scheduled for Spacelab experiments. But since the consumables loaded were lasting longer than expected, it was decided to extend the flight by one day to help finish some of the experiments that were running behind and to perform additional experiments with the equipment on board. On the eighth day of the flight, the Spacelab was hot soaked to conduct solar-observation experiments, with *Columbia*'s cargo bay pointed toward the sun. After a day of this, the bay was again pointed toward Earth to finish up the Earth-observation experiments. The final day on orbit before landing consisted of additional engineering checkouts of the lab complex. It performed better than expected.

Columbia itself had some problems on its return to Earth. During a hot-fire test of the shuttle's reaction control system in preparation for reentry, two of the shuttle's general-purpose computers crashed due to a ball of loose solder inside causing a short. One general-purpose computer was restarted successfully; after a delay of a few hours to verify that it would work properly, *Columbia* executed a normal reentry and landed safely. However, two additional problems were discovered after landing.

First, leaking hydrazine from one of the shuttle's auxiliary power units produced a small fire about fifteen minutes after touchdown. A second and potentially more serious problem, which wasn't reported at the time, involved *Columbia*'s left-hand orbital maneuvering system (OMS) pod. The pod experienced a breach in its thermal protection during reentry as a hole was burned through it down to the graphite structure. Just behind the skin at the front of the pod sat the hypergolic fuel and oxidizer tanks for the OMS engine and thrusters on that side. If those tanks had been breached during reentry, the entire pod could have caught fire and perhaps exploded, potentially dooming the shuttle with critical damage.

Wind tunnel analysis after the flight determined that ice buildup on the shuttle's waste-water discharge chutes might have caused a disturbance to the shuttle's plasma flow on reentry, which resulted in a localized hot spot on the front of the left OMS pod. To remedy this, engineers replaced the white tiles in spots on the front of both OMS pods with higher-temperature black ones for future shuttle missions.

Overall, Spacelab performed better than expected. The engineering evaluation of the lab complex achieved 100 percent of its test objectives. Science goals in the various disciplines studied on orbit ranged from 65 percent to over 90 percent depending on the experiments themselves. The low percentage on some of the mission goals were due to some small glitches and equipment failures, but this was to be expected, given that it was the first flight of a new system. NASA judged the mission an overwhelming success.

Spacelab 3, the Second Spacelab Mission

Due to delays in the development of the IPS, a major piece of equipment needed for the pallet-only Spacelab 2 mission, it was decided to fly the second pressurized laboratory mission next. This flight would be a NASA-only mission with no ESA participation, although the French agency CNES would have some experiments flying on board. While Spacelab 1 was a jack-of-all-trades mission to test out the shuttle and Spacelab in many different disciplines, Spacelab 3 would focus more heavily on only five specific fields: materials science, life science, fluid mechanics, and atmospheric and astronomical observation.

MOL veteran Robert Overmyer was in command of the mission, with Fred Gregory as pilot. Class of 1967 scientist-astronauts Don Lind and Dr. William Thornton would fly as mission specialists. Lind had a nineteen-year wait after astronaut selection before flying, while Thornton had flown on a previous shuttle mission. They were joined by another mission specialist, Dr. Norman Thagard. Thagard was a medical doctor and a physicist. Thagard also served as a fighter pilot in the Marine Corps Reserve, and his résumé included combat missions flying F-4s during the Vietnam War. He was selected as part of the first class of shuttle astronauts in 1978.

Joining them for the flight would be payload specialists Taylor Wang and Lodewijk van den Berg. Wang was the first Chinese American astronaut, as he was born in mainland China in 1940 before his family moved to Taiwan in 1952. After achieving his doctorate in low-temperature physics, or superfluid and solid-state physics, Wang joined Caltech's Jet Propulsion Laboratory (a center operated in partnership with NASA) in 1972 as a senior scientist and became a U.S. citizen in 1975. For Spacelab 3 he performed several experiments in the behavior of fluid spheres in a zero-g environment.

Lodewijk van den Berg at first glance did not look like the typical astronaut as he was fifty-three years old (making him the oldest space rookie at the time), had a rather skinny build, and wore thick glasses. Van den Berg held both a master's degree and a PhD in applied science, and he was the first astronaut born in the Netherlands to fly in space, although residents of the country consider Wubbo Ockels to be their first astronaut since van den Berg was a naturalized U.S. citizen at the time of his flight.

While working for U.S. defense contractor EG&G Energy Measure-

36. The Spacelab 3 crew inside a training mock-up. Standing (*back row, left to right*) are Lind, Wang, Thagard, Thornton, and van den Berg. Sitting (*in front*) are Overmyer and Gregory. Courtesy NASA.

ments, van den Berg created a vapor-crystal growth experiment that caught the eye of NASA. He was asked to select eight scientists who were familiar with the experiment to become candidates for payload specialist to run the experiment on orbit. It was felt that it would be easier to train a scientist familiar with crystal growth to become an astronaut than to do the reverse. Van den Berg could only come up with seven candidates; so at the urging of a colleague, he put himself on the list as well to fill out the roster. No one, not even van den Berg himself, figured he would be picked. Surprisingly, van den Berg along with one other EG&G scientist made the cut. After months of training, van den Berg was told that he was a prime crewmember for Spacelab 3.

In addition to the seven astronauts, Spacelab 3's science payload included cages filled with twelve squirrel monkeys and twenty-four rats to measure their adaptation to space. This live cargo was loaded the day before launch, using an elaborate chair rig erected vertically in the lab at the launchpad. This mission would last seven days, with six of them devoted to science.

Spacelab 3 launched into orbit aboard the space shuttle *Challenger* for mission STS-51B on 29 April 1985. Again the mission was very successful, with most of its scientific goals being achieved. Some highlights included using a laser spectrometer to analyze Earth's ozone layer. Taylor Wang's fluid-drop experiments got off to a slow start due to equipment failures, but he lobbied NASA to try to fix his equipment and was successful in repairing the problems. Van den Berg's crystal growth experiments seemed to perform quite well also.

The monkeys and rats stole most of the headlines, though, as the crew spent a lot of their time cleaning up food particles and feces that leaked from the animal cages when their supply troughs were opened for feeding and cleaning. Two of the monkeys got sick once they reached orbit. One recovered after a day, but the second one wasn't doing so well and would not eat. William Thornton successfully nursed it back to health by hand-feeding it during his off-duty time until it was healthy again. Problems were also encountered with the lab's scientific airlock. And the primary experiment computer crashed, but the backup computer worked without any problems. The space shuttle *Challenger* itself also performed with no problems and landed successfully with the 213,000-pound Spacelab in its cargo bay, encountering none of the reentry problems experienced by *Columbia* on STS-9.

After analyzing the data, NASA judged the science gathering during the Spacelab 3 mission to be a total success. But due to the problems encountered with the animal cages on this flight, it was decided not to fly any more live animals above a certain size, because the cages were apparently not up to keeping the contents inside. Part of the problem apparently was due to the animals' somersaults inside the cages, causing turbulence that overwhelmed the system's containment fans. The astronauts were also not too keen on having to spend hours cleaning up monkey and rat poop. From that point on, typically only small biological samples or insects would fly on life science missions.

Spacelab 2

The first Spacelab flight to feature an all-pallet configuration with the long-delayed Instrument Pointing System was finally ready to fly in July of 1985 aboard space shuttle *Challenger* as part of shuttle mission STS-51F. The IPS was delivered in November of 1984 with a number of tasks still left unfinished (mostly software related). The IPS had an X-ray telescope mounted in it. The other two pallets were outfitted with plasma detectors, a helium-cooled infrared telescope, and a superfluid helium experiment.

A plasma diagnostics package that could be grappled by the shuttle's RMS (remote manipulator system) was also fitted. This package had originally flown on STS-3. By using the RMS, the shuttle could move the package around the orbiter to measure the plasma interactions between the orbiter and the trace gases surrounding it. An electron-beam gun was also fitted to fire arcs at the diagnostics package as well to see how the energy wave would react in space. The crew would also conduct their fair share of biomedical experiments on the shuttle's middeck.

STS-51F's crew consisted of STS-3 veteran Gordon Fullerton as commander and rookie pilot Roy Bridges Jr. Joining them were mission specialists Story Musgrave and Karl Henize from the 1967 XS-11 class, along with rookie mission specialist Anthony England. The payload specialists were Loren Acton and John-David Bartoe. As with the other Spacelab flights, the crew would be divided into red and blue teams, but with Fullerton being available for any major changes to the shuttle's orientation. This would be a bit different from the other Spacelab flights, though, given that the flight deck and middeck would both be utilized for experiments. The medical data collection was scheduled for when both sets of crews were awake so as not to interfere with scheduled sleep periods.

The trip to orbit was anything but smooth. On *Challenger*'s first launch attempt, on 12 July, the shuttle's ignited main engines shut down three seconds before scheduled launch. Fullerton later told reporters, "It was the longest three seconds I've ever experienced." Checkout of the shuttle and replacement of equipment could reveal no major fault, so launch was rescheduled for late July. Prior to the second launch attempt, one of the control computers in the pallet's igloo failed during a prelaunch checkout. Since the igloo could not be serviced at the pad, it was decided to continue launch preparations using the igloo's two redundant computers instead. But if one of the remaining computers failed, science-gathering goals would be crippled even though engineering evaluation of the pallet-only Spacelab configuration could still take place.

On 29 July 1985 *Challenger* finally got off the pad at 17:00 EDT, but its troubles were far from over. Six minutes into the flight, one of the shuttle's main engines prematurely shut down due to the failure of two of the engine's temperature sensors. A second engine was very close to shutting down from the same problem when an alert controller in Houston made the decision it was a sensor problem and not the engine itself. The sensor readings were bypassed, and *Challenger* managed to limp into a lower than planned orbit on two engines. This was the first and only abort to orbit (ATO) of the shuttle program.

Even with the lower-than-planned orbit, the flight was able to perform most of its science experiments and tested out the IPS. The IPS developed some software-related problems during testing, and it took several days of troubleshooting by the crew as well as the engineers on the ground to get it to work properly. But the sensors mounted to the IPS returned some excellent data, including better images of the sun than were achieved from *Skylab*'s ATM only a decade before. The Spacelab 2 mission returned mountains of data, with only a small portion of the scientific objectives affected by the lower-than-planned orbit. The shuttle's return to Earth on 6 August was much less dramatic than its launch, as *Challenger* landed safely.

Cola Wars in Space

One cargo that threatened to overshadow the rest of the STS-51F mission in the public spotlight was the flight of two experimental soda cans developed by the Coca-Cola and Pepsi-Cola companies. The Coke can was developed first as the company had been conducting a serious engineering study into how to make a drink container that could dispense a carbonated beverage to give it the characteristic fizz in a weightless environment. A normal cola can opened in zero gravity ran the risk of rocketing itself across a compartment, resulting in a very sticky mess of cola droplets to clean up. The challenge was to develop a can that could release its fizzy beverage only on command. The Coke dispenser was a fully approved NASA experiment, but when PepsiCo heard about it, they exerted some political pressure to get NASA to fly a Pepsi dispenser as well.

During the 1980s the cola wars were in full swing in the United States, as each company tried their best to capture the market share. Flying beverage products in space was considered a big selling point. From NASA's standpoint, there wasn't much they could do to oppose this pressure, as their charter prevents them from publicly endorsing commercial products.

Pepsi's can was developed in a much quicker period, and it didn't look all that different from a whipped-cream dispenser, yet it seemed to work fine.

So the drink experiments were conducted. Since the shuttle had no refrigerator, the contents were "warm and frothy," according to the astronauts, and not very enjoyable. Carbonated beverages will likely never fly in space regularly until a form of artificial gravity is developed. The reason is that while the drink can be consumed in zero gravity, built-up gas bubbles in an astronaut's stomach can be painful and tough to get rid of. When one burps in freefall, it is not a "dry" burp as on Earth, where the gas is lighter than the liquid. Since this mission, Pepsi has not flown any other dispensers. Coke flew another version of their can to space station *Mir* in 1991, but the company seems to have had greater success flying Coke syrup dispensers to flavor normal water-filled drink pouches without the carbonation. They tested such a system during two shuttle flights in 1995 and 1996.

Other Spacelab Flights

The next flight for the Spacelab was STS-61A, aboard the space shuttle *Challenger*, which was the first flight of the second Spacelab pressurized module. This mission, known as Spacelab D-1 (*D* for "Deutschland"), was the first Spacelab flight funded entirely by West Germany, and it had two German payload specialists: Reinhard Furrer and Ernst Messerschmid. Wubbo Ockels would also fly as an ESA representative on his first space mission. The pilot-astronauts assigned were MOL veteran Hank Hartsfield as commander and Steven Nagel as pilot. The mission specialists for the mission were Bonnie Dunbar, James Buchli, and Guion Bluford. Bluford had previously flown as the first African American astronaut, on STS-8 in 1983. This was the first and only time the shuttle had flown with eight crewmembers on board.

For this flight, instead of the PCCs in Houston or Huntsville, the German Space Operations Center near Munich, Germany, handled communications with the Spacelab crew. This meant that at least some of the crew had to have their sleep cycles adjusted to European time. As before, the crew was divided up into two teams for around-the-clock science gathering.

German prelaunch media coverage for the mission was a bit biased against NASA, as there were still some bruised feelings that Germany had to pay NASA to fly "Germany's Spacelab." Even Ulf Merbold, Germany's first astronaut, seemed to echo those comments in the press. Regardless of the public perception, NASA, the mission crew, and their German customers got along very well with one another. Spacelab D-1 launched on 30 October 1985 and returned on 6 November. The crew conducted seventy-six experiments on orbit, with a large portion being microgravity studies in biological and materials science, as well as tests of the astronauts' vestibular systems.

Unfortunately, this mission would be the last successful one for the space shuttle *Challenger*, as it was lost with all hands on its next mission, STS-51L, on 28 January 1986. The loss was primarily due to hot exhaust gases leaking from an aft field joint on one of the shuttle's two SRBs, due in part to improper sealing by the joint's O-rings. *Challenger* was launched in very cold temperatures that morning, and contractor representatives from Morton Thiokol, producers of the shuttle's SRBs, had conducted a conference call with NASA the night before, urging them to postpone the launch due to concerns with how the O-rings would behave at low temperatures. NASA managers responsible for the SRBs wouldn't listen and pressed ahead with the countdown anyway.

The resulting loss of the shuttle and its seven-person crew would cause a thirty-two-month delay in the program before flights resumed in September 1988. The safety stand-down revealed many flaws in NASA's management of the shuttle program, and many changes were made. The results would fundamentally change how the shuttle was utilized.

No longer would shuttle missions launch commercial satellites, and no attempt would be made to try to ramp up the shuttle's flight cycle to twelve missions a year. Planned shuttle launches of DOD payloads into polar orbits from Vandenberg AFB were also canceled. There would be other missions to fly, such as launching and repairing the Hubble Telescope; sending probes to study Jupiter, Venus, and the sun; and performing EVAs to test out construction techniques for a proposed space station. But from that point on, most shuttle missions would involve science gathering in one form or another with Spacelab hardware being used for a large percentage of it.

The first flight of a Spacelab after *Challenger*'s loss was STS-35. It was called ASTRO-1, and it used two pallets and the IPS as a combined X-ray and ultraviolet telescope system. The payload flew aboard the space shuttle *Columbia* from 2 December to 10 December 1990, with shuttle commander Vance Brand and mission specialist Bob Parker among the seven-person crew, flying their final trips into space. ASTRO-I would be the first in a long line of twenty science missions out of twenty-four shuttle flights that used Spacelab hardware. The vast majority of these flights involved the pressurized lab modules, but many others flew in the pallet-only configuration (some with the igloo, some without). Germany also sponsored and flew a second dedicated Spacelab mission (STS-55, Spacelab D2), and Japan even sponsored a Spacelab flight of its own (STS-47, Spacelab J).

Funny enough, the one customer that never got to fully utilize Spacelab was the ESA. Shuttle missions can be expensive; after the decade spent in design and building the laboratory, the ESA had no desire to fund dedicated Spacelab missions, opting to focus instead on other ventures after the *Challenger* disaster, such as their commercial launch activities with the Ariane launch vehicles. Success with the Ariane has made the ESA one of the premier providers of commercial space launch services in the world today.

Ultimately the ESA would take part in a support capacity and benefit from the science gathered on the flights. But it is ironic that the agency that built the Spacelab wouldn't be one of its major users. Looking at the bigger picture, though, the production and managerial experience gained by the ESA during the Spacelab program did prove invaluable when the time came for Europe to take part in NASA's next big project, the International Space Station.

9. Soviet Space Station *Mir*

The Death of a Designer

By the early 1980s there wasn't all that much for Vladimir Chelomei to look forward to. The manned Almaz station program was canceled after Chelomei's best ally in the military, Marshal Grechko, died from a heart attack in 1976. A state decree in 1979 called for the Almaz OPS program to fully merge with the DOS program under the control of NPO Energia. Chelomei's bureau was still doing first-class work on their military projects, and Chelomei himself had also won a state prize in 1982. But without Almaz, his bureau's participation in manned spaceflight activities would only be as a subcontractor at best. As a consolation prize, the TKS spacecraft finally got to fly in the late 1970s, and refined versions of the ship successfully docked with both *Salyut 6* and *Salyut 7*. They worked as advertised, even if they never carried a crew into orbit or back.

There was a glimmer of hope in 1984. Dmitry Ustinov had contracted pneumonia in late October and was hospitalized. He was unable to make any public appearances to fulfill his duties as minister of defense. In November he underwent emergency surgery to correct an aortic aneurysm. His deteriorating health continued through early December. There are rumors that when Chelomei heard of this, he was elated that his longtime nemesis was finally on his apparent deathbed.

Unfortunately, Chelomei would not live to even see Ustinov's death. Thanks to his prestige, the designer managed to acquire a Mercedes, which he drove on a regular basis. One day in early December, he pulled his car out of the garage (some reports say out a gate) and stopped it with the engine running to get out and close the garage door. While Chelomei was closing the door, the car rolled forward and crashed into him, crushing his leg between the car bumper and the door.

Chelomei was rushed to the hospital; even with the broken leg, he was reportedly in good spirits, still talking about grand plans for the future of his design bureau without Ustinov's interference. But it wasn't meant to be. On 8 December 1984 Vladimir Chelomei died suddenly in the hospital while on the phone with his wife. This was the result of a blood clot from the broken leg. The clot broke loose and lodged in his lungs, causing a pulmonary embolism. Ustinov would ultimately die of heart failure in the hospital on 24 December 1984. It is unfortunate that Chelomei would not outlive Ustinov, since he might have finally been able to take public credit for his design work.

Mir's Launch

In early 1986, agencies and individuals following the Soviet space program noted increased activity hinting at the imminent launch of another space station. All indications were that it would be another Salyut station and not a TKS or other type of vehicle that required a Proton booster. This was unusual, considering *Salyut 7*, with the TKS spacecraft Cosmos 1686, was still "operational" with the last crew abruptly leaving it in mid-November of 1985.

Suspicions were confirmed when a Proton rocket lifted off from Baikonur early on the morning of 20 February 1986. The booster worked perfectly and injected its payload into nearly the same orbit as *Salyut 7*. But rather than being called *Salyut 8* as many Western analysts had guessed, this space station's name was revealed by Soviet state news as *Mir*. The word *mir* in the Russian language has multiple meanings depending on its context. It can mean "peace," "new world," or "community." Most Western news agencies reported it as meaning "peace." Given that the launch occurred less than a month after the loss of the space shuttle *Challenger*, it is likely that the Soviets were trying to capitalize on a stumble by NASA to grab the world's spotlight once more. With *Mir*, they would do so in a big way.

Like its Salyut predecessors, *Mir* was a DOS space station (DOS-7) that both inwardly and outwardly resembled its second generation cousins in general layout, but there were some key differences. When launched, the station only had two solar arrays, with one on each side, although these new arrays had almost twice the surface area of their predecessors. There were provisions for a third array in the normal dorsal location, even though none was fitted at launch. There was no large camera or telescope mount on the bottom of the

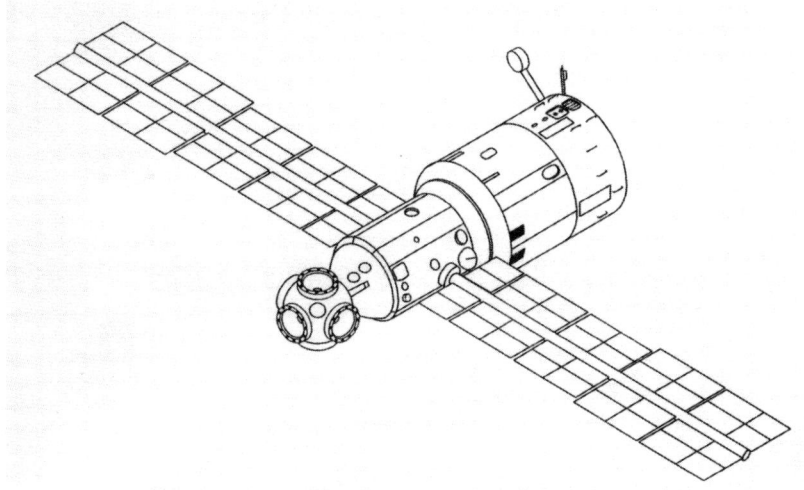

37. The *Mir* core module featured four radial docking ports for other modules. Courtesy NASA.

station. In its place there was a small scientific airlock. Even with the exterior differences, the internal layout of *Mir* was almost identical to *Salyut 7*.

The docking system was upgraded. *Mir* had the old Igla-based rendezvous-and-docking system, but in addition to that, it also had a new system called Kurs. Unlike Igla, which required a narrow line of sight between antennae for a craft to line up with the ports, Kurs transmits signals from multiple antennae. The receiver system on the spacecraft uses the Kurs signals to interpret its position from anywhere around the station, automatically orienting itself with a proper docking port.

The most important change *Mir* had was a multiple-port docking node at the front of the station that had one axial port and four radial ones set apart at ninety-degree angles. Taking full advantage of the lessons learned from the Almaz program, these new docking ports could accommodate heavy modules based on the TKS spacecraft. Design and construction of this docking node took some time to sort out since berthing something heavy on a radial docking port can produce high-torque loads, especially during orbital maneuvers such as reboosting.

The first crew to fly to *Mir* launched aboard *Soyuz T-15* on 13 March 1986 and consisted of cosmonaut veterans Leonid Kizim and Vladimir Solovyov, who both set EVA records during their repairs of *Salyut 7*'s fuel system. The

Soviets had high hopes for the mission, announcing the crew before launch and covering the liftoff live on state television. The crew docked with the new station the next day and got right to work unpacking equipment and setting up the new complex.

Salyut 7's Final Visit

Mir's first crew conducted several months of normal operations, with two Progress resupply visits taking place to deliver equipment. At the end of this period, the crew returned the station to its automatic control and prepared to leave. But they weren't returning to Earth. Instead, they had another destination in mind, *Salyut 7*. The plan was to undock from *Mir*, rendezvous and dock with *Salyut 7*, spend a few weeks there, and then rerendezvous and dock with *Mir*, to transfer over salvaged hardware from the older station. This would be the most ambitious feat of orbital rendezvous ever undertaken by any space program to date, and it would be the only time that a single crew would set up residence in two stations during one mission.

Phase one of the plan had the docked *Progress 26* cargo craft fire its engines on May 5 to drop *Mir*'s orbit slightly below that of *Salyut 7*'s in order to begin reducing the range from three thousand kilometers to a more manageable distance. The crew then loaded five hundred kilograms of supplies aboard the Soyuz, boarded it, and closed the hatch before undocking a few hours later. After a day of independent flight, they caught sight of the Salyut. After closing the distance farther, Kizim brought the Soyuz in for a successful manual docking.

For the next week, the crew spent time restoring *Salyut 7* to operational flight, reconditioning its systems, and then resuming *Salyut 7*'s Earth-observation program. At the end of the month, they donned the new Orlan suits that had been delivered by the Cosmos 1669 spacecraft the previous year to replace ones damaged by the freeze, and then they ventured out for a space walk. Their prime task was to test an experimental, folding girder segment that could expand to fifteen meters in length. The segment, also delivered by Cosmos 1669, was slated for testing by the previous crew before their mission got cut short. Next, the pair retrieved some experiment cassettes. The EVA tasks took about four hours to complete. Portions of the space walk were broadcast on television.

Two days later, the crew conducted another space walk to perform additional girder experiments, including the use of a low-power laser to measure for any flex or distortion to the girder's shape. A variation of the welding experiment used by Svetlana Savitskaya was also performed. The girder was jettisoned from the Salyut after the welding tests were completed. For the next month, the crew conducted a normal scientific program, which included plant cultivation, production of different compounds with the station's furnaces, and Earth-observation studies. The Earth studies included measuring the environmental impact of the Chernobyl nuclear reactor disaster that had taken place in April.

Toward the end of June, the crew dismantled several pieces of equipment from *Salyut 7* to take back to *Mir*. Included in the haul were a mapping camera, a low-light camera, two furnaces, an electrophoresis apparatus, biomedical gear, video equipment, unexposed and exposed film canisters, several spectrometers, and the two Orlan space suits. Their haul accounted for four hundred kilograms of equipment. On 25 June the crew finished "mothballing" the station and returned it to automatic flight before boarding the Soyuz for their return to *Mir*, which was now about one thousand kilometers in front of the Salyut. The crew spent a total of fifty days aboard *Salyut 7* as its final resident crew.

After a day of free flight, the crew again docked with *Mir* and off-loaded their gear. They would only spend a little over two weeks on board before returning home. It had been announced when the station was launched that *Mir* was intended for continuous occupation, but minor delays with the new Soyuz TM (*TM* standing for "Transport Modified") design meant that it was not ready for a manned flight just yet. The first Soyuz TM had flown unmanned in May to test out the new systems, and it achieved a successful docking with *Mir* using the new Kurs system while the *Soyuz T-15* crew was aboard *Salyut 7*. After a couple of days of docked operations, *Soyuz TM-1* returned home successfully. Even with this success, it would take some time to analyze the flight data before allowing a crew to fly.

Mir was not set up exactly like the previous Salyut stations. It was designed from the outset to be modular, expanding its capabilities as new modules were added. Part of the reason for this methodology was an attempt to reduce internal clutter. Deep into their missions, the walls of the Salyut stations were cluttered with bags of supplies and equipment racks. There was less room for cosmonauts to get around in something that began to resemble a storage closet more than a habitable living space. *Mir*'s plan was to reduce the clutter in the core module's working compartment by placing the scientific equipment in add-on modules.

Salyut 7's End

After *Soyuz T-15*'s visit, no additional crews were sent to *Salyut 7*. Previous practice would have had the old station perform a controlled deorbit to eventually burn up harmlessly over an ocean, but controllers on the ground decided to use this opportunity to monitor the complex in order to study the degradation of its systems. In August 1986 ground controllers used the remaining propellant aboard Cosmos 1686 to boost the orbit of the complex to over 450 kilometers in altitude.

Salyut 7 spent the next five years in orbit while its systems were monitored before the orbit decayed. The station reentered Earth's atmosphere on 7 February 1991. The initial impact projections predicted the complex would land in the southern Pacific Ocean, west of South America. But it overshot that area, and several debris fragments rained down over the city of Capitán Bermúdez in eastern Argentina. Fortunately, nobody on the ground was injured by the debris. While *Salyut 7* did not enjoy the success of its illustrious predecessor *Salyut 6*, it still taught a lot of lessons that would be applied to future missions to *Mir*.

Kvant 1

It would be almost a year after *Mir*'s launch before the next crew would occupy the station. On 5 February 1987 the brand new *Soyuz TM-2* spacecraft with veteran cosmonaut Yuri Romanenko and rookie Aleksandr Laveykin lifted off from Baikonur. As had been done with *Soyuz T-15*, the launch date and crew of *Soyuz TM-2* were announced ahead of time. The crew docked with the *Mir* complex after flying a two-day rendezvous, encountering no problems with their new spacecraft. Laveykin apparently encountered some problems adapting to space, but after a few days, he was fine. For these flights to *Mir*, since long duration and continuous occupation were the key, the Soviets came up with a new system for mission designation. This was the second expedition to *Mir*, so it was referred to as Mir EO-2 (*EO* being a Russian acronym meaning "Principle Investigation").

On 31 March a Proton rocket lifted off from Baikonur carrying a new module called Kvant 1 (a Russian name meaning "Quantum"). It was a variation of the TKS spacecraft. The plan was for the TKS tug to dock its payload with *Mir*'s aft docking port, which to that point had only been used by the Progress resupply tugs. Once the new module was firmly docked, the tug would undock itself to expose a new aft docking port on the Kvant 1, which would be used for future spacecraft visits. This meant that the main engines in *Mir*'s core module would no longer be useable, but the Progress system proved itself to be more than capable of conducting orbital reboosts. The Kvant 1 had propellant lines to the *Mir* core, so Progress craft could still refill *Mir*'s thruster fuel supply.

After five days, Kvant 1 caught up with *Mir* and began the final phase of its docking maneuvers using the Igla system. All went well until it lost transponder lock two hundred meters out. The attempt was aborted, but the slowly rolling TKS continued to approach. Luckily, it did not impact the station, although it came very close as it was seen slowly drifting by one of *Mir*'s portholes. After telemetry analysis discovered and corrected a software error, another docking attempt was made on 9 April. This time, the Kvant 1 managed to achieve a successful soft dock. However, the capture latches would not engage when the command was given to perform a hard dock. It seemed like something was obstructing it.

Two days later, Romanenko and Laveykin performed an EVA using the Orlan suits brought over from *Salyut 7*. They climbed down the length of the core module (also known as the base block) to get to the aft port. Laveykin peered down toward the *Mir* docking port and saw something white fouling the port. The ground commanded Kvant 1's probe to extend, opening up the gap so he could remove the problem. It turned out to be a cloth trash bag that had somehow worked its way loose and gotten stuck on *Mir*'s aft docking port when the previous Progress spacecraft departed. After Laveykin successfully removed the bag, ground controllers commanded the probe to retract again, and Kvant 1 achieved a successful hard dock with the *Mir* base block. Later that day, the TKS tug was released to fly free, exposing the new aft docking port for use by future Soyuz and Progress craft.

Kvant 1 was only about one-third of the length of the *Mir* base block, but it was well-outfitted. The module was equipped with X-ray and ultraviolet telescopes, a wide-angle camera, additional X-ray experiments, and the "Svetlana" electrophoresis experiment unit. It also included an additional

Elektron oxygen-generating system to supplement the one found in the *Mir* base block, an additional carbon dioxide scrubber system, and six additional gyrodynes to help control the orientation of the station. The module was also packed with supplies, including a folded solar array to place on the top of the *Mir* base block. The new array was installed during an EVA in June.

A Syrian Visit and Crew Swap

Soyuz TM-3 lifted off from Baikonur on 22 July 1987 and docked two days later on *Mir*'s new aft Kvant 1 port. The crew commanded by Aleksandr Viktorenko would perform a Soyuz ferry swap. While the new TM series Soyuz craft had the potential of staying docked for six months, it was not cleared for a full mission just yet, because the configuration was new and the Soviets were still evaluating it. Aleksandr Aleksandrov, a veteran of *Salyut 7*'s fourth expedition, was the flight engineer. His assignment was to replace Laveykin on orbit as part of the EO-2 crew. During the EVA conducted to remove the trash bag that fouled Kvant 1's docking port, Laveykin developed an irregular heartbeat detected by a new biomedical apparatus he was testing, and he was temporarily banned from EVAs until it cleared up. While it eventually did, allowing him to perform a solar array attachment EVA in June, doctors on the ground were concerned enough to have him return home sooner than expected.

Muhammed Faris would be the research cosmonaut on this flight. He was a military pilot from Syria and became the second Arab to fly in space. As with previous Intercosmos flights, the research cosmonaut would perform some short-duration experiments and use onboard instruments to photograph his home country. After nearly seven days on orbit, the crew boarded the *TM-2* ferry and returned home. High winds almost blew the Soyuz descent module onto the top of a farmhouse, but it landed safely otherwise. Laveykin was taken to a hospital to get looked over by top Soviet cardiac specialists. They gave him a clean bill of health and pronounced him fit to fly, but he would never fly in space again and retired from the cosmonaut program in 1994.

Muhammed Faris continued his career in the Syrian military after his spaceflight. In 2011 a civil war broke out between Syrian president Bashar al-Assad's government and opposition forces who wanted to see the dictator ousted. The Syrian uprising was inspired by the Arab Spring, which had taken place in other Middle Eastern countries during the previous years. The city of Allepo (Faris's birthplace and hometown) became the front line for some of the fiercest fighting in the conflict, with thousands of refugees fleeing the area. Faris managed to defect to Turkey after several attempts and joined the Free Syrian Army in August of 2012.

The new EO-2 crew of Romanenko and Aleksandrov docked the ferry at a different port a few days after *TM-2*'s departure and settled down for an additional six-month tour of duty with heavy photographic and spectral studies using Kvant 1's equipment. Romanenko would be pushing the limit of human endurance on this flight to over three hundred days in space. It was originally planned for him to do it with Laveykin, but having Aleksandrov on board became a blessing, as the fresher cosmonaut ended up doing most of the heavy lifting toward the end since Romanenko was growing fatigued, even with the strict exercise regime.

Orbital Handoff

Relief was on the way as *Soyuz TM-4* launched on 21 December 1987 with a two-member Mir EO-3 crew of Vladimir Titov and Musa Manarov. Joining them was test pilot Anatoli Levchenko, one of the cosmonaut test pilots from the Soviet Buran shuttle program. As with Igor Volk's mission to *Salyut 7*, the plan was to see if a short-duration spaceflight would degrade a cosmonaut's flying skills, especially for the critical landing phase. Volk and Levchenko were assigned to be the commanders of the first manned Buran flights, and their Soyuz trips were necessary due to the Soviet mandate of having at least one crewmember on a mission with previous spaceflight experience.

Soyuz TM-4 docked two days later, and the EO-3 crew performed an orderly handoff with the EO-2 crew over the next week. This was the first time that a scheduled handoff between two long-duration crews had taken place in orbit. The EO-2 crew and Levchenko undocked and returned to Earth in *Soyuz TM-3* on 29 December. All told, Yuri Romanenko had spent 326 days in orbit and set a new record in the process. Combined with his time aboard *Salyut 7*, Romanenko had accumulated a total of almost 431 days in space. Aleksandrov walked to the waiting helicopter unaided while Romanenko was carried by the recovery crew. While Aleksandrov had only spent 160 days in orbit on this mission, when combined with his *Salyut 7* time, he had spent over 300 days in space total.

The Buran shuttle program would ultimately have only one unmanned test flight, which lifted off into dark and snowy skies on 15 November 1988 aboard a massive Energia booster. Buran carried out a two-orbit flight test and then conducted a fully automated reentry, approach, and landing with no major problems. There were plans to fly a manned mission in 1991 using a second Buran shuttle, with eventual flights to *Mir*, but the project was a big drain on the Soviet economy. Ultimately, no manned Buran missions would ever take place, even though the decision to cancel Buran would not come for many years yet.

Sadly, Anatoli Levchenko would not live to even see Buran fly. Shortly after his return from orbit, Levchenko was diagnosed with a brain tumor and died less than a year later on 6 August 1988. Born in the Ukraine, he was only forty-seven years old and married with one child. None of the other cosmonaut test pilots selected for the Buran program would ever get to fly in orbit, leaving Igor Volk as the only surviving member of the group with any spaceflight experience.

A Year in Space!

Medical tests of Romanenko didn't show anything unexpected, so the green light was given for Titov and Manarov to try to set a new record of spending a year in space aboard *Mir*. This mission was packed with many objectives; for the first six months, the crew conducted numerous scientific experiments in the Kvant 1 module and made use of new equipment delivered by several Progress ferries. For this mission, the crew asked for and was granted a request to have several music cassettes sent up to help pass the time. Cosmonauts have always considered missions of this type to be marathons. Care is typically exercised to not load down a crew with too many tasks too early. Still, the crew got plenty of work done in those early days and conducted a space walk in February to replace a panel on the dorsal solar array with a more efficient one.

At the end of six months, all had gone well in orbit. Regular Progress supply flights had kept *Mir*'s consumables topped off. While there were only minor failures of equipment, including a particle detector on Kvant 1 that would require an EVA to fix it, *Mir* was performing fine. Even with these issues, the crew was in good spirits, so the go-ahead was given for a six-month extension to their tour of duty.

The EO-3 crew would host visitors on 9 June 1988 as *Soyuz TM-5* docked at the station's rear port with the three-person EP-2 short-duration crew of Soyuz commander Anatoly Solovyov (no relation to Vladimir Solovyov), who was on his first spaceflight, and Salyut veteran Viktor Savinykh. Joining them was a Bulgarian Intercosmos cosmonaut, Aleksandr Panayotov Aleksandrov. While the Bulgarian shared almost exactly the same name with Soviet cosmonaut who had flown on the previous *Mir* expedition (even down to the same middle initial, *P*), the two men are not related. The Bulgarian Aleksandrov was the backup for fellow Bulgarian cosmonaut Georgi Ivanov, who launched into space aboard *Soyuz 33* yet was unable to visit *Salyut 6* due to a Soyuz engine malfunction.

With this flight, Bulgaria concluded the Soviet Intercosmos program on a high note. The mission was a typical research cosmonaut flight with Aleksandrov performing numerous biomedical experiments and photographic studies of his home country. As with previous Intercosmos flights, the week of docked operations was packed with activity for the new crewmember. The visit of the three newcomers was welcomed by the original crew as it helped to have a change of pace, even if only for a short while. The EP-2 crew returned home on 17 June in *Soyuz TM-4*, leaving the *Soyuz TM-5* to act as a fresh lifeboat.

Titov and Manarov transferred their new Soyuz to the front docking port the next day and underwent preparations to replace the defective detector on Kvant 1's X-ray telescope with a new unit. In a showcase of the warming of relations between the Soviet Union and Western Europe, the original detector and its replacement had been built by scientists from England and Belgium working for the ESA. The European specialists were on hand in the Kaliningrad control center during the 30 June space walk to act as advisors. The crew performed the EVA, although they ran into difficulties with some bolts that were very difficult to break loose with their gloved hands. The operation took longer than expected, and they ultimately had to be cut short when a special tool they were using broke, leaving the pair unable to fully extract the faulty detector. They would finish the task once a beefed-up tool was delivered on a Progress, but they still managed to get a lot of work done.

A Progress flight delivered more supplies in late July, including a pair of brand new Orlan DM space suits. Unlike the suits designed earlier, which the cosmonauts had been using, these new units had a fully self-contained electrical system and communications gear, so they did not require a tethered umbilical from which to draw power. Instead, the new suits would use a smaller safety tether line. This line could be clipped on to handholds as needed during EVAs, allowing the cosmonauts to access new parts of *Mir*'s exterior.

New Visitors and a House Call

Soyuz TM-6 was the next spacecraft to visit *Mir*, docking with the complex on 31 August 1988. Veteran cosmonaut Vladimir Lyakhov was in command of the EP-3 mission. The flight engineer position was occupied by Valery Polyakov, a medical doctor. It was originally planned for Polyakov to fly to *Mir* during Romanenko's long-duration visit and act as a medical specialist on hand to observe the spaceflight record attempt, but a last-minute crew change had Buran pilot Levchenko flying in his place instead.

The third crew couch was occupied by Abdul Mohmand, a research cosmonaut from Afghanistan. The Soviet military was in the beginning stages of a planned pullout from the long war in Afghanistan, with the first major troop withdrawal taking place in mid-1988. The Communist Party wanted to capitalize on a bit of positive propaganda by having a citizen from that country flying in space before Soviet forces completely withdrew by early 1989. Mohmand's cosmonaut training only lasted six months. This period was much shorter than the two-year training periods normally given to Intercosmos candidates. But Mohmand was a Soviet-trained fighter pilot in the Afghan Air Force, and he could speak Russian fluently. These factors aided his cosmonaut training.

The EP-3 mission had Mohmand doing standard photography and biomedical experiments in orbit, although these used existing equipment since there wasn't time to fly experiments designed in Afghanistan. This mission would be a little different, as Polyakov would join the EO-3 crew for the final few months of its flight and then remain in orbit with the EO-4 crew. Lyakhov and Mohmand would return home on the *Soyuz TM-5*, swapping it for the new ferry.

"Stuck in the Middle with You"

Soyuz TM-5 undocked on 6 September. To help provide a better fuel cushion, Soyuz TM missions to that point had the orbital module being jettisoned before retrofire, so less fuel would be needed to slow down the spacecraft for a proper deorbit. The procedure for the descent called for *TM-5* to fire the deorbit engine just after entering daylight for a morning landing in Kazakhstan. But just prior to the engine firing, orbital sunrise temporarily blinded the horizon sensors on the Soyuz craft. It took a few minutes to find out what the problem was, and by then the Soyuz was too far down range from the recovery site to execute a proper deorbit. Two orbits later, a second attempt was made, but the engine cut off after only six seconds of burn time. On an attempt to relight it, it again cut off after fifty seconds, well short of the nearly four minutes required for a proper deorbit. It was decided by Kaliningrad that the crew would remain in orbit for a whole day while engineers on the ground analyzed the problem. That was easier said than done.

Without the Soyuz orbital module, Lyakhov and Mohmand would have to spend almost twenty-four hours in orbit wearing their Sokol pressure suits. They couldn't stretch out since the orbital module had already been jettisoned. Plus, the spacecraft's toilet was aboard the orbital module as well. So they pretty much had to sit cramped in the fetal position in a slightly cold Soyuz descent module with no fresh food and no toilet while awaiting the next deorbit window. Analysis of telemetry from the craft's Argon computer revealed that it had selected the wrong program for its engine firing and that the engine itself was okay. Sure enough, the engine fired properly the next day, and the crew returned home safely.

While Abdul Mohmand was given the usual state medals and hailed as a hero for his flight, he didn't remain in Afghanistan for long. After the Soviet forces completed their withdrawal in 1989, Afghanistan fell back into a civil war between government forces and rebel factions who had Western support during the Soviet occupation years. The Afghan Communist government fell in 1992, and Mohmand fled to Germany, knowing that his life was in danger if he remained behind, since he was a symbol of Afghanistan's old regime. He applied for and was granted asylum, eventually becoming a German citizen in 2003.

Going for the Record

With *Soyuz TM-5* safely on the ground, the crew got back down to business with Polyakov performing his duties as a doctor and flight engineer in support of Titov and Manarov. *Soyuz TM-6* was redocked on the front port, and the next Progress to arrive carried among its cargo the redesigned tool needed to help with replacement of the X-ray detector on Kvant 1. After making preparations to finish the job, Titov and Manarov got down to business during an EVA on 20 October. This time with the new equipment, the crew got the job done ahead of schedule and replaced the faulty detector with a new one. Kvant 1 was fully operational once again.

Dr. Polyakov had the crew on an increased exercise regime to help combat the effects of weightlessness and prepare them for a return to Earth. This did the trick, as Titov and Manarov didn't seem to suffer from the same weakness problems Romanenko encountered. The two men broke Romanenko's 326 day record on 12 November and Polyakov gave them a go to spend "a year and a day" in space.

First European Space Walk

A few days after Buran's successful two-orbit, unmanned flight test in November, *Soyuz TM-7* lifted off from Baikonur with Soviet cosmonauts Aleksandr Volkov and space rookie Sergei Krikalev on 28 November. Joining them would be an old visitor, French CNES astronaut Jean-Loup Chrétien, making his second spaceflight. After Chrétien's first spaceflight, he trained as a NASA payload specialist backup to fellow French astronaut Patrick Baudry for shuttle mission STS-51G, which flew in June of 1985. Chrétien was the first guest cosmonaut to be granted a second spaceflight, and this mission would last almost a month.

This was the first time *Mir* had hosted six visitors at once, and things were a bit cramped in those days. With only the Kvant 1 module docked to the base block, *Mir* wasn't all that much bigger than a standard Salyut, but the crew got down to business in an organized fashion. Chrétien had nearly six hundred kilograms of equipment at his disposal and conducted a full range of technology and medical experiments. This was something unique for the French, given that NASA's shuttle was only capable of sending somebody up for a little over a week at most. Some of the French experiments tested equipment planned for the ESA Hermes space shuttle program.

The highlight of Chrétien's visit, though, was an EVA to deploy a French-designed folding truss called the Era structure. It was part of a French study to see if this specific design could be used for a radio antenna on a future spacecraft. Volkov and Chrétien performed the space walk on 9 December. Chrétien also mounted a couple of experiment cassettes on the outside of *Mir* for exposure, before the two men hooked up the truss to an external power port on *Mir*'s base block. Once they finished their task, Krikalev, inside *Mir*, flipped a switch to command the structure to unfurl. It didn't budge. A few nudges by Chrétien did nothing. Finally, Volkov gave it a few kicks with one of his legs, and the structure unfurled fully. After a few minutes of deployment where sensors on board collected data on the truss to determine how well it worked, it was jettisoned. The space walk took six hours to complete and set a new record. It was also the first space walk ever conducted by a person not of Soviet or U.S. nationality.

Handoff

By 15 December Titov and Manarov had surpassed the previous record by 10 percent, making it an official space record to the International Astronautics Foundation. The handoff went well, with the new crew settling in while the old crew got ready to return home with their French guest. Polyakov would remain in orbit with Volkov and Krikalev to potentially try to set a record on his own. By 21 December, everything was ready for Titov, Manarov, and Chrétien to return home aboard *Soyuz TM-6*.

After the difficulties encountered by *Soyuz TM-5*, the deorbit procedures were revised so that the orbital module would not be jettisoned until after retrofire took place. It was a good thing, too, as during the first attempt to perform the deorbit burn, the computer became overloaded and canceled the sequence. After another orbit, the engine fired, and *TM-6* began its journey home. Reentry and landing were normal. The recovery forces were on hand to greet the two men who had spent a year in space; while they took time to adjust back to Earth's gravity, they were no worse off than crews who had flown for a typical six-month *Mir* expedition. Further medical testing came to the same conclusion, showing that the cosmonauts had not suffered any additional demineralization of their bone structure than someone who had spent half that time in zero gravity. After a few days of physical therapy, which included rehabilitation exercise in a swimming pool to help stimulate muscles made long dormant by weightlessness, the crew were up and walking around in almost no time. The increased exercise regime on orbit seemed to do the job.

New Year, New Challenges

When Volkov and Krikalev joined Polyakov on orbit as part of the Mir EO-4, it was planned for *Mir* to receive at least one and possibly two new modules to expand the complex in 1989. Both men were trained in the tasks required to activate the new systems. Both modules were built from TKS spacecraft and weighed about twenty tons each, almost as much as the *Mir* base block. The first module, Kvant 2, was originally planned for flight during Titov and Manarov's time in orbit, until it experienced delays. This also meant that the launch date for the second module, Kristall, would be delayed as well.

After Chrétien's return, the Soviets announced that they wouldn't be offering anymore "free rides" to guest cosmonauts. Future short-duration flights to *Mir* would require payment of a fee for the services, so most future Soyuz flights in the near term would only have two crewmembers. With the certified design life of six months on orbit for a Soyuz TM module, there wouldn't be a need for dedicated Soyuz ferry swap missions either. As for future long-duration expeditions, with one pair having successfully set a record of a year in space, there wasn't really a need to push the envelope any further with a full *Mir* crew. Future endurance records would be attempted by a single cosmonaut only.

Changes at Energia

During the delay of the next two *Mir* modules, Energia's chief designer Valentin Glushko died of natural causes on 10 January 1989. As a member of the first group of chief designers from the old days of the Soviet space program, his loss was a great one. During his tenure at Korolev's old bureau, Glushko was mostly involved with the development of the LHX-LOX-powered Energia rocket. This was a reversal of his previous stance, as he resisted calls to work on hydrogen-powered rocket engines when Korolev was alive. While the Energia rocket had successfully flown two test flights, including Buran's launch, there was no money to develop it further. The Soviet Union's economy was slowly beginning to crumble from the weight of decades of economic stagnation. While the Energia rocket itself would not fly again, its strap-on boosters would form the basis of a very successful satellite launch vehicle called Zenit (a Russian term meaning "Zenith").

Glushko had given stability to the bureau, something that many have said it hadn't had since Korolev's death. Yuri Semenov, the deputy designer who first came to prominence in his management of the first Salyut station project two decades earlier, became Energia's next leader. Semenov had spent many years managing the DOS program; since it had become the primary focus of the Soviet manned space efforts, he was considered to be the best man to take the bureau into the next decade.

A Brief Pause

By April of 1989, delays in the development of the Kvant 2 and Kristall modules for *Mir* meant that they would not be ready to fly by the time the Mir EO-4 was scheduled to return home. One module could perhaps have been launched sooner, and it could have been left docked on *Mir*'s forward axial docking port, but this meant that the station wouldn't be able to accept any Progress supply flights with a Soyuz occupying the aft port. The next module would have to be moved to one of the radial docking ports as soon as possible to free up the front port.

The docking of a twenty-ton module to the radial port meant that the station would be L shaped with an asymmetric mass configuration and the center of mass outside of the station's thrust line. If a reboost burn were conducted, the station would likely end up tumbling end over end. Small orbital maneuvers for stellar or planetary observations would be limited as well due to the strain on the station's gyrodynes and attitude-control computer. To compensate for this, it was better to dock two modules of the same mass to opposite radial ports, giving the station a T shape. This would give *Mir* the proper center of mass.

It was decided to bring the Mir EO-4 crew home at their scheduled time, rather than having them wait on orbit for who knew how long. *Mir* had received several new pieces of scientific apparatus on multiple Progress flights during their stay, but they were useless until the new modules were docked and just took up space inside the station. EO-4's return without another crew to relieve them meant that Polyakov would have to cut short his attempt at a record. The next expedition would be ready to fly when the next module was.

So the crew loaded up the *Soyuz TM-7* with the results of their experiments and returned *Mir* to automatic flight, undocking from it on 27

April 1989. They performed a successful deorbit and landed the same day, with Volkov and Krikalev having spent 151 days in space. Polyakov's total was 240 days. While it was planned for *Mir* to be continuously occupied, the station was still in excellent shape, and no problems were encountered during this unmanned period. The X-ray experiments on Kvant 1 could be operated autonomously from the ground and continued to collect data. All three cosmonauts were none the worse for wear upon their return.

Reoccupation

Prior to the launch of the next Soyuz, a new variant of the Progress craft was launched in late August. This spacecraft, known as *Progress M-1*, was an updated version of the craft. Like the Soyuz TM series, it contained upgraded systems, which included replacement of the Igla rendezvous system with Kurs. It docked with the front port of *Mir* two days after launch. *Mir*'s front port was set up with fuel and water transfer lines, same as the aft port, but it only had the Kurs system and couldn't guide in Igla-equipped Progress modules. Kurs-equipped Progress-Ms could now replenish the station's consumables from either end, although the aft port was still preferred so that Progress craft could conduct periodic orbit reboosts. The Progress-M design could also supplement *Mir*'s electrical power with its own solar arrays while docked.

On 8 September 1989, *Soyuz TM-8* lifted off with Mir EO-5 crewmembers Aleksandr Viktorenko and Aleksandr Serebrov. It docked with *Mir*'s aft port on Kvant 1 two days later. Both cosmonauts were veterans of the Salyut program with extensive experience. They were trained to perform many of the tasks originally assigned to the previous crew, preparing the station for Kvant 2's arrival.

While the crew got down to business as usual unloading the Progress and conducting experiments, a solar flare in late September added some excitement to the mission, as radiation from the flare set off a detector in the *Mir* base block. While the levels of radiation weren't high enough to warrant an immediate evacuation, the crew spent the next few days sleeping in the Kvant 1 module since it was more heavily shielded than the base block or the Soyuz. The Soviets announced that the crew only received a radiation exposure from the flare that was equivalent to two normal weeks in space. Neither Viktorenko nor Serebrov seemed to suffer any ill effects from the encounter.

The Kvant 2 would experience one more delay before its launch as a computer fault was detected during its final checkout. The fault was traced to a batch of computer chips in the Kurs rendezvous-and-docking system. So as a precaution, engineers on the ground opted to replace all the suspect computer chips with newer ones. This would delay Kvant 2's launch by about forty days from mid-October to late November. The original plans were for the crew to receive the Kristall module in early 1990 before their return home. But with the slip in the schedule, they only received the first module.

Kvant 2 Arrives

The Proton rocket carrying the Kvant 2 finally got airborne on 26 November, and the module entered orbit normally. But a problem was detected during the opening of its solar arrays as one of the arrays would not fully unfurl. The module could still dock with *Mir*, but it was decided to make an attempt to unfurl the panel before Kvant 2 docked with the station, as an unsecured solar array could cause problems. The array was also needed to deliver critical electrical power to the module both before and after docking. Engineers on the ground commanded the module to begin a slow roll and simultaneously commanded the array's rotation motor (used normally to track the sun) to move back and forth to try to jostle it. The hope was that centrifugal force and the jostling would coax the panel to open. The process worked, and the panel opened fully.

After six days of free flight on a trajectory set to conserve fuel consumption, the Kvant 2 module became visible to the *Mir* occupants. The Progress had been jettisoned from the front port a couple of days earlier to prepare for this moment. The Kurs docking system guided the Kvant 2 to within twenty kilometers of *Mir* before aborting the rendezvous after sensing the approach was too fast. Coincidentally, the station's Argon computer also went down, knocking out the station's gyrodynes. The crew got *Mir* stabilized, and controllers in Kaliningrad revised the approach procedures for the module. Four days later, they tried again. Viktorenko used the thrusters on the Soyuz ferry at the aft port to hold the station's attitude stable while the Kvant 2 continued its approach. The module made a smooth hard dock on the front axial port with no problems.

The next part was something new. Kvant 2 extended a pivoted, blunt structure known as the Lyappa arm into one of two receptacles on *Mir*'s front hub. These openings were offset by 45 degrees each from *Mir*'s vertical

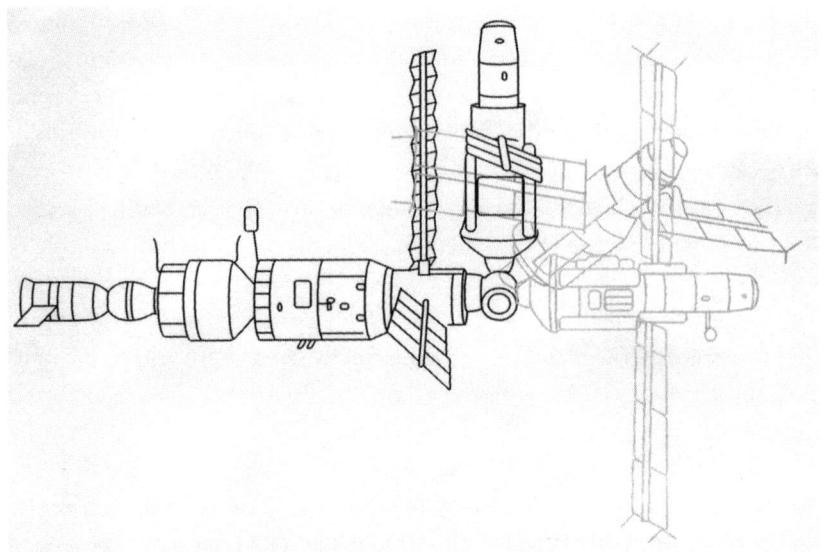

38. Illustration showing how the Kvant 2 module was moved with its Lyappa arm after docking. Courtesy NASA.

axis. The arm and the receptacle resembled a smaller version of a docking probe and drogue. Once the arm was fully seated, the module was lifted out of the axial docking port by the arm and did a half-pivot and half-twist maneuver until it ended up fully seated in the top docking port, 90 degrees offset from the axial port. The operation took an hour to perform.

Kvant 2 added some new capabilities to *Mir* and augmented some of its older equipment. It contained six new gyrodynes to supplement those on the base block and Kvant 1; it featured two additional water storage tanks and an improved water-recycling system. Internally, the module was divided into three compartments, with a new airlock for EVAs being mounted at the far end, a second instrument and cargo compartment in the middle that could also double as an emergency airlock, and an inner compartment that was equipped with a variation of the crew shower first flown on the Salyut stations. The shower proved problematic, though, so later crews turned it into a steam room before it was finally dismantled in the mid-1990s, with crews preferring to take "baths" by wiping themselves down with wet towels and soap containing a biocide. The module also contained a new main computer designed to replace *Mir*'s four-year-old Argon 16B system.

Being designed as a multipurpose module, Kvant 2 was loaded with equipment for science gathering. It contained no less than six different spectrometers in various wavelengths and also included a new version of the venerable MKF-6 Earth resources camera system, known as MKF-6MA. Also on board was a bird egg incubator designed to help hatch quail eggs so that cosmonauts could study the development of embryos in zero gravity. Kvant 2 also contained a protein-crystal growth experiment from the American company Payload Systems Incorporated. The company was founded by STS-9 veteran and payload specialist Bryon Lichtenberg. Due to U.S. restrictions on transferring technology to the Soviet Union, this particular experiment was sealed and only had a single power switch.

Some new EVA equipment was also delivered, which included two brand-new, third-generation Orlan-DMA space suits and the YMK (a Russian acronym meaning "Cosmonaut Maneuvering Unit") jet pack. The YMK was a free-flight pack that worked by firing jets of compressed nitrogen gas to control orientation and translation of the cosmonaut wearing it on a space walk. It was inspired by NASA's Manned Maneuvering Unit, first tested on shuttle mission STS-41B by astronaut Bruce McCandless. The YMK was larger than the Manned Maneuvering Unit but performed the same functions.

The rest of December was spent checking out the Kvant 2's systems after it was secured in its new location and off-loading a Progress module packed with additional supplies. Viktorenko and Serebrov would not have to wait long before trying out their new EVA equipment. In mid-January they conducted two EVAs to attach a pair of star-tracker ports to Kvant 1 and then switched out some exposure cassettes. The second space walk was a combination EVA and IVA. Inside the depressurized docking adaptor, the crew removed the docking drogue assembly from Kvant 2's hatch and placed it on the bottom docking port in preparation for the Kristall module.

On 28 January the crew conducted a third space walk, this time from Kvant 2's airlock to place more experiment cassettes outside. The highlight of this third EVA was to test out the brand new YMK jet pack. Serebrov tested the unit first, putting it through a series of maneuvers. Unlike McCandless, who was completely free from the shuttle on his flight, Serebrov stayed attached to *Mir* with a thin nylon tether line. If he ran into problems, Viktorenko could reel him back in. The pack performed as advertised. Three days later, Viktorenko repeated the YMK tests. He also took along an X-ray and ultraviolet scanning apparatus attached to his chest and pointed it at *Mir* from forty-five meters out. The gear was used to measure radiation levels generated as the station passed through Earth's ionosphere. After the test was complete, Viktorenko docked the YMK on an exterior cradle on Kvant 2. The pack only had a limited supply of fuel, so it was not used again after these tests.

In early February the Mir EO-5 crew was nearing the end of its tour of duty. The Mir EO-6 crew of Anatoly Solovyov and Aleksandr Balandin launched on *Soyuz TM-9* and docked with *Mir* on 13 February. As the Soyuz closed with the station, the cosmonauts on board noticed that their periscope was partially obscured, and the docking camera on Kvant 1 detected several loose objects on the side of the Soyuz. But even with the problem, docking was made as normal. After six days of joint operations, Viktorenko and Serebrov loaded up *Soyuz TM-8* with their experiment results (including the sealed American crystal growth experiment) and undocked.

Before returning home, *Soyuz TM-8* performed a visual inspection of *Soyuz TM-9* and noted that three of the insulation blankets on the descent module had come loose at the bottom but remained attached at the top. There were some concerns that the loose blankets would expose the descent module's structure directly to freezing temperatures and perhaps cause condensation to build up inside the craft, creating risk of an electrical short. Managers on the ground said there wasn't much cause for concern, but Solovyov and Balandin would eventually need to conduct a space walk to reanchor the dislodged blankets to keep them from interfering with the craft's horizon scanners during their return in a few months. With the inspection complete, the Mir EO-5 crew used their thrusters to leave the vicinity of *Mir* and experienced a normal deorbit, reentry, and landing.

The next month was mostly occupied by standard *Mir* scientific experiments. The crew installed and integrated the station's new Salyut 5B computer into *Mir*'s control system to replace the original Argon system. On 17 March a quail chick hatched in the incubator. It came from a batch of fertilized quail eggs delivered by a Progress resupply flight. Controllers on the ground announced that the quail eggs and embryos were part of a study into potentially using quail as a fresh food source for crewmembers on missions to other planets.

Kristall's Arrival

The long-delayed Kristall (a Russian word meaning "Crystal") module was finally launched into orbit via a Proton booster on 31 May 1990 and began its approach to *Mir* on 6 June. However, a thruster malfunction caused the Kurs system to abort the approach in a situation similar to what happened with Kvant 2's delivery. Four days later, on 10 June, a second attempt was made, using the Kristall's backup thrusters; docking was achieved successfully. The next day, the command was given for Kristall to extend its Lyappa arm into the lower receptacle on *Mir*'s docking adaptor, and the next hour was spent slowly moving the module to *Mir*'s ventral port. Now in a T configuration, *Mir* had a stable center of mass, and its orbit could be rebooosted once again by the Progress modules.

The Kristall module's primary function was materials processing. It was set up with several processing furnaces for growing semiconductors and an electrophoresis separation unit. It also contained several astrophysics and magnetic spectrometer experiments in addition to a set of Earth resources cameras. The most noticeable external features on the module were two APAS-89 docking ports, which were directly based on the APAS-75 common docking adaptor developed for ASTP. One was located on the end of the module, and the second one pointed to *Mir*'s rear. The purpose for the first adaptor was to accommodate a Buran shuttle. The second adaptor was intended to host an X-ray telescope intended for delivery by the Buran. Due to the tight-clearance issues with Buran's payload bay, the solar arrays on Kristall could be folded as needed and redeployed later. While the docking port would never host a Buran, it would come in handy a few years later.

Solovyov and Balandin spent the next few weeks commissioning Kristall's systems, but they also conducted a space walk in mid-July to inspect their Soyuz module's heat shield and to patch the loose insulation blankets. Media reports alleged that the crew was "stranded" due to the damage to the blankets, but managers at Kaliningrad downgraded the danger. Still, there was a Soyuz craft standing by with a single cosmonaut ready to launch in case any damage was detected with *Soyuz TM-9*'s heat shield.

This would not be an easy space walk; in order to reach the descent module, they needed to secure a ladder to the orbital module so that one cosmonaut could climb down to tuck in the blankets. In the crew's haste to

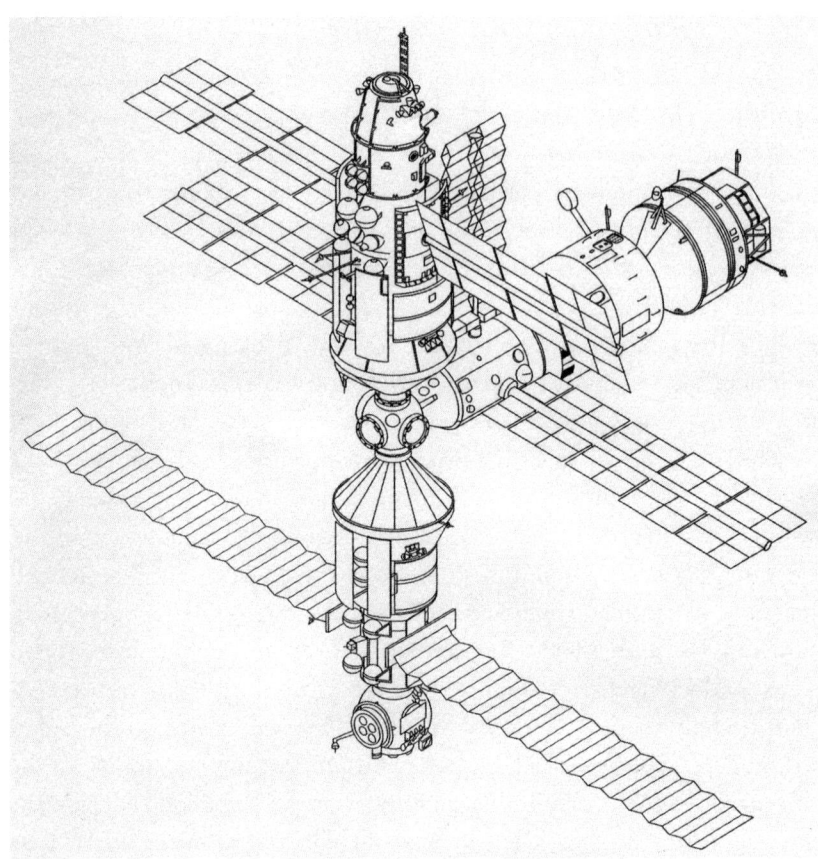

39. *Mir* in 1990 with Kvant 1, Kvant 2, and Kristall modules docked. Courtesy NASA.

get started, though, they opened Kvant 2's airlock hatch before the cabin was fully depressurized, and the pressure differential caused the hatch to blow open quickly, bending the hinge. After conducting a grueling space walk that lasted over seven hours, the crew didn't discover the problem until they climbed inside and the hatch would not completely seal shut. Thankfully, Kvant 2's middle compartment could be used as a second airlock, which the crew used to finish their record-setting EVA. Another space walk to remove the ladder from the Soyuz was conducted a few days later, and television images from the cosmonauts relayed the hatch damage to the ground. Spare parts were needed in order to fix it.

Late summer of 1990 brought about another crew change as *Soyuz TM-10* arrived at *Mir* on 3 August with Gennadi Manakov and Gennadi Strekalov on board. Solovyov and Balandin departed a few days later, with *Soyuz TM-9* conducting a normal reentry and recovery with no problems. The new crew was only going to be aboard the station through December, and their major task would be maintenance, which included rerouting power cables in *Mir*'s core module and performing repairs to the Kvant 2 airlock hatch. While the crew was unable to repair all the damage to the hatch's hinge mechanism, they were able to replace some of the parts and got it to close fully, allowing the normal airlock to be repressurized. The rest of the hatch's repairs would take place with the next crew.

To help with returning more materials-processing samples from Kristall's furnaces, *Progress M-5* contained a new item, a Raduga return capsule. This capsule, measuring one and a half meters long, looked like a truncated cylinder. It could hold 150 kilograms of cargo. The Raduga was designed to be installed in place of the Progress module's hatch prior to undocking from the station. During Progress deorbit, the Raduga would be ejected, enter Earth's atmosphere on its own, and deploy a parachute for a soft landing. While the Raduga would reduce a Progress's cargo capacity somewhat due to their size and weight, it would mean sample returns would be less dependent on cramped Soyuz descent modules. Over the next few years, ten Raduga capsules would be used, with nine successfully returning their cargos to Earth.

A Japanese Visitor

Soyuz TM-11 lifted off from Baikonur on 2 December, bringing year-in-space veteran Musa Manarov back to *Mir* along with Soyuz commander Viktor Afanasyev for his first visit. Accompanying them was Japanese television journalist Toyohiro Akiyama, *Mir*'s first paying customer. His flight was funded by the Tokyo Broadcasting System (TBS) of Japan. Part of the reason for Manakov and Strekalov's rewiring efforts on the base block of *Mir* was to accommodate Akiyama's television and camera equipment, as he was there to document life in space for television viewers and to conduct a handful of Japanese-funded research experiments. Akiyama's time in orbit wasn't necessarily a great one, though, as he came down with space adaptation syndrome very early in the flight and never quite recovered before returning home with Manakov and Strekalov on 10 December.

Afanasyev and Manarov conducted an EVA on 7 January 1991 to finish repairs on Kvant 2's busted hatch. In late January they installed a hand-operated boom crane known as Strela on Kvant 2 during a second EVA. This crane could extend up to forty-six feet in length with telescoping segments. A cosmonaut would ride on the crane while the second one operated the hand cranks. The system would allow a cosmonaut on the end of it to have access to many new areas of the *Mir* complex. It was a versatile piece of hardware, and *Mir* was eventually equipped with two of them.

The rest of Afanasyev and Manarov's *Mir* stint also included troubleshooting Kvant 1's Kurs docking system, after *Progress M-7* had failed to dock on two separate attempts, aborting its approach both times at the last moment. The crew boarded *Soyuz TM-11* and flew a Kurs approach; sure enough, a problem developed, forcing them to dock at the aft port manually. The fault was eventually traced to an antenna that was out of alignment. The Progress was then successfully docked on the vacated front port. Other than that, the rest of 1991 was shaping up to be business as usual.

Soyuz TM-12 lifted off from Baikonur on 18 May 1991 with cosmonauts Anatoly Artsebarsky and Sergei Krikalev heading up for a planned five-month stay in orbit to perform scheduled repairs and updates to the five-year-old station. British passenger Helen Sharman was also along for the ride. To celebrate the English visitor's flight to *Mir*, the R-7 rocket for the mission had a large Union Jack painted on the spacecraft's launch shroud.

Sharman's path to orbit was thanks to a privately funded British venture known as Project Juno. The plan was to use funding from various private companies and a lottery to come up with enough cash to send a UK resident into orbit for a week of experiments on *Mir*. Born in Sheffield, England, Sharman held both a bachelor's of science degree and a PhD from two British universities and was working as a chemist for the Mars candy corporation when she heard an advertisement on British radio in 1989 that said, "Astronaut wanted. No experience necessary." A total of thirteen thousand applications were received, and the selection process whittled the candidates down to four British residents who were sent to Star City, Russia, to begin training. Ultimately, the four candidates were pared down to just Sharman and her backup, Timothy Mace, a major in the British Army.

Project Juno had plans to fly several microgravity experiments for Sharman to carry out; in the end, they lacked enough funding to fly the ambitious science program. It has been said that they also fell short of the required funding to even fly Sharman until Soviet president Mikhail Gorbachev apparently mandated that the flight should go ahead anyway. Once the crew docked with *Mir*, Sharman still had plenty to do, becoming a guinea pig to study her adaptation to weightlessness and conducting experiments with available equipment on board, primarily with the station's agricultural cultivator.

Unlike her Japanese predecessor, Helen Sharman adapted to space very well and spent much of her time enjoying the view, while also taking time to talk with school children in England via a ham radio. Compared to her overworked predecessors with packed flight plans designed to maximize the return in orbit, Sharman seemed to have a much better time on orbit as England's first citizen in space, due to the more open schedule. Her cosmonaut colleagues acted like gentlemen during her stay. During a light moment, Krikalev even wore a tie during one of the dinners the crew had together, although one of the other cosmonauts was reported to have said, "A woman's place is in the kitchen, not the cosmos."

Eventually, after a week in space, the time came for Sharman to return home with Afanasyev and Manarov. The Soyuz undocked and landed successfully on 26 May, just four days before Helen's twenty-eighth birthday. Sharman became a celebrity in England after her flight. In both 1992 and 1998, Sharman's name was submitted by UK officials as a candidate for joining the ESA astronaut ranks as a mission specialist, but she was not selected on either occasion. Today she works as a lecturer and broadcaster specializing in space education while also working for the National Physical Laboratory in the United Kingdom.

Artsebarsky and Krikalev's time in orbit for the next five months was primarily devoted to EVA repair work and construction. They deployed a cosmic-ray detector from the University of California on Kvant 2, replaced an antenna for the Kurs docking system on Kvant 1, and erected a girder called Sofora from Kvant 1 as well, during a total of six space walks. To signal the completion of Sofora's assembly, which towered fourteen meters over the station, the crew unfurled the Soviet flag from the top to acknowledge this achievement in space construction. Little did the crew know at the time that the Soviet Union's days as a country were numbered.

Winds of Change

Back on Earth, a fundamental change in how the Soviets conducted their business was well underway. Since taking office in 1985, General Secretary Mikhail Gorbachev had been trying to steer the Soviet Union's government and economy on a path that conducted more open business with the West and didn't suppress opposing opinions internally. The Soviet Union under Leonid Brezhnev had become more hard-line with increased expenditures in military hardware and less in social programs until his death in 1982. The two Soviet leaders who followed after Brezhnev didn't last long due to their advanced age, as Yuri Andropov died in early 1984 and Konstantin Chernenko was already in poor health when he held office for a little less than a year before his own death in 1985.

The Soviet economy was in bad shape, as many Soviets were living well below the poverty level by Western standards. It was hoped that by offering social and economic reforms, the Soviet way of governing could still be preserved. While the intentions were good, the reforms led to increased calls for nationalism and self-governing among the many ethnic territories and once-independent countries that the Soviets had ruled with an iron hand for decades. This call for change extended to allied countries as well.

By 1989 there were movements underway in Eastern Europe to abandon communism and go with freely elected governments. In years past, such moves would have been squashed with the Soviet military moving in, as had been done in Czechoslovakia in 1968 and in Afghanistan a decade later. But Gorbachev reversed this policy in part due to the drain of Soviet resources brought about by the long war in Afghanistan.

In 1989, revolutions broke out in most of the Eastern European nations of the Warsaw Pact. These revolutions were inspired in part by a pro-democracy demonstration in China that ended with a well-publicized hard-line crackdown against student protesters in Tiananmen Square. Compared to the Chinese experience, most of the European revolutions took place without a shot being fired and ended with peaceful changes of government and the implementation of free elections within a year or two. Countries with once-closed borders began to open them, allowing their citizens to travel freely to Western Europe. The most visible image of this was the tearing down of the Berlin Wall in November of 1989, which ultimately led to reunification of the once-divided Germany by 1991.

The spirit of revolution gained root in many of the Soviet Union's republics as well. Competitive elections in 1990, which were allowed by Gorbachev's reforms, had the CPSU (Communist Party of the Soviet Union) losing many seats in government to candidates from opposition parties. During 1991, in an attempt to help stave off the collapse of the Soviet Union, Gorbachev hoped to re-form the Soviet territories into a confederation of sorts where the Soviet government would be more decentralized, but it would still have a common president, a common army, one economy, and the same general interests. In early August, Gorbachev was only a few days away from signing a treaty to create this new USSR when a coup was staged by Communist Party hard-liners and members of the military who wanted to see the old ways continue. Gorbachev was kept under house arrest during the coup.

The coup ran into public opposition, primarily led by newly elected Russian president Boris Yeltsin, who favored Russia moving completely to a free-market economy and totally abolishing the USSR. After three days, the coup collapsed, and Gorbachev was freed. Upon learning that it was Yeltsin's people who broke the coup and not CPSU loyalists, Gorbachev knew the future of Russia and the other republics would no longer be a Soviet-led one. By the end of 1991 the Soviet Union was dissolved, with most of its territory becoming independent countries.

These world events would alter the Mir program at a fundamental level and change how the space program did things, as it had to answer to a new leadership, a freely elected Russian one. The results of this alteration would ultimately affect the future of manned spaceflight in ways that neither Russia nor the United States could possibly imagine. For better or for worse, space station *Mir* would become the center stage for what came next.

10. The Odd Couple

Russia took over management of the former Soviet space program once the Soviet Union was dissolved. By government decree, the Russian Federal Space Agency (RKA), later to be known as Roscosmos, was formed on 25 February 1992. A space probe design engineer from the NPO Lavochkin design bureau, Yuri Koptev, was placed in charge. By all rights, this agency should have had final say over space management issues. But with the manned program, things were a bit more complicated than that.

Under the Soviet system, it was the design bureaus that were in charge of their own projects with approval from the Soviet government. While the Russian Space Agency was mandated to be in charge of the entire space program, NPO Energia was still the primary driving force behind the manned program since it supplied spacecraft, rockets, and technical personnel. This rendered the fledgling RKA all but powerless after its formation.

In the new economy, Energia became a privately traded company known as Russian Space Corporation (RSC) Energia. This was a necessary change as Energia would have to generate its own revenue to help keep the projects going to support *Mir*'s operation in the face of major budget cuts. Many of the supporting bureaus that supplied hardware also became private companies as well. Unfortunately for Energia and RKA, not all the suppliers were based in Russia anymore. Prices for many of the goods and services these companies provided increased as Russian currency devalued heavily during the 1990s.

Mir for Hire

Even before the Soviet Union fell, Energia turned toward flying paying customers to *Mir* by advertising it as an orbiting laboratory for hire. Frequently, guest cosmonauts would take rides on the Soyuz to *Mir* for about

$35 million a seat, but some rides to *Mir* were provided as favors. *Soyuz TM-13*, which launched to *Mir* on 2 October 1991, was an example of both since Aleksandr Volkov was the only career cosmonaut on board. The other two seats were occupied by Austrian research cosmonaut Franz Viehböck, who had a privately funded seat, and Toktar Aubakirov, who was flying as a citizen of the soon to be independent Kazakhstan as a favor to that country's government. The flight for a Kazakh citizen was considered important since the Baikonur launch complex was located in Kazakhstan.

This meant that Sergei Krikalev, who was scheduled to return home with Anatoly Artsebarsky at the end of a normal five-month tour, would have to remain in orbit for at least another five months with Volkov. Krikalev didn't mind the extension since he had flown with Volkov before and the two men got along well with one another, but the situation brought rampant speculation in the media that Krikalev was stranded in orbit with no ride home, when technically he had access to a Soyuz in an emergency. Ultimately, Sergei Krikalev would spend ten months aboard *Mir* before returning home in March of 1992 with Volkov. Krikalev holds the unique distinction of having launched into space as a Soviet citizen and returning to Earth as a Russian citizen.

During the next few years, *Mir* would host many research cosmonauts from other countries. Klaus-Dietrich Flade from a newly unified Germany flew in March of 1992. French astronaut Michel Tognini went up next as part of a program where CNES astronauts would fly to the complex once every one to two years to conduct ongoing experiments. Jean-Pierre Haigneré from France followed Tognini into orbit in 1993. Even Ulf Merbold flew to *Mir* in late 1994 to conduct scientific work for the ESA and Germany. *Mir* itself seemed to be functioning well with equipment upgrades as everything appeared stable to the outside world. While Russia was suffering from an economic crisis, the space program seemed to outside observers to be one bright spot that people could point to as a tangible success story. To observers, the Russians were flying for less money than what NASA was spending on continued space shuttle operations.

However, the monetary troubles were beginning to manifest themselves. Conflicts with agencies that built flight hardware were beginning to cause work stoppages as some of the contractors would not deliver their goods until the bills were paid. Kazakhstan also cut power to the Baikonur Cosmodrome a couple of times when the electric bill wasn't paid. In 1993 the Buran program was officially canceled. This resulted in the loss of one quarter of Russia's entire space workforce, which numbered four hundred thousand strong in the late 1980s.

Since independent states now also controlled many of the former Soviet sea ports, many vessels in the tracking network fleet were unable to set sail until port fees were paid. During the 1980s Russia had launched a series of satellites that could allow for continuous coverage of *Mir* in much the same way as NASA's TDRS network. But some of these satellites were beginning to break down, and others needed the tracking ships to relay signals back to Kaliningrad. Without the ships, *Mir*'s communications periods were limited for the most part to when it passed directly over the tracking stations in mainland Russia.

The cosmonauts themselves also had to resort to peddling commercial products in orbit by filming television commercials for the advertising revenue. However, this income along with the revenue from the flights of foreign researchers wouldn't solve all the problems. Infrastructure at Star City, the TSUP at Kaliningrad, and Baikonur were all beginning to show signs of age and decay. Because of these problems, both Roscosmos and Energia began to look for additional opportunities to drum up business. They found it from a seemingly unlikely partner, the United States.

Space Station *Freedom*

At the time of Russia's space budget woes, NASA was having problems justifying its own future in manned spaceflight. After the early success of the shuttle program in the 1980s, then NASA administrator James Beggs sold the Reagan administration on the idea that NASA's "next logical step" should be the construction of a space station. In 1984 President Reagan made a public announcement to endorse the project. The new program would be called Space Station Freedom. The loss of the space shuttle *Challenger* in 1986 slowed those plans. But they weren't canceled outright, and NASA spent many years refining the design of the new station and its capabilities. All of this took money, and a portion of NASA's budget during subsequent years was channeled to *Freedom* for early design and testing work.

Once the shuttle fleet was flying again in 1988, NASA expanded its focus on Space Station Freedom, and they had a more pressing need to do so. As a result of the *Challenger* disaster, the planned frequency of shuttle flights was heavily curtailed. Since the shuttle program would no longer be launching commercial satellites, NASA would no longer have a source of revenue to help offset the costs. The DOD also scaled back its planned use of shuttle after the handful of classified shuttle missions had been flown, moving its future payloads back to unmanned rockets. NASA looked to the space station *Freedom* to become the shuttle's destination and the purpose for the manned space program's existence.

Even with NASA's post-*Challenger* situation, Reagan never abandoned his support for the station. But it would be up to his successor, George H. W. Bush, to take the next step when Reagan concluded his second term in office. On 20 July 1989 during the twentieth anniversary of the Apollo 11 moon landing, President Bush announced a "Space Exploration Initiative" with plans to build the space station *Freedom*, a return to the moon within ten years, and an ultimate destination of Mars within about twenty years. NASA's administrator Richard Truly, who was appointed after the *Challenger* disaster, had a ninety-day study conducted to see how much these goals would cost. The resulting figure was an astronomical sum of $500 billion, spent over the next two to three decades. This resulted in a bad case of sticker shock for the president and Congress. The lunar return and Martian plans were soon curtailed but not abandoned entirely.

Freedom continued to receive funding to support design work, although the program was underfunded from the start and suffering from cost overruns. By 1993 *Freedom* had gone through no fewer than seven major design changes since 1984; each time, its capabilities were pared down in an attempt to bring the budget in check. International partners from the ESA and Japan were brought in to help build some of the hardware. But even bringing in these partners would not take care of all the problems, and there were very vocal calls in Washington DC to cancel the program outright.

The Road to Russian Partnership

NASA's new administrator, Daniel Goldin, was appointed near the end of Bush's first term in office to help refocus NASA. While Richard Truly's experience as an MOL and shuttle astronaut had made him a very good fit for the agency after *Challenger*, the Bush administration felt that there was a disconnect between the White House and NASA. The astronomical dollar figure quoted for the space initiative did not sit well with Bush's advisors, and it was felt that *Freedom*'s cost overruns were not being reigned in properly. Bush wanted somebody who was more in tune with what the White House wanted yet still had the charisma to both rally NASA and inspire the general public.

Dan Goldin worked as an engineer for NASA in the early 1960s, but he left the agency to work for TRW, eventually becoming its general manager during a twenty-five-year career. Part of the reason he was nominated to become the NASA administrator was his success at TRW in controlling the costs of space hardware development while still maintaining a high standard of quality. Dan's mantra became known as "Faster, Better, Cheaper" during his tenure at NASA. What helped Goldin to remain as NASA's administrator after Bush lost the 1992 election was his political party affiliation and his philosophy, which meshed well with the administration of newly elected president Bill Clinton.

Cowboy Diplomacy

One of Goldin's first duties as administrator was to restart a project that was agreed to, at least in principal, by President Bush and Mikhail Gorbachev during a summit meeting: an exchange of space travelers to revive the spirit of cooperation from ASTP. The plan was for the Soviets to fly a cosmonaut on the shuttle while a NASA astronaut would fly to *Mir*. The plan fell by the wayside when the Soviet Union was dissolved, but Goldin felt that it would be a public and foreign relations coup if the same agreement were made with the Russians. It took some serious arm-twisting to get members of the State Department to go along with it. But Goldin was successful, and an agreement was made for the crew swap during a summit meeting between Bush and Yeltsin.

The early discussions were done with RKA and not Energia, which meant that Energia wasn't exactly a willing participant when a NASA delegation went to Moscow to hash out the details of the exchange. To Energia, *Mir* was theirs; if NASA wanted to deal, they had to negotiate directly, which would require hard currency. The attitude of many from Energia toward the NASA representatives was less than inviting in a number of ways. For instance, Energia's lead negotiator, flight director Valery Ryumin—the big, burly, stereotypical Russian cosmonaut who had accumulated a total of 362 days in space—was not exactly cordial to the American astronauts who were part of NASA's negotiating teams. To Ryumin and other cosmonauts who had spent hundreds of days in orbit, the Americans were little more than rank amateurs. To Americans like astronaut Brian O'Connor, who was the lead NASA negotiator on the first team, Ryumin wasn't exactly somebody who oozed confidence either. As O'Connor said in Bryan Burrough's book *Dragonfly*, "I never trusted him, and he never appreciated me. He would bad-mouth people and complain about everything—and drink vodka. He was just a jerk."

There were concerns from many at NASA that having an astronaut on *Mir* wouldn't be much more than a publicity stunt with no practical space-flight value. But the decision was a political one, and it was already made at the top. So NASA had to follow through with it, assuming Energia would play ball. Energia's stonewalling on the issue was largely put to rest when Dan Goldin used his characteristic bluntness during a reception to explain to Energia's leader, Yuri Semenov, in no uncertain terms that this cooperative space effort was going to happen since the presidents of both of their nations had already agreed to it and Semenov did not have the power to veto it. To drive home his point, from that moment on, Goldin dealt primarily with Yuri Koptev and the RKA. If Semenov needed something from the NASA administrator, he had to go through the Russian space agency, as opposed to going through direct channels.

Goldin's decision is said to have given Roscosmos legitimacy, although many observers felt that his blunt approach and breach of international protocol might have doomed things from the start. But for those who had knowledge of how the Russians did things, sometimes being blunt and rude was how you got things done if nothing else worked. For better or for worse, there would at least be one exchange of an astronaut and a cosmonaut on each other's vehicles. But this was only the beginning.

The Shotgun Wedding with a Russian Bride

When President Clinton took office, things were a bit bleak for *Freedom*'s future, as few politicians in Washington wanted anything to do with it. Leon Panetta, who back then served as the head of the White House's Office of Management and Budget (OMB), informed Goldin that NASA's budget was going to be cut by 20 percent in the coming fiscal year and that the station was on the chopping block. Goldin asked for time to study the situation further before the budget was presented to Congress and was granted a "stay of execution" until he could come up with possible alternatives. Goldin analyzed the problem and consulted veteran NASA people such as George Abbey and John Young among others. Goldin's handpicked team came up with some ideas. One of the ideas involved a partnership with the Russians to build the new station.

In theory, having the Russians participate was a sound idea. Part of the reason for the cost overruns with *Freedom* had to do with NASA's and its contractors' relative lack of experience in these matters. Sure, there was the experience of the Skylab project to draw from. But Skylab was a program with a beginning and an end. By comparison, the Russians had a lot more experience in designing stations for long-term occupancy and open-ended missions. Over the course of five Salyut stations and *Mir*, they had done simple maintenance, performed life-extending repairs, and even saved their stations from potential mission-ending damage on at least two occasions, all while keeping the stations resupplied with a gradually maturing fleet of Soyuz ferries and automated Progress cargo vehicles. *Mir* had also given the Russians experience with the building block approach to station construction, something that NASA had yet to accomplish as a standard practice.

The idea appealed to President Clinton, as he had another potential problem brewing with an international security matter. The Russians had been in negotiations to supply rocket booster technology to India, for use in a new satellite launcher design. India had succeeded in launching a couple of small research satellites on their own, but they had no capability to launch anything bigger. But a rocket booster capable of sending a satellite into geosynchronous orbit can just as easily carry a nuclear warhead. Given the long-simmering tensions between India and its neighbor Pakistan, the idea of India with a heavy-rocket capability did not sit well with Clinton's security advisors. There were also concerns that if the Russians were successful in selling heavy-booster technology to India, then they might be inclined to do the same with nations potentially more hostile to the United States.

There was talk of imposing sanctions if the Russians went ahead with the deal, but Clinton decided to go for a combination of the carrot and the stick approaches. If the Russians could be encouraged to drop their deal with the Indians and instead have their efforts focused on a partnership to construct an international space station with NASA for potentially more money, it would be a win-win situation for both sides. So with the White House's blessing, NASA made the pitch.

It turns out that Energia was thinking along the same lines concerning a space station partnership. A lot of details would need to be hashed out, but it seemed like the two countries were on the same page. After NASA, Roscosmos, the international partners of the ESA, and Japan's National Space Development Agency had negotiated the details, all parties involved would cooperate to build the International Space Station, also known as the ISS. Ultimately, the Indian rocket deal wasn't canceled, but the Russians agreed to only sell built rocket stages and not provide India with manufacturing equipment.

Genesis of the Shuttle-Mir Program

As part of the deal made with the Russians for their commitment to the ISS, it was decided to get American astronauts as much flight experience as possible on the *Mir* space station; so the first joint program, with one cosmonaut flying on a shuttle and one astronaut flying to *Mir*, was expanded to a total of seven astronauts visiting *Mir* over a three-and-a-half-year period. Most would be delivered by a shuttle, which would dock with the complex by using the APAS-89 docking adaptor built for Buran. On one flight, cosmonauts from *Mir* would also return to Earth on the shuttle, instead of on a Soyuz vehicle. The collaboration between NASA and *Mir* officially became known as ISS Phase One, while actual construction of the International Space Station would be known as ISS Phase Two. To most everyone who worked on ISS Phase One, it simply became known as the Shuttle-Mir Program.

Before a shuttle could dock with *Mir*, the APAS-89 docking system had to be tested. *Soyuz TM-16* with an APAS docking collar was launched on 24 January 1993 and executed a proper docking with the Kristall module's Buran port. When *Soyuz TM-17* arrived six months later, it shot some unique photos of *Mir* with *TM-16* on Kristall, a Progress module docked with the station's Kvant 1 port, and a second Progress backing away from *Mir*'s front port to make room for *TM-17* to dock. It gave *Mir* the appearance of a busy airport in space.

The shuttle would deliver logistical supplies to the station and additional experiments as well. NASA would also supply funding to finish *Mir*'s final two modules: Spektr (a Russian term meaning "Spectrum") and Priroda (a Russian term meaning "Nature"). These two modules, based on the TKS spacecraft like the Kvant 2 and Kristall modules, sat mothballed in the factory, as there were insufficient funds to complete them. Their internal systems would be heavily reconfigured to act as a pair of scientific modules, and Spektr would also serve as living quarters for the NASA astronauts. If the Apollo-Soyuz Test Project was anything to go by, the Shuttle-Mir Program should have gone off without any major hitches, but events would transpire to make Shuttle-Mir more than what either side bargained for. It would be an experience that many of the people involved would not soon forget.

Astronaut Selection

When the announcement was made about the Shuttle-Mir Program, many members of the astronaut office were less than enthusiastic about taking part. But there were some, both veterans and rookies, who felt that a tour of duty on *Mir* just might be a good opportunity. The numbers of volunteers were not large, though. Rather than being able to select the best candidates for the Shuttle-Mir missions from a sea of applicants while taking into account psychological temperament and experience, NASA didn't have much choice except to take all those who volunteered for the Mir program.

Norm Thagard, a veteran of the first 1978 class of shuttle astronauts, would be the first visitor as part of the original pre-ISS agreement. He would fly to *Mir* aboard a Soyuz craft and, after several months in orbit, return home with his cosmonaut crewmates aboard a shuttle that ferried up two other cosmonauts to take over the next increment. Fellow 1978 veteran Shannon Lucid would fly the first official ISS Phase One mission a few months later. John Blaha would fly next, exchanging positions with Lucid. He would be followed by newcomer astronaut Jerry Linenger, who joined the astronaut office in 1992. Michael Foale, a veteran astronaut of dual U.S. and UK citizenship, would fly next. Taking over for him would be Wendy Lawrence, while David Wolf would perform the final increment before ISS Phase One was completed. Selected for backup astronaut roles were Bonnie Dunbar for Thagard's mission and Andy Thomas for the program's other nonflying backup. Scott Parazynski had also been selected for the Mir program, but he was ultimately scrubbed from the assignment since he was a little too tall to safely fit in a Soyuz TM crew couch, as each astronaut would have to be fitted for a Sokol pressure suit and custom seat liner in case they needed to use the Soyuz in an emergency.

The Language Barrier

While the ASTP astronauts and their cosmonaut colleagues had many years to train and integrate as a combined crew, that wasn't necessarily an option open to the Shuttle-Mir astronauts. Each training period would last about two years, but they wouldn't be able to integrate with specific cosmonaut crews due to the complications brought about by launch delays. The language barrier would be the largest hurdle to overcome, so each astronaut and many of their support crewmembers on the ground would have to take what was akin to a crash course in Russian in order to communicate with their cosmonaut crewmates and instructors. Many of the Russians spoke English, but it was not their native language. And unlike ASTP, where the crews would speak each other's languages, the cosmonauts had no mandate to improve their English unless they were flying on shuttle missions. The instructors at Star City would conduct training sessions in Russian as well, both spoken and written. While language interpreters helped, they couldn't be sent into orbit of course.

Norm Thagard began taking classes in Russian before he was officially announced in order to help give him a leg up on training. Some of the other Shuttle-Mir candidates followed his lead. To help with the language, the first astronauts assigned to the Mir program were sent to the Defense Language Institute in Monterrey, California, for Russian study. The DLI was a school established in World War II to teach members of the military and intelligence community foreign languages. During the Cold War its role was expanded to train members of other federal agencies, and the language base was expanded to include Russian.

Instructors at DLI said that at minimum it would take eighteen months to learn Russian with intense study, compared to six months needed to become reasonably fluent in Spanish or French. While it wasn't the hardest language of all to learn, it would certainly not be an easy one. According to Jerry Linenger's account, the commanding officer of the DLI reinforced this to him in their first meeting by saying there were "no shortcuts." Each astronaut had only five weeks to learn Russian because somebody at NASA had explained to the DLI that their candidates were willing to learn and smarter than average people. The DLI itself at least welcomed the work, albeit reluctantly. So five weeks it would be, for better or for worse.

One might liken the training approach used by the DLI for a five-week Russian-language class to taking somebody with only a rudimentary knowledge of swimming and throwing them into the middle of a lake. The student would either sink or quickly learn to swim. At the end of five weeks, after showing at least a basic understanding of the language (which, depending on who was asked, is the subject of debate), it was off to Russia for training at Star City on Mir and Soyuz systems in the Russian language. After the Russian-language classes, training in a post-Soviet-era Russia likely had many astronauts wondering if they had gone off the deep end.

Culture Shock

Compared to Star City of the early 1970s, Star City of the 1990s was both similar in many ways and very different. By 1994 the years had not been kind. Hyperinflation had caused the Russian ruble to devalue to the point where eight rubles would barely buy a pack of chewing gum. U.S. dollars became the preferred currency for most purchases. Retired Russian military officers and officials living at Star City who had worked most of their lives for the state saw their pensions become nearly worthless, to the point that some members of their families were working jobs in Moscow just to make ends meet. During the training period for the first three Mir increments, astronauts were staying at the old three-story hotel built for ASTP, which was known by its rather unflattering name, the Prophylactorium. According to eyewitness accounts, it had seen better days.

Part of the economic package from the United States was to fund the building of brand new duplex facilities to house astronauts and their wives (or husbands), but the work remained mostly unfinished until Jerry Linenger and Michael Foale's training periods. There have been reports that some of that money was funneled away to build lavish houses for some members of the Energia staff and members of the military. There were also rumors of Russian mafia corruption.

Jerry Linenger's book *Off the Planet* told of times when the Chevy vans purchased by NASA for official use would end up getting stolen from Star City and never be seen again. On another occasion, the duplex apartment of Michael Foale was vandalized and had valuables stolen from it while he was back in Houston. Linenger wrote that security didn't really improve until he planted a rumor to some Russian colleagues that he was going to apply for a gun permit to protect his wife. She was staying with him in Russia while she worked for a NASA contractor and was pregnant with their first child at the time. After that, Russian security dramatically improved, and the thefts dropped to almost nothing.

Russian training on space systems, at a glance, is similar to training in the United States. But while NASA tends to favor training in specific skill sets, Russia focuses primarily on theory to make their cosmonauts jacks-of-all-trades. Crewmembers had to take periodic exams to see if they absorbed the information they were being taught, and a crew would not fly if they failed their final exam on the eve of a spaceflight.

What made things difficult for the Americans, though, was that a lot of the information being taught was not written down, and people were discouraged from taking what few training manuals they could find back to their apartments to study. This practice didn't just extend to the Americans, though, as research cosmonauts and commercial fliers from other countries have reported similar behavior by the Russians toward them. The Soyuz simulators were up to the task, but there were concerns that the *Mir* simulators were not completely representative of the nearly decade-old station with years of clutter and modified systems. Still, everyone made do.

Camaraderie and close relationships were formed between many of the NASA, Roscosmos, U.S. contractor, and Energia people as time went on, but it didn't start out that way. Some Americans bonded more openly with their Russian counterparts than others. Of the astronauts who lived and trained in Star City, Michael Foale became the most "Russianized," as he tried to totally immerse himself in Russian culture, speaking Russian whenever possible during his off-duty times and opening his apartment to his Russian friends in Star City. Dave Wolf apparently also succeeded in drinking some of his Russian colleagues under the table during some legendary wild parties on the weekends. To the Russians, the ability to hold liquor can command just as much respect as anything else, and Dave could do it as well as anyone.

Both NASA and elements of its contractor workforce set up offices at both Star City and the TSUP in Kaliningrad. In addition to the instruction on Soyuz and Mir systems, it would be up to these support teams to come up with and approve the scientific experiments that would fly as well as to provide on-orbit support since once the shuttle had undocked, there would be almost no direct support from Houston. It has been said that NASA didn't necessarily send their best support people to Russia. Ideally, personnel from NASA's Mission Operations Directorate should have been sent to manage the work that would take place on orbit, since that was their purpose for shuttle missions. The Mission Operations Directorate had their hands full, though, working on the ISS, and only got involved with the shuttle-related tasks. But assurances were made from NASA management that the Russians had things well in hand. In those days, NASA would mostly be just along for the ride.

On 3 February 1994 the space shuttle *Discovery* launched from KSC on mission STS-60, which would be the first ISS Phase One mission. While not scheduled to rendezvous or dock with *Mir*, the shuttle carried Russian cosmonaut Sergei Krikalev into orbit as part of a six-person crew for a nine-day scientific mission that involved biomedical and materials-processing experiments in the newly developed Spacehab module. As part of a three-way communications link, Sergei Krikalev contacted his crewmates on *Mir*, and their exchange was seen on live television in the United States. The next time *Mir* and a shuttle would be in communication with one another, they would be a lot closer.

Shuttle Preparations

As NASA's side of the ISS Phase One program progressed, the role of the shuttle was refined somewhat. While the shuttle had been designed to support a space station, operations in a close proximity to *Mir* were something new and required precision maneuvers that the shuttle had not originally been intended to fly. The Russians also had concerns about exhaust from the shuttle's hypergolic thrusters potentially contaminating *Mir*'s solar arrays, since shuttle thrusters were much larger than those found on Russian spacecraft.

To help limit the amount of thruster firings needed for rendezvous and docking, the shuttle would fly a plus R-bar approach (*R* standing for "radius"). Rather than having the shuttle approach *Mir* from in front or behind and using thrusters for braking, it would approach the station from below, increasing orbital altitude as it closed the distance. This approach would act as a natural form of braking due to how the orbit was set up, so thrusters would not need to be fired close to the station before dock-ing. NASA successfully tested this approach method using the space shut-tle *Atlantis* on STS-66 with a free-flying satellite payload.

As a full dress rehearsal to the first shuttle-docking mission, *Discovery* lifted off on mission STS-63 in early February 1995. The launch window for this mission was only five minutes long in order to target *Mir*'s 51.6-degree orbital inclination. Joining the NASA crew was cosmonaut Vladimir Titov, only the second Russian to fly on shuttle. While STS-63 would conduct some experiments directly related to future ISS construction on this flight, along with some joint Russian and U.S. experiments, the highlight would be the rendezvous and close approach to *Mir* midway through the mission. But it almost didn't happen.

Just after launch, two of *Discovery*'s thrusters developed slow leaks and were trailing fuel particles. The Russians were very concerned about having the orbiter make a close approach to the station with such a leak. But after a day in orbit, the leaks abated, and the go was given for close approach. On *Discovery*, shuttle commander Jim Weatherbee and female shuttle pilot Eileen Collins approached to within eleven meters of *Mir*.

Aboard *Mir*, cosmonauts Aleksandr Viktorenko, Valery Polyakov, and Yelena Kondakova monitored the approach, shooting video and still photo-graphs of the event. Polyakov was on board setting a new space endurance record of over a year in space. During his mission, he spent 437 days in orbit before returning home in March with the TM-20 crew. Yelena Kondakova was only the third female cosmonaut to fly in orbit after Tereshkova and Savitskaya. She was married to Valery Ryumin and served a normal five-month tour of duty on *Mir* before any of the Americans (including Shan-non Lucid) got to visit the station.

During the point of closest approach, Weatherbee commemorated the event by saying, "As we are bringing our spaceships closer together, we are bringing our nations closer together. The next time we approach, we will shake your hand, and together we will lead our world into the next millen-nium." From *Mir*, Viktorenko responded, "We are one. We are human." After closest approach, *Discovery* backed away and spent a portion of the next orbit carefully circling *Mir* from a safe distance, shooting photographs of the complex before leaving. The next time a shuttle visited, it would link up with the complex.

Norm Thagard's Mission

On 14 March 1995 the Soyuz rocket carrying Mir EO-18 was ready to blast into the heavens from Baikonur. Aboard *Soyuz TM-21*, Vladimir Dezhurov and Gennadi Strekalov were joined by NASA astronaut Norman Thagard. It was the first time an American had ever flown on a rocket of Russian design. Launch proceeded just fine; after two days, the trio firmly docked at *Mir* and performed the normal handover from the Mir EO-17 crew of Viktorenko, Polyakov, and Kondakova, who returned home soon after.

The schedule called for Thagard to conduct most of his scientific research in the Spektr module, which was supposed to be delivered early in the mission. But Spektr's delivery date slipped, and NASA's science team led by Peggy Whitson had to scale back their plans and use equipment that could either be delivered by a Progress module or was already aboard *Mir*. As Whitson explained, "We found out that the Spektr science module was going to be delayed until near the end of that expedition. So over the course of a weekend, we had to go through and cut and reprioritize. So it was very much pulling things together that we could get on a Progress, make it happen, and still meet as many of our science objectives as possible."

Part of the data gathering Thagard was assigned to do included collection of blood and tissue samples from both himself and his cosmonaut crewmates. Without access to Spektr's hardware, Thagard was placing the samples in a small freezer left on *Mir* by ESA astronaut Ulf Merbold, who had visited the station five months earlier. But the freezer was having trouble maintaining low temperatures during Thagard's mission due to a buildup of ice on its condenser, and it needed to be defrosted. Unfortunately, the attempts to defrost the freezer proved futile, and many of the biological samples had to be scrapped.

As for other experiments Thagard did, they primarily involved diet and metabolic measurements. The original plan was for Norm to eat a structured diet from bar-coded food containers with known nutritional contents. Supplemental food could be eaten, but a diary would need to be kept for scientists on the ground. Unfortunately, *Mir* didn't exactly have a convenient stash of paper to take notes on, so Thagard stuck to the prearranged food and didn't eat anything else that would require note taking. Midway through the mission, the regular body mass measurements were showing that Thagard was losing weight since the caloric intake was not keeping up with his daily nutritional requirements. Finally, it took a direct order from doctors on the ground before Thagard would eat the snack foods on board to get his caloric intake back to normal levels.

Spektr Arrives

The Spektr module finally was launched in late May and docked with *Mir* on 1 June 1995. In preparation for Spektr's arrival, the Kristall module had its solar arrays retracted and was moved using the Lyappa arm and drogue system to the right-side docking port. Dezhurov and Strekalov relocated one of Kristall's two solar arrays to a new mounting point on Kvant 1 during an EVA in late May, and the second Kristall array was folded up to prevent any clearance issues with a docking shuttle. Spektr executed a perfect docking and then performed the now-standard Lyappa arm transfer to finally be berthed in Kristall's old location of the bottom docking port.

Spektr contained about seven hundred kilograms of American equipment for experiments in addition to many of the normal systems it was equipped with prior to its mothballing. Spektr was originally outfitted with systems of a military nature, as it was intended to carry surveillance equipment like the Almaz and the Cosmos 1686 spacecraft. When Spektr was repurposed, the surveillance equipment located at the front of the module was replaced with a cone-shaped section containing two solar arrays mounted diagonally, in addition to its normal arrays mounted midway down on each side of the module. Some of the scientific equipment it contained included a binocular radiometer, a lidar system for atmospheric study, an interstellar gas detector, and several different spectrometers. It also contained a scientific airlock for exposing various experiments to space. The four new solar arrays would help to supplement *Mir*'s power supply.

For his final month on *Mir*, Norm Thagard spent his time unpacking the Spektr module and getting the equipment ready for the next American astronaut to visit the station, even though that visit wouldn't occur for a few months yet. He also got a chance to conduct some additional science as well. Given that Thagard had spent much of his time not really having any experiments to perform for most of May due to the busted freezer and the problems with the food-intake experiment, it was a welcome change. All he could really do before Spektr arrived was perform his normal exercise regime. Thagard's presence on *Mir* was almost like that of a research cosmonaut on a short-duration mission. Although he was fully trained in *Mir*'s systems, controllers and his cosmonaut crewmates apparently would not allow him to take part in normal station-maintenance chores or aid in preparation for their space walks. Having a lot of downtime, for an astronaut who has trained hard and knows that every minute spent on orbit can be precious due to its cost, can be almost torture.

Shuttle Docking

The space shuttle *Atlantis* lifted off from the Kennedy Space Center on 27 June on shuttle mission STS-71. In its payload bay, it was fitted with the pressurized Spacelab module and the APAS-89 docking adaptor. The Spacelab would be used to conduct joint American and Russian life sciences experiments both while docked to *Mir* and after undocking. Joining the five-person American crew were Russian cosmonauts Anatoly Solovyev on his fourth spaceflight and Nikolai Budarin on his first. Both men would swap places with the Mir EO-18 crew of Dezhurov, Strekalov, and Thagard.

The EO-18 crew had transferred Kristall back to *Mir*'s front docking port to provide *Atlantis* enough clearance from the solar arrays on *Mir*'s base block for a safe docking. *Mir* would temporarily stretch almost thirty meters in length from the aft docking port on the Kvant I to the APAS docking port at the extreme opposite end of Kristall. After the shuttle departed, Kristall would be moved back to make room for Soyuz and Progress craft.

Docking day for STS-71 came on 29 June. At the controls were veteran shuttle commander Robert "Hoot" Gibson and pilot Charlie Precourt. At 08:00 Houston time, *Atlantis* achieved a soft dock with *Mir* as the pair orbited over central Russia. By all accounts, it was a very smooth docking with almost no misalignment issues to speak of or any jolt during the actual linkup. Once hard dock was achieved, the pair formed the largest combined spacecraft in orbit, weighing almost half a million pounds in the process. After a few hours, when the seal checks showed that everything was satisfactory, the hatches were opened, and Gibson shook hands with cosmonaut Dezhurov, paying homage to a similar handshake conducted on ASTP between Stafford and Leonov.

For the next five days, the combined ten-person crew conducted transfer of materials between *Mir* and the shuttle as well as joint experiments in

Spacelab to measure the crew's adaptation to space. Logistically, the shuttle could carry many more supplies than a Progress module, and its fuel cells also came in handy for delivery of fresh water to *Mir*'s tanks. The shuttle's ability to haul supplies during the coming years would be very important for *Mir*'s continued use.

To take advantage of a rare photo opportunity at the end of the five-day period, Solovyev and Budarin boarded the Soyuz and undocked from *Mir*, taking up position several meters to the left of the complex in order to shoot photographs of the two docked spacecraft. Once *Atlantis* had undocked, the Soyuz would redock with the station once again. Unfortunately, *Mir*'s attitude-control computer picked this time to go off-line, so Solovyev and Budarin had to redock manually as soon as possible to stabilize the station's attitude before the solar arrays stopped tracking the sun and the batteries began to drain. This incident would be a hint of things to come.

After a few more days of solo operations, *Atlantis* was ready to deorbit and land. Since Dezhurov, Strekalov, and Thagard had spent over one hundred days in orbit and shuttle crews normally sit upright during reentry, there were concerns about the long-duration crewmembers being prone to blackouts from the g-forces caused by blood pooling in their legs. To help ease the strain on their bodies, *Atlantis* was fitted with three reclined crew couches on the middeck. The crewmembers coming from *Mir* would sit with their backs on the floor and their legs elevated in a similar position to Soyuz seats so that their hearts were at the same level as their heads. This seating arrangement would become standard for crewmembers of *Mir* and the ISS when returning to earth on the shuttle. *Atlantis* landed successfully at KSC on 7 July. Thagard survived his ordeal none the worse for wear, as did the cosmonauts. The Americans and the Russians hailed the entire mission publicly as a success, and the way was cleared for the next missions to *Mir* by U.S. astronauts.

The next visit of a shuttle to *Mir* occurred on 15 November 1995. For this mission, the shuttle carried a new Russian-built docking module for *Mir*. This barrel-shaped module was equipped with APAS-89 docking collars on each end with a cylindrical section in the middle. On orbit, the module was attached to the shuttle's airlock; upon undocking, the module would remain attached to *Mir*. This allowed for future shuttles to dock with the Kristall module at its normal location since the new module provided the

40. Space shuttle *Atlantis* docked with *Mir* on STS-71. This image was taken by Solovyev and Budarin aboard *Soyuz TM-21*. Courtesy NASA.

proper clearance between the shuttle and *Mir*'s solar arrays. The docking module also contained two new solar arrays that would be installed on the station's Kvant 1 module during two planned EVAS.

No crewmembers were exchanged on this mission, as the purpose of this flight was to load the station with additional supplies. The mission was a short one by shuttle standards, as *Atlantis* was only docked to *Mir* for four days out of the eight it spent in orbit. One of the crewmembers on this flight was Chris Hadfield, a mission specialist from the Canadian Space Agency. Chris became the only Canadian astronaut to visit *Mir*.

Shannon Lucid's Mission

It would be about a year after Thagard's launch before the next American would call *Mir* home. Astronaut Shannon Lucid was part of the first group of women to join the NASA astronaut ranks in 1978. Her mission would be the first of a continuous American presence on *Mir* for the next two years. While Lucid's flight was packed with experiments, many of which were originally intended for use on Thagard's mission, some problems with scheduling of the training meant that Lucid didn't have access to any decent documentation for the science she would be performing until very close to launch time. Fortunately, Lucid had one of the better support groups of the ISS Phase One program, as Bill Gerstenmaier, an engineer from NASA's Mission Operations Directorate, was the primary lead and did what was necessary to get her all the support she needed. Many consider Lucid's increment with *Mir* to be one of the smoothest running of the entire Shuttle-Mir Program.

When the space shuttle *Atlantis* arrived on 24 March 1996 during mission STS-76, Lucid's cosmonaut crewmates, Yuri Onufriyenko and Yuri Usachev, had already been aboard *Mir* for over a month, after arriving aboard *Soyuz TM-23*. The combined crews transferred 1,500 pounds of water to the station and two tons of additional equipment and supplies. A highlight of the mission was an EVA in which shuttle astronauts Michael Clifford and Linda Goodwin conducted a six-hour space walk and placed four racks of materials known as the Mir Environmental Effects Payload (MEEP) on the docking module. These racks were exposed to space to evaluate materials being considered for use on the ISS. It was the first time American astronauts had performed a space walk around a Russian station. The samples were retrieved during an EVA on STS-86 almost eighteen months later.

After five days of joint operations, *Atlantis* undocked and left Lucid to begin what was originally planned to be a mission lasting four and a half months. Ultimately, due to a couple of delays in processing *Atlantis* for the STS-79 mission stemming from concerns about the SRBs, Shannon Lucid would end up spending 188 days in space and setting a new single-mission endurance record for both American and female astronauts. There were some concerns prior to the flight as to how her Russian crewmates might treat her. But Onufriyenko and Usachev reportedly acted like gentlemen, calling their new crewmate "Ms. Shannon," and the trio got along very well together.

Priroda's Arrival

Mir's final module, Priroda launched on 23 April 1996. Unlike the other *Mir* modules, it was not equipped with solar arrays. Instead, it was outfitted with a large load of storage batteries to provide electricity until it could

41. The complete *Mir* complex. The radial modules (*clockwise from the upper left*) are Kvant 2, Priroda, Spektr, and Kristall with shuttle docking module. A Soyuz is docked in front, and a Progress is docked to Kvant 1 in back. Courtesy NASA.

be docked with the station. Three days later, the new module approached *Mir*'s docking port. During those three days, a problem with one of Priroda's electrical connectors cut its available power supply in half. So Priroda would only have enough power to perform one docking attempt. This was not a good development, as most of the other *Mir* modules failed to dock on their first attempt. But Priroda performed a successful docking on *Mir*'s front port and was moved to the vacant axial docking port soon after with no problems. *Mir* was now complete, and the complex measured 27.5 meters tall and 31 meters across, weighing almost 130 metric tons.

Priroda's primary scientific load was for studies of Earth's atmosphere and surface features, using a combination of microwave systems and infrared-detection equipment that measured the near, the passive, and the active infrared spectrum. It also contained a lidar that could be used to measure cloud-top heights in Earth's atmosphere. Priroda's most visible feature was a large parabolic antenna for a synthetic aperture radar system called Travers. Similar radar systems had been flown on shuttle flights, but this was the first time such a system had been designed for *Mir*. All told, twelve countries supplied experiments for use on Priroda, making *Mir* a true international space station. Much of the science from European countries would take place in Priroda, while most of the American work would be conducted in Spektr.

Onufriyenko and Usachev conducted a total of six EVAs during their five-month stay on *Mir*. Much of the work revolved around preparing Kvant 1 to receive its two new solar arrays brought up with the shuttle-docking module. They also manually deployed the Travers radar antenna after it had failed to fully unfurl during its automatic deployment sequence. On one space walk, the pair also performed a rather unusual role in an advertising gimmick as they unfurled and inflated a giant Pepsi-Cola can and filmed it for use in a commercial that ultimately never aired. It was yet another use of *Mir* by Energia to generate revenue from product endorsements to help offset the station's operating costs.

Due to the delays in Shannon Lucid's return shuttle mission, the next Russian crew arrived at *Mir* before the shuttle did, as cosmonauts Valery Korzun and Aleksandr "Sasha" Kaleri arrived aboard *Soyuz TM-24* in mid-August. Joining them was France's first female astronaut, Claudie André-Deshays, who conducted a short-duration mission to the station during the crew handover period as part of a series of joint Russian and French experiments. Onufriyenko, Usachev, and André-Deshays returned to Earth aboard *Soyuz TM-23* safely on 2 September.

John Blaha's Mission

STS-79 finally got off the ground on 16 September 1996 and docked with *Mir* a couple of days later. Shannon Lucid was very happy to see her NASA colleagues, as she was seen floating alone in *Mir*'s docking compartment while awaiting the hatches to be opened between the two spacecraft. After nearly five days of joint operations, Lucid returned home aboard *Atlantis* while her replacement, John Blaha, took up residence for his own four-and-a-half-month tour of duty aboard the station.

As a veteran shuttle commander, John Blaha was unique among the NASA crewmembers in the Mir program, since every other astronaut in the Mir program was a mission specialist. Blaha came up through the test pilot ranks in the U.S. Air Force and was one of the last graduates of Edwards AFB's ARPS program before it was eliminated from the Test Pilots School's curriculum. By all indications, Blaha was a very smart individual. It comes as a bit of a surprise that Blaha's time on *Mir* as documented in the book *Dragonfly* seems to paint a picture of an astronaut that didn't seem to have a firm grasp of what he was doing on board. In an interview conducted after *Dragonfly* was published, Blaha considers the sections written about him to be a work of fiction done at his expense to sell books.

Part of Blaha's problems seemed to stem from an overly ambitious scientific program that was tightly scheduled on a daily basis by a support crew that seemed to not have a firm grasp of the equipment on orbit or the state of *Mir*'s systems. The daily schedules, known as "Form 24s," were sent up to the station on a regular basis and had the astronaut's entire day mapped out from wake-up to sleep period. The updates might call for Blaha to work with one specific piece of scientific apparatus with a specific time to perform the experiment, but the schedules didn't take into account that Blaha would have to likely spend hours collecting pieces stored all over *Mir* and assembling them to perform an experiment. In a couple of cases, specific pieces of equipment couldn't be found at all.

These schedules had been agreed to ahead of time between NASA and flight directors at the TSUP and couldn't be amended easily after their approval. So if an experiment wasn't conducted for some reason, it might pop up on the schedule again in a few days, starting the whole time-wasting process all over again. The usual result typically was frustration between Blaha on orbit and his American support crew on the ground.

Blaha also apparently clashed a little bit with *Mir* commander Valery Korzun, an air force officer who joined the cosmonaut ranks as a test pilot. Korzun was flying his first space mission and apparently tried to compensate for his relative lack of space experience by micromanaging everything. Blaha, being a career U.S. Air Force officer with command experience, didn't seem to appreciate that, and the two apparently butted heads on a few occasions. Of all the cosmonauts, Korzun was considered to be the most Westernized of the group, and he typically got along well with his astronaut counterparts. The fact that Blaha and Korzun didn't train together likely contributed to the problems. Korzun's crew was originally the backup to the prime crew of Gennadi Manakov and Pavel Vinogradov. When Manakov developed heart problems and had to retire from the cosmonaut ranks, Korzun and Kaleri were assigned to fly instead. In contrast to Blaha, Jerry Linenger seemed to get along very well with Korzun, as they had trained together.

When Jerry Linenger arrived on *Mir* aboard *Atlantis* on STS-81 in early January of 1997, Blaha seemed to be rather exhausted and glad to be returning home. Blaha explained to Linenger many things, including his frustration with the scheduling. Linenger tried his best to make the transition a smooth one. At the time, nobody really had any idea what the next five months would have in store for the most junior astronaut in NASA's Mir program.

Jerry Linenger's "Eventful" Mission

Jerry Linenger hadn't been an astronaut very long when he was selected to take part in the Shuttle-Mir Program. Linenger had been a medical doctor with a bachelor's degree in bioscience and a U.S. Navy flight surgeon, and his desire to take part in a long-duration space mission is one of the factors that led to his astronaut selection in 1992. To get spaceflight experience prior to his trip to *Mir*, Linenger flew aboard the space shuttle *Discovery* on STS-64 in 1994.

It has been said that in every story of actual events told by two people, there are three versions: what one person said, what the other person said, and the truth. Jerry Linenger's mission to *Mir* has been documented in print both in *Dragonfly* and in his autobiography, *Off the Planet*. Both have similarities, but they also paint rather different pictures of some events. What they have in common is that Linenger experienced some of the same frustrations that Blaha did, at least in the early days. Linenger was also irritated that scheduled time periods where he was supposed to be in contact with his support crew on the ground or having personal communications with his wife (who worked for one of the NASA contractors in Russia) were usually interrupted by official business considered to be more pressing by the TSUP flight controllers. Even Linenger's cosmonaut crewmates found this irritating since they knew that voice time scheduled for personal matters was not something that was usually interrupted since it was considered important for crew morale.

Eventually, Linenger opted to cut off all voice communications with his support team except during very important periods, preferring instead to just use the email system on his laptop computer, since he considered the communications likely to get interrupted anyway. To the Russian doctors and psychologists on the ground, this painted a picture that Linenger was not being a team player and that there was possible strife on orbit between the American and his Russian colleagues. Linenger's own account denies that was ever the case.

Fire in Space!

The communications and psychological concerns didn't really begin to manifest themselves all that much, though, during the first month of the mission. Indeed, Korzun pulled rank on the American support team at one point on Linenger's behalf when he noticed how they were scheduling the astronaut, front-loading him with a lot of scientific work at a fast pace as if it were a shuttle mission. Experienced cosmonauts knew that long-term spaceflight was a marathon, not a sprint. Trying to do too much too soon risked hitting the wall and burning out. The support team backed off.

On 12 February 1997 *Soyuz TM-25* docked to *Mir* with crewmembers Vasily Tsibliyev and Aleksandr "Sasha" Lazutkin on board. Joining them was German physicist Dr. Reinhold Ewald, who would be conducting a ten-day scientific program during the handover. Ewald would return home with Korzun and Kaleri. Having six crewmembers on board meant that *Mir*'s life-support systems were a bit overtaxed. Of the two primary Elektron oxygen-generating systems located in *Mir*'s base block and Kvant 1, only one was working and not at full efficiency. During docked shuttle operations, the shuttle would carry the life-support load for both itself and *Mir*. But without the shuttle, something else was needed to generate the required oxygen for six people.

Mir's supplemental oxygen generation came in the form of canisters known to the Russians as Vika (a Russian term meaning "Pressure" or "Power") and to NASA as SFOG (solid-fuel oxygen generation). Each Vika canister contains a quantity of lithium perchlorate and an internal heating element. When the heating element is activated, it heats up to four hundred degrees Celsius, and the resulting chemical reaction releases fresh oxygen. It is a very efficient system, given its compact size; the Russians also use them in submarines. One canister can provide enough oxygen for one person for twenty-four hours. With more than three people on *Mir*, a canister needed to be activated every few hours.

The cosmonauts referred to the canisters as "candles," and the duty of activating them typically fell to a flight engineer, in this case Lazutkin. The apparatus that housed an activated canister was located in Kvant 1. A few days after the new crew arrived, the old crew decided to have a party for their new crewmates on 24 February 1997. As Lazutkin described it, they opened and consumed some of the comfort foods, such as red caviar, normally only reserved for very special occasions due to the cost. During dinner, a warning light indicating low oxygen turned on, so it was time to activate a canister. Lazutkin excused himself to do so, as he had done several times on the mission already. He switched out the spent canister for a fresh one and activated it by pulling its activator pin.

When an SFOG canister is operating properly, it feels slightly warm to the touch due to the heating element, and a breeze of "fresh" oxygen can be sensed. But this time, the canister literally almost exploded, as a blowtorch-like flame measuring about a foot across and two to three feet long erupted from the canister. Flames in zero gravity normally form a ball, but this one was straight, like a rocket's exhaust plume. The fire was being fed by freshly generated oxygen under pressure. Molten droplets of metal from the burning canister were splattering on the far bulkhead according to Linenger's accounts of the fire. It was like a fireworks display gone horribly wrong.

Dense smoke began to fill *Mir* as the crews reached for their emergency oxygen respirators. The first one Linenger reached for didn't work, but the second one did. Lazutkin, in an interview for a television program about *Mir*, described what happened next: "When I saw [*Mir*] was full of smoke, my natural reaction was to want to open a window. And then I was truly afraid for the first time. You can't escape the smoke. You can't just open a window to ventilate the room."

After an attempt to smother the blaze with a blanket had little effect, Korzun grabbed a fire extinguisher and tried to put it out with the foam setting, but the foam didn't do much good. He switched to water to cool the fire area down. This was critical since if the fire didn't get under control, the flame might breach the far wall of the Kvant module and open it to space, killing everyone in the process. Linenger acted as part of the fire-fighting team, stabilizing Korzun by the waist as the cosmonaut worked to spray water on the fire. Linenger checked Korzun periodically with tugs on the legs to make sure he was still conscious, since he couldn't see his friend easily in the dense smoke or hear him through the sound of the inferno and the respirator mask. With each tug, Korzun would move his legs to acknowledge he was still conscious.

At the same time, Kaleri switched off the circulation fans and began printing out critical orbit data and checklists in the *Mir* base block that would be needed to help undock and deorbit the Soyuz modules if the station had to be abandoned. The three newcomers did whatever they could to support the firefighting crew by grabbing more extinguishers and making preparations in the Soyuz that was still accessible. The fire raging in Kvant 1 meant that half the crew would be unable to get to their spacecraft until the fire was put out, as the passageway to the second Soyuz was completely blocked by the flame.

At least three extinguishers were used on the fire, and Linenger has said that they didn't really make a dent in it. But they helped to prevent secondary fires from braking out. A few weeks earlier, Kvant 1 was packed with full trash bags before they were all loaded into a Progress module and jettisoned to make ready for the next Soyuz. If the trash bags had still been present, they would have easily caught fire.

Finally, after what seemed like too long of a time, the fire was out. But the crew was still in danger, as they still had a station filled with smoke. Energia claimed that *Mir*'s air-purification systems made short work of the smoke, but the crew on board was pretty sure that it was the damp conditions and cold metal that absorbed the smoke. The crew tried to do as little activity as they could for the next hour, attempting to let their emergency respirators last as long as possible while the smoke cleared. After about an hour, Korzun removed his respirator first and felt that the air was breathable enough. Accounts of how long the fire burned vary between who was asked, as officially the Russians say it only lasted about ninety seconds, while Jerry Linenger has said it burned for closer to fourteen minutes. While flight controllers on the ground downplayed to the media and to NASA the whole situation and just how dangerous it was, everyone in orbit knew that they had dodged a serious bullet. At least the crew got to share a laugh as everyone complimented Kaleri on how calm he was, doing orderly printouts of critical data while chaos was going on around him.

As to what caused the fire, nobody is really certain, since the oxygen canister was almost completely destroyed. The likely cause is that the heating element somehow got contaminated by an organic compound, such as oil residue, probably during its manufacture. Flight controllers at TSUP initially accused Lazutkin of contaminating the canister himself, but those accusations were ultimately dismissed. For a while, though, the crew was forbidden from activating any canisters from the same production batch. Ultimately, they had to do so when supplies ran low, and no problems developed with any of the other canisters. When Korzun, Kaleri, and Ewald returned home on *Soyuz TM-24*, the strain on *Mir*'s systems was reduced, and things returned to normal for the remaining three crewmembers.

Other Problems

While things got back to a "normal" routine after the fire, *Mir* was certainly not quite the same vessel that had been launched eleven years earlier. Frequent leaks were developing in the station's ethylene glycol coolant loops, and droplets of the green fluid could sometimes be found floating free in the cabin. On past stations, a glycol leak would have resulted in abandonment of the station, but the TSUP was seemingly doing everything it could to keep the station manned and going at all costs. Tsibliyev and Lazutkin were spending much of their time trying to look for and plug coolant fluid leaks in the maze of pipes located behind equipment panels in *Mir*'s base block and Kvant 1 modules.

Ethylene glycol, the same stuff used in automobile radiator systems, can cause lethal kidney damage if ingested, and nobody exactly knew what breathing glycol fumes would do to the crew on board. Linenger described that whenever a coolant line was pressurized with air to look for a leak, his nose would immediately get stuffed up and he would begin to cough, even if he was on the other side of the station. Russian flight controllers again downplayed the concerns much to the frustration of the crew on orbit, and Linenger did his best as a trained medical doctor to periodically check his crewmates for signs of lingering glycol effects regardless of what the TSUP controllers were saying.

That wasn't the only problem, as the station's guidance computer was also shutting down periodically. When that happened, *Mir* would drift out of alignment, and its solar arrays would no longer get efficient light from the sun. On one occasion, not enough systems were shut down before the batteries on *Mir* drained and the station was plunged into darkness. When it happened, it took hours to realign the station manually to generate power and days before the batteries were charged enough to resume normal operations.

Near Miss

During a normal docking with *Mir*, the Kurs rendezvous-and-docking system would handle the whole process. As a backup to the Kurs system on the Progress modules, *Mir* was outfitted with a remote control station known as TORU (a Russian acronym meaning "Tele-operated Mode of [Spacecraft] Control"). It featured a CRT display screen and a control panel with two control sticks. It would allow a cosmonaut to take over control of a Progress module and dock it manually like a Soyuz if a problem developed with the Kurs. However, the Russians had more ambitious plans for TORU.

On the ground a crisis was brewing. The Kurs system was manufactured not in Russia but rather in the Ukraine by the Kiev Radio Factory. When the Soviet Union was dissolved, Ukraine became an independent country and formed its own space agency, which continued to manufacture the Kurs system for the Russians, but for a price. While Russia struggled through its economic problems, the price of the Kurs had gone up to the point that Energia began to look for alternatives. Officials felt that if cosmonauts aboard *Mir* could fly the rendezvous-and-docking sequence of a Progress with TORU manually, Russia would not have to buy the Kurs system anymore. Therefore, they could fit the Progress with a simpler system to supply range and rate data for docking and load it with more cargo.

While a close-proximity docking test had taken place with TORU before, Tsibliyev as command cosmonaut would perform a much more ambitious test with it. *Progress M-33* arrived in early February, containing fresh supplies. Progress equipment is not loosely loaded on board. Instead, it is hard bolted and strapped into a web truss structure to prevent shifting during thruster burns, and it had to be loaded in a certain way so as not to upset the craft's center of mass. Linenger recalled how his crew would begin to stow equipment on an outgoing Progress based on a manifest radioed up from the ground. Then, new orders would be issued by TSUP requiring the crew to remove things in order to stick in additional equipment during a very time-consuming process. While the weight of each piece of equipment was accounted for at launch, there was no effective way to weigh the trash and spent equipment in zero gravity. So no one knew exactly how much stuff was loaded into a Progress before jettisoning it.

The attempt to dock manually was made on 4 March as the Progress, which had been undocked a week prior, was commanded by the ground to begin a burn that would send it back to the station. The plan was deceptively simple. When the craft was in range, Tsibliyev would receive a black-and-white television feed from the Progress via the TORU, along with range and rate data from its Kurs system to judge the ship's distance from the station; then he would fire the braking rockets on the Progress to slow it down and guide it in for a docking. Everything would take place in orbit with no help from the ground, since the station was out of communications range after the Progress rendezvous burn command was sent.

Problems began almost immediately. At the appointed time, the Progress was not transmitting a television picture. Linenger and Lazutkin tried to look for the craft from *Mir*'s portholes, but that was easier said than done, as the craft was coming in at an angle from which they couldn't easily see it. As minutes ticked by, the concerns grew among the crew as they still had no television signal or a visual in the portholes.

Finally Lazutkin spotted it coming in way faster than it should have. He tried to instruct his commander in what direction to send the craft, but that didn't work too well. Without a common point of reference of whether the craft was right side up or upside down in addition to a reversed perspective, up, down, left, and right might be very different. Unable to get a television picture, Tsibliyev began moving to the portholes, looking for the Progress himself. When he saw it, he immediately moved back to TORU and fired the translation thrusters in a last-ditch attempt to force the Progress to miss, as it appeared to be on a collision course. Everyone braced for an inevitable collision. When it didn't come, everyone let out a sigh of relief. They had literally dodged a bullet as the Progress went sailing on by.

At the time of the attempted docking, the Russians downplayed the danger to NASA, saying that it was "routine" and had been done before. No one in Houston had any idea of just how dangerous it was. One of the mission control rooms in Houston was set up with a reduced staff to monitor the docking. But apparently controllers at TSUP had told them the docking attempt had already been made, so the U.S. controllers called it a night. No attempts were made by NASA to examine the "docking test" further.

One last milestone on Linenger's mission involved a space walk he performed with Tsibliyev to fit some experiment modules to the outside of *Mir*. It was the first time an American astronaut had performed a space walk in a Russian Orlan suit. For Linenger, this would be his first space walk; although he had trained in the Hydrolab at Star City, he had not conducted any similar training in Houston's own facilities. The space walk went well, although Linenger felt an almost-overwhelming sense of falling when he was positioned on *Mir*'s Strela crane arm, and it took all his concentration to overcome a sense of fear and near panic from his subconscious. Prior to that space walk, no other astronauts had admitted to having similar fears, but some came forward afterward to express that they had encountered the same thing when perched on very high vantage points, outside the relative safety of the shuttle's cargo bay.

At the end of Linenger's nearly five months in orbit, he was very ready to come home and expressed his feelings to managers in private emails that he didn't think *Mir* was safe for astronauts to occupy due to its almost-constant need for repairs. But he still managed to finish all his science and photography goals. Given that he didn't seem to bond with the crew according to Tsibliyev's and Lazutkin's accounts in *Dragonfly*, it seems that his crewmates were as ready to get rid of Linenger as he was to leave the station. Linenger paints a bit of a different picture in his own account, saying that while he didn't work hand in hand with the crew on certain repair tasks (ones that only a single person could best do anyway), he still made himself available to conduct some of the Russian research by changing out materials experiments in the furnaces to give his crewmates time to devote to the repairs to *Mir*'s glycol systems, which they were best equipped to deal with in the first place. Ultimately, after returning home and fully recovering from his long-term spaceflight, Linenger was asked if he wanted to fly another mission. He declined the offer, as he felt that his time on *Mir* was enough and left the astronaut corps soon after.

Michael Foale's Mission

On 17 May 1997 *Atlantis* arrived on mission STS-84 to exchange Linenger with Michael Foale. Foale was born in Louth, England, to an English father and an American mother, which gave him dual citizenship in both countries. Foale wanted to become an astronaut since an early age and learned to fly planes on his own time in addition to his schooling as an astrophysicist. Foale began working for NASA upon joining the Mission Operations Directorate in 1983, but he wasn't accepted to the astronaut corps until 1987.

The language training and acclimation efforts in Russia paid off, as by all accounts, Foale was a team player with the Russian crews he flew with. They treated one another as equals and got along together rather well compared to some of the things Tsibliyev and Lazutkin continued to say about Linenger after his departure. Given the stress that the cosmonauts were under with the coolant leaks and the near miss by the Progress, not to mention the general state of decay that *Mir*'s systems were experiencing, the comments were likely more intended as a form of blowing off steam as opposed to being genuine criticisms centered on their former astronaut crewmate.

About a month into Foale's mission, plans were made at the TSUP to try another TORU docking with *Progress M-34*. As before, NASA wasn't in the loop on this decision; while Linenger had likely mentioned *M-33*'s near collision in debriefings, efforts were slow on the part of NASA managers to look into things. While Tsibliyev expressed some concerns about trying another docking to his crewmates, he was still a proud military pilot, and a successful docking meant that he would get an additional docking bonus added to his paycheck. Russian cosmonauts would get bonuses of about one thousand dollars each if they performed special tasks such as EVAs or achieved successes on dockings. Indeed, it has been said that some cosmonauts would override the Kurs system on Soyuz dockings with *Mir* just so they could get an extra grand for an "unscheduled" docking. In a post-Soviet-era Russia, where a cosmonaut made a fraction of what NASA astronauts typically did, every little bit helped.

While it was unrelated to the TORU docking attempts, Tsibliyev had a close call on a previous mission. On 14 January 1994 at the conclusion of a standard five-month tour, Tsibliyev, Aleksandr Serebrov, and French CNES research cosmonaut Jean-Pierre Haigneré undocked *Soyuz TM-17* in preparation to come home. Before the crew left *Mir*'s vicinity, controllers on the ground wanted them to station keep with the Kristall module and take photographs of the APAS docking collar, which *Soyuz TM-16* had previously used. Tsibliyev was handling the controls of the Soyuz from the descent module while Serebrov was in the orbital module preparing to take the photographs.

Tsibliyev reported that the Soyuz was handling sluggishly. The hand controller that governed acceleration and braking was not working. He still had some control with the second stick, which he used to keep the Soyuz from colliding with *Mir*'s solar arrays. Ten minutes after undocking, *Soyuz TM-17* had a glancing collision with the Kristall module near its attachment point with the base block, but there was no real damage. It turns out that somebody had not set a switch properly in the Soyuz orbital module, which has a duplicate set of control sticks. Because the orbital module's acceleration and braking controller was activated, it locked out the one Tsibliyev was using in the descent module. Everyone considered the incident "no harm, no foul," and Tsibliyev was not punished. Little did anyone realize three years and five months later what was about to happen.

Russian controllers had analyzed the data from the previous aborted TORU docking attempt and determined that the radar signal from the Kurs had jammed the television picture. So for this attempt, there would be no radar data coming from the Progress to tell its velocity or distance. Tsibliyev would only have the Progress television feed and a grid scale on the TORU monitor to help judge how far out from *Mir* it was during the approach.

Lazutkin also set up a camera to record the docking and had it focused on the TORU station. As before, ground controllers sent instructions to the Progress to begin its approach, before the station drifted out of range of the ground stations. At the proper time, *Mir* got a good camera signal from the Progress. The station looked tiny on the monitor as clouds and ocean drifted underneath the station. The Progress was coming in from above the complex and slightly behind it. *Mir* itself in relation to Earth was upside down, with Kvant 2 pointed down toward the surface and Spektr pointed upward. The plan was for the Progress to dock with *Mir*'s Kvant 1 docking port.

Michael Foale in a television interview for NOVA picked up the story from there: "I had no idea what they were doing. No one at NASA knew what the Russians were doing. I'm still thinking everything is normal, because nobody has told me anything about what systems were turned off or on."

The picture in the TORU monitor was not the greatest since the low-contrast, black-and-white image meant that the station was about the same color as the clouds below it. The plan was for Lazutkin and Foale to try to acquire the Progress visually in a porthole and use hand-held laser range finders to provide a more accurate measure of distance, but they couldn't see it in order to do so.

The Progress was coming in very fast, and attempts by Tsibliyev to fire the braking thrusters were not working. If anything, the ship almost appeared to speed up a little bit. Lazutkin finally saw it out the bottom porthole of the *Mir* base block, which was pointed up toward space: "Through the porthole, I could see the cargo ship gliding below us." Lazutkin stood erect, looked toward Foale, and shouted, "Michael, va korabyl!" (*korabyl* is Russian for "spacecraft"), which essentially meant, "Get to the Soyuz!" Michael Foale got up from his spot next to Tsibliyev and floated at a high speed toward the front of *Mir*, since he understood something bad was about to happen. Tsibliyev tried one last-ditch attempt to fire the Progress's translation thrusters to try to have it miss *Mir*.

The Progress, which had slowed down somewhat but had not stopped, filled Lazutkin's porthole. The craft's front was visible, with its nose slightly angled toward the rear of the station as it approached. It looked almost like a train coming to a railroad crossing with *Mir* stuck on the tracks. Lazutkin described it in similar terms: "It was full of menace, like a shark. I watched this black body, covered in spots, sliding past below me. I looked closer and at that point, there was a great thump and the whole station shook." There were actually at least two thumps. The orbital module of the Progress collided with *Mir*'s base block at an angle, which was fortunate since that part of the Progress was a smooth ball with no sharp protrusions. The Progress bounced off and sideways before a back corner of its propulsion section collided with the Spektr module near its centrally mounted solar arrays, tearing a jagged hole in one of them.

The decompression alarms went off, indicating that there was a leak in the station. Initially Foale thought that the source of the leak was in the base block, since that is where the Progress had impacted first, but Lazutkin had seen the Progress hit the Spektr module and correctly guessed that the leak was coming from there. It was a slow leak, which gave them time to try isolating it before abandoning the station. Tsibliyev monitored the pressure drop from his command station while Lazutkin and Foale worked to unhook cables between Spektr and the base block. Pressure was steadily dropping when *Mir* came back into communications range and controllers at the TSUP were informed of what happened.

One last power cable frustrated Lazutkin since he couldn't find its plug, and he began trying to cut it with a kitchen knife. Sparks erupted from the

42. Visible damage to Spektr after the Progress collision. Courtesy NASA.

cable, and Foale indicated it was not a good idea to do that. Eventually the two men traced the cable to its source and unplugged it. By this time, the pressure differential was producing suction, and they were unable to close Spektr's hatch since it had to be pulled shut against the airflow. Finally, the pair got ahold of one of the hatch plug assemblies *Mir* had been equipped with when it launched, and they positioned that in the hatchway, sealing off Spektr and the leak successfully.

But their ordeal was not over yet. The Progress collision and the leak had knocked *Mir* into free drift and a slow spin, so the remaining solar arrays stopped tracking the sun. The low-voltage alarms came on as *Mir* slowly lost power and the interior was plunged into darkness. Foale recalled, "For the first time I experienced a totally silent, still space station, where there are no fans moving, there is no light on, nothing is alive. Just our breathing is causing any sound." Eventually, it was Foale who came up with an idea to stabilize *Mir*'s attitude. Using the Soyuz, the crew would fire its thrusters to arrest *Mir*'s slow spin and then orient it properly so that its solar arrays could begin the process of recharging the batteries. With Spektr isolated, power would be at more of a premium, given that the four newest solar arrays (one of them damaged) were now useless in a dead module.

After a few days of steady recovery, during some housecleaning to remove some of the clutter and stow cables that were no longer hooked to Spektr, Lazutkin accidentally unplugged the power cord to *Mir*'s main computer, and the station drifted out of alignment again, plunging it into darkness as the batteries drained. Again the crew had to work to restabilize everything and recover systems during the next few days. It was not a promising future for the station or NASA's involvement. Foale's experiments and personal effects, including his laptop computer, were inside the dead Spektr module. But in the days following the collision, there wasn't much time to do scientific work anyway.

The stress caused to the crew during the collision and its aftermath took a heavy mental toll on them. Reportedly, Lazutkin apparently made the decision to commit suicide ten days after the collision, likely because of the second power loss caused by unplugging the wrong power cable on top of everything else. But ultimately, he changed his mind and went back to work after having a chance to sleep on it.

Aftermath of the *Mir* Collision

Back on Earth, all parties involved with the program, especially on the NASA side, were shocked at how the events transpired to allow such a collision to occur in the first place. It seems that a double standard in safety had evolved where although NASA maintained a very tight reign on hardware and procedures they directly controlled, they pretty much just took Russia's word for granted regarding *Mir*'s condition. The safety oversight was not there, since NASA-built hardware wasn't involved and since most of the work in Houston was concentrating on the ISS. Energia, on the other hand, seemed to treat their "partners" like little more than customers who had almost no say in how things were done on *Mir*, except for when a shuttle would dock.

In the wake of what happened in June of 1997, plus the events that transpired on Jerry Linenger's *Mir* stay, there were a lot of calls for NASA to abandon the Mir program on the justification of safety alone. Congress held hearings, while interviews of participants were conducted to find out what had happened. Linenger had become a very vocal critic, saying that the next astronaut to fly after Michael Foale should not be allowed to stay on the station due to the danger involved.

In the weeks after the incident, Tsibliyev and Foale were scheduled to conduct an internal space walk. The crew would depressurize the station and enter Spektr, where they would retrieve some of Foale's personal belongings and route the module's power cables through an airtight hatch plug. They would also try to find the air leak and patch it. However, doctors on the ground detected that Tsibliyev had developed an irregular heartbeat, likely due to all the stress related to the mission; so the ground control decided to delay the first of the repair space walks until after the next crew had arrived. Tsibliyev and Lazutkin returned home after cosmonauts Anatoly Solovyev and Pavel Vinogradov arrived in early August.

It would be up to the two newcomers to try to salvage the station. Returning Spektr's solar arrays to the electrical grid was the most important task. Priroda had no solar arrays of its own and needed the power that Spektr's arrays provided if any semblance of a science mission were to be salvaged. Solovyev was up to the task as he had already performed space walks on previous missions. Along with normal EVA space walks, the internal space walk was successful in returning Spektr's solar arrays to *Mir*'s power grid, and Michael Foale was able to resume a scientific program with equipment in Priroda.

Over the course of their five-month stay in orbit, Solovyev and Vinogradov accomplished many of the repair goals. While Spektr was lost, the rest of the station seemed stable enough. Michael Foale even conducted a space walk with Solovyev and also provided invaluable experience when the station's attitude-control computer dropped off-line a couple of more times. The crew this time was able to shut off enough critical systems before the ground told them to do so, preventing another complete power failure from occurring.

Back home, NASA made the decision to continue the Shuttle-Mir Program, provided the station could be given a clean bill of health. In spite of all the criticism, justification was made that unplanned repair tasks might occur on future missions to the planets when an escape was not possible, so the experience was potentially invaluable. To many observers, it seemed like tenuous justification, at best, with NASA dodging accusations that it was ignoring its own safety protocols. However, NASA knew that even though they were about a year or two away from the first module of the ISS flying into orbit, pulling out from the Mir program at that time might cause Russian involvement in the ISS to crumble like a house of cards.

There were improvements in the dialog among NASA, Roscosmos, and Energia after the collision. NASA was able to flex its muscle in getting the Russians to delay a visit by Léopold Eyharts to the station on the *Soyuz TM-25* flight with Solovyev and Vinogradov until after the more critical problems had been dealt with. The Russians wanted to fly the French mission since they needed the money. But they also needed support from shuttle's heavy-lift capability to send up new equipment and more supplies. So at this point, the partnership was finally becoming a two-way street, with both teams working together as opposed to the Russians dictating the terms to the Americans. The Russians began to discuss their problems more openly, and concessions were made to the Americans. In exchange, NASA allowed the Russians to save face.

The Final Shuttle-Mir Missions

There would be a change as to which astronaut would next visit *Mir*, since that person needed to be capable of conducting a space walk if the need arose. Wendy Lawrence, the next astronaut scheduled to fly to *Mir*, was too small to fit in the Orlan suit and hadn't trained for a space walk, so the decision was made to replace her with Dave Wolf, who had EVA experience. This meant that Andy Thomas, originally scheduled to be a non-flying backup, would serve the last tour of duty on *Mir*. Dave Wolf, also being a bit of an inventor and multitasker with a medical degree and an engineering background, seemed perfect for the assignment in helping to get the NASA-Mir science program back on track. Some of the equipment used on *Mir* was of his own design.

Arriving on the station during STS-86, Dave Wolf seemed to have a good attitude for his mission. He had to get used to the Russian way of doing things as well when his crewmates didn't hand out praise or a simple "thank you" as easily as Americans do. Wolf also had to cram a lot of training with an Orlan suit into a five-week period, since although he was EVA qualified in an EMU, there were initially no plans for an American to conduct an Orlan EVA on *Mir* after Linenger's mission. Dave Wolf was up to the task, though, and relished it. On orbit at the end of a space walk that he conducted with Anatoly Solovyev, the veteran of sixteen space walks told him, "Good job," and Dave felt that it was all worth it. He was a fully accepted member of the team.

Unfortunately, nothing could be done to repressurize the Spektr module. Attempts were made to pump air through it from a sealed canister with a tracer material to highlight the leak, but the cosmonauts were unable to find the source. A special cap fitting was delivered on STS-86 by cosmonaut Vladimir Titov and astronaut Scott Parazynski during an EVA, the first conducted by a Russian in an EMU suit. This cap was installed around Spektr's damaged solar array during a *Mir* space walk to try to seal a possible leak in that location, but it didn't work either. Except for the module's reconnected solar arrays and the equipment salvaged from it, Spektr was a dead and useless module.

At the end of Dave Wolf's mission, the space shuttle *Endeavour* moved to dock with *Mir* on STS-89. On board, Australian-born astronaut Andy Thomas was ready to conduct the final American stay on *Mir*. Like Dave Wolf, he took steps to bond with his crew as much as possible. One benefit of Thomas's training was that he did not have to visit the Defense Language Institute to learn Russian. Instead, he had access to a very good tutor at JSC as preparations were underway to offer Russian-language classes for astronauts and support engineers closer to home before they had to make trips to Russia for ISS missions.

While there were the usual mechanical breakdowns and problems with *Mir*'s systems, Andy Thomas's stay on the station was relatively uneventful. Solovyov and Vinogradov returned home not long after he arrived, and the Mir EO-25 crew of Talgat Musabayev and Nikolai Budarin took over. French research astronaut Léopold Eyharts also conducted his delayed short visit to *Mir* to continue France's research program, and he returned to Earth with the previous Russian crew.

But there was one more incident that reminded everyone involved as to just how dangerous spaceflight can be. *Mir*'s carbon dioxide scrubbers used a chemical bed of charcoal, and periodically that bed needed to be changed out for a second one. To purify the used bed, its sealed tray would be opened to space and heated to red-hot conditions to bake it clean for reuse. During one particular baking session, the valves weren't configured right, so the red-hot charcoal bed was still open to *Mir*'s internal atmosphere and caught fire, pumping dense smoke and carbon monoxide into the cabin before corrective action was taken. Andy took readings with a carbon monoxide detector and reported his findings to the ground, but controllers dismissed the readings as being suspect since they were awfully high. Thomas also collected an air sample. When the sample was analyzed after he returned to Earth, scientists at JSC were shocked as to how high the carbon monoxide readings were. All Andy Thomas could do in the days after the fire was to take medication for his carbon monoxide–induced headache while the air quality improved, and he suffered no lingering effects from the incident.

On 4 June 1998 the space shuttle *Discovery* on mission STS-91 docked to bring Andy Thomas home. This would also be the final docking between *Mir* and a shuttle. After four days of supply transfers, the two craft separated. When *Discovery* landed on 12 June 1998, ISS Phase One was officially concluded.

As NASA's administrator Daniel Goldin predicted, NASA had learned a lot about space station operations from the Russians and certainly more than they ever wanted to. The Russians, on the other hand, learned a lot about their American counterparts. They learned enough to know that astronauts were as capable of hard work as their own cosmonauts. While the relationship between the two countries' space agencies wasn't always a harmonious one, it was still productive and delivered results. Even though the political and economic situation would throw a few curve balls at the partnership in the coming years, the agencies were finally ready to commence construction of the International Space Station.

11. The International Space Station

Five months after Andy Thomas returned home from the last NASA mission to *Mir*, the next phase of space station development was ready to get underway. On 20 November 1998 a Proton rocket launched into orbit carrying the first element of the ISS. It was an FGB node, based on Chelomei's TKS spacecraft design. Prior to the flight, it was given the name Zarya (a Russian word meaning "Sunrise"). It seemed like a very fitting name since it was, for both the United States and Russia, the sunrise of a new project. To space workers with experience in the Salyut program, the name of this first module likely gave them a sense of pride. Three decades after a space station with *Zarya* painted on the side flew as the first Salyut, a direct descendant would finally take the name officially into orbit.

A total of eighteen member nations were involved with the International Space Station project, spread across five major space agencies. NASA and Roscosmos were the highest-profile members of the project, but the ESA, the Canadian Space Agency, and the National Space Development Agency of Japan would also be heavily involved in development of hardware and providing crewmembers for the station. In terms of on-orbit construction projects, the ISS was massive and would require close coordination among all involved. During the next few years, over 150 space walks would be carried out in direct support of ISS assembly tasks.

Russia's modules would provide the backbone for the station in the form of the Russian Service Module (RSM). The RSM started life as DOS-8, the engineering backup module for the *Mir* station. It was originally planned to become the *Mir 2* space station before the collapse of the Soviet Union. From here, a three-person crew would begin regular occupancy of a module that at a glance didn't look all that much different from the previous DOS stations. Andy Thomas visited the RSM on one of the early shuttle missions to the ISS and said he had a sense of déjà vu as to how much it looked like *Mir* on the inside in terms of its equipment layout.

But just because the RSM looked like *Mir* on the inside, that doesn't mean it was a totally off-the-shelf design. Thanks to the lessons learned from *Mir*'s time in space, extensive changes were made in regards to how the station's systems were set up. Some problems with *Mir* revolved around how its modules were hooked together for power and the glycol coolant loops. Critical time due to the leak was spent in unhooking hoses and power lines between Spektr and the *Mir* base block after the Progress collision. A future depressurization event on the ISS might prove to be even more time critical, so every attempt was made in the RSM's redesign to move coolant lines and power hookups to outside the pressurized areas so that a hatch could be sealed off at a moment's notice.

The cooling system used by the ISS would be ammonia based, with the primary coolant lines running completely outside the habitat areas. Ammonia is a very good coolant and more efficient than the glycol-based coolants used on *Salyut* and *Mir*, which is why it is used in industrial refrigeration units on Earth. However, its vapors can also be toxic to astronauts in smaller amounts than glycol. For the internal cooling loops inside the modules, a nontoxic fluid was used instead, and redesign work was done to the coolant lines to help safeguard them against corrosion over the service life of the station.

Zarya

The decision to use an FGB module for the ISS came about because of delays in development of the RSM. The Zarya FGB included its own solar arrays along with guidance and propulsion systems, so it could act as the nucleus of a station until the RSM was ready to fly. Even though Energia was in a little healthier state financially by the late 1990s, work delays and bureaucracy were still the order of the day, leading to slippages in work schedules. Thankfully, the Zarya module was not under Energia's direct control, as it was built by the Khrunichev State Research and Production Space Center.

Khrunichev was a merger of two of the older Soviet factories that had ties to Chelomei's OKB-52. The former Myasishchev factory in Fili, which became part of TSKBM and later NPO Energia in the late 1970s when the DOS and OPS Salyut programs were combined, became an independent group in 1988 known as the Salyut design bureau with responsibilities over module development for the Mir program. The second group was the Khrunichev factory, which manufactured Proton boosters while also maintaining its manufacturing responsibilities for Energia and other firms. The two companies merged together to become the Khrunichev State Research and Production Space Center in 1993.

While Energia's problems were causing so many headaches for NASA during the 1990s, Khrunichev was having more success. The four-stage version of Chelomei's Proton booster had become a very popular launch vehicle, winning several contracts for lofting communications satellites. The fall of the Soviet Union made it possible for Western companies to deal directly with providers of launch services in Russia, and Proton was the most capable booster in their arsenal. Today, Khrunichev continues to build on its successes as part of the International Launch Services (ILS) partnership with Lockheed Martin in the United States.

While technically Khrunichev was a major subcontractor on the Russian Service Module, that particular project was managed by Energia. On the other hand, Khrunichev was the prime contractor for the Zarya, and the group coordinated its work with both Boeing, which was the prime U.S. contractor for the American side of the ISS, and NASA directly. So while the RSM fell behind schedule, the Zarya FGB was delivered on time and on budget. All parties involved considered the partnership between Khrunichev and the Americans to be a harmonious one compared to some aspects of the Energia partnership.

As part of the revised construction sequence of the ISS, the Zarya module was launched first and would be joined in orbit by the first American segment, the Unity node, on shuttle mission STS-88. The original plan was for the Zarya and Unity complex to independently operate from six to eight months before the arrival of the RSM. But ultimately, the RSM would not be ready to fly for nearly two years. Thankfully, Khrunichev built their module well, as the FGB performed its job as the early brain of the ISS with very few problems.

Unity Launches!

On 4 December 1998 the space shuttle *Endeavour* sat on the launchpad at KSC in preparation for launch on mission STS-88, the first shuttle mission dedicated to construction of the ISS. At 03:35 EST the exhaust plumes from

Endeavour's SRBs turned night into day as the shuttle climbed into orbit. In its payload bay was the first U.S. module, the Unity node of the ISS. A node module is essentially an intersection module, featuring an attachment port on both ends as well as four radial ports. Each port is known as a Common Berthing Mechanism (CBM) and would allow for other station modules designed in the United States, Europe, and Japan to attach to the node like a giant construction toy.

Being more than a simple docking port, CBMs also allow a module to be plugged in to the station's power and cooling grids, as they include electrical and support system connections that are linked inside after docked modules are firmly secured together. By allowing the modules to dock this way, the connecting hatch passageways are free of obstructions, and the hatches themselves can be shut quickly in the event of an emergency. The CBMs themselves are round, but the transfer hatches are much larger and square in shape compared to a typical round docking hatch. This was done in part to allow large equipment racks to be floated through the station's hatchways as needed.

To dock with the Russian Zarya module and shuttle, each end of the Unity node was equipped with a Pressurized Mating Adaptor (PMA). A PMA essentially turns a CBM into a docking port capable of accommodating a spacecraft equipped with an APAS-style docking port, such as the shuttle, a properly equipped Soyuz, or other spacecraft that were still on the drawing board in 1998. The reason for not using a CBM as a standard docking port was related to the size of the hatch and the requirement for extremely tight tolerances. An RMS arm is needed to position a module on a CBM. A PMA also has some built-in shock-absorbing capability, should it see a rougher-than-normal docking.

In cross section, a PMA looks sort of like an offset cone with a squared-off skinny end. The offsetting was done to provide the proper clearance issues with the space shuttle's docking port. One PMA was permanently attached to allow Unity to join with a docking port at the end of the Zarya module. The second PMA would only be attached temporarily for shuttle dockings, and it would be relocated to make room for additional modules on later flights. Since PMAs can be docked with other CBMs, they can either be used as active docking ports or stored temporarily out of the way as needed.

After reaching orbit, the Unity module was attached to *Endeavour's* docking port with the shuttle's RMS; after three days of flight, the shuttle caught up with the Zarya module. Instead of trying to dock something as massive as a shuttle with a module to a relatively lightweight FGB, the shuttle RMS was used to grasp Zarya and berth it to the PMA on top of Unity. The two modules were docked with one another early on the morning of 7 December. The reason for the unusual hours of operation was so that the two modules could be docked over Russia's tracking stations during daylight hours.

At docking, the new ISS weighed thirty-five tons and measured seventy-six feet long (twenty-three meters) from one end to the other. From that point on, it would only get bigger. After docking, shuttle mission specialists Jerry Ross and James Newman conducted space walks over the next two days to hook up power and data cables between Zarya, the PMA, and Unity. These tasks would become commonplace in the coming years of ISS construction, along with other tasks such as hooking up ammonia coolant lines. On the second space walk, the pair removed launch restraints and covers from Unity's exposed CBMs.

On 11 December, hatches between the shuttle and the ISS were opened for the first time by mission commander Bob Cabana and Russian mission specialist Sergei Krikalev. For the next few days until undocking, the shuttle crew spent their time unstowing equipment inside Unity from prelaunch storage containers. Some equipment was changed out as the two modules were given a thorough checkout in preparation for the first crew. In Krikalev's case, he would be revisiting the ISS as a crewmember on Expedition 1, once the RSM was ready to fly. A third space walk was conducted a day later to install additional communications and docking antennae on the outside of both modules. After six days of joint flight, *Endeavour* undocked from the ISS, allowing it to fly free. Its new communications capabilities and activated systems would allow controllers on the ground in both Houston and Moscow to monitor its systems and operate them remotely as needed.

In addition to the normal equipment transfers, astronaut Bob Cabana also carried up a couple of interesting items of significance. The first was a logbook for the station. This book was the brainchild of fellow pilot-astronaut Jeff Ashby, who felt that having a logbook in the finest tradition of seafaring ships of old would be a way to help record its history. Unlike the great creations found on planet Earth, such as the pyramids of Egypt, one day the ISS will eventually be deorbited, leaving no other permanent record for future generations to look back on. Ashby approached the National Naval Aviation Museum in Pensacola, Florida, and recruited them to custom make and donate a logbook to the cause, which they did entirely on their own at no cost to NASA. On the front of its metal cover were printed the words "International Space Station One" (which in hindsight Ashby says is not entirely accurate, since *Mir* was technically the first international space station). On the back cover was printed an appropriate quote by English poet T. S. Eliot: "We shall not cease from exploration, and the end of all our exploring will be to arrive where we started and know the place for the first time."

The second item carried up by Cabana was a ship's bell. This was another idea from Ashby, with the intention of starting a second tradition on the ISS with its roots in nautical history. Its intent was to show good order and mutual respect. The bell could be rung each time there was a return of an ISS crewmember or a change of command. But what it is perhaps best known for is that each time a shuttle would arrive or depart, the bell would be rung twice to signify the event. Again, Ashby enlisted the assistance of the National Naval Aviation Museum in Pensacola to create the bell. To give the bell a connection with previous spaceflight endeavors, a small space was hollowed out in part of the bell, and inside it were placed shavings from the Naval Academy class ring of Mercury astronaut Alan Shepard; shavings from the ship's bell of the nuclear aircraft carrier USS *Enterprise*; a screw from *Mir*; and by some mysterious circumstances, shavings from Ashby's own naval aviator wings.

The logbook was briefly brought back for refurbishment in 2006 before being returned to the ISS. However, the bell was replaced after a couple of years since the tone of its ring didn't sound so good in orbit. It was replaced with a different bell that had a better sound. The original ISS bell is currently housed at the National Naval Aviation Museum.

Like *Mir*, the orbital inclination of the ISS is 51.6 degrees, and this was needed due to how far north the Baikonur Cosmodrome is from Earth's equator. Rockets launching from Baikonur can't launch into lower-inclination orbits. There was some criticism that the orbit selected would prevent the ISS from being used as an effective launching platform for probes to the moon or other planets, but the orbit would also allow for much of Earth's surface to be surveyed on an ongoing basis.

Another problem with the orbit is that it limited the amount of cargo that can be carried up. In a lower-inclination orbit, a spacecraft can gain additional velocity due to the speed of Earth's rotation. To compensate for a lack of this velocity assistance, NASA put the space shuttles *Discovery*, *Atlantis*, and *Endeavour* on a diet during their scheduled refits to help save a few hundred more pounds of weight for cargo. More powerful main engines had also been developed for the shuttles to give them a little more kick on the climb to orbit.

The launch window from KSC to intersect with the orbit of the ISS is only about five minutes long, so shuttle flights to the ISS would require precise launch timing. During each day, the time that the ISS orbit passes over a certain point of the earth advances by about thirty-seven minutes. A missed launch opportunity means that a spacecraft launching from KSC has to wait about twenty-three and a quarter hours before it can make another launch attempt. Technically, the ISS orbit passes over Florida twice each day due to its ground track with one orbit being an ascending node trajectory (its path being from southwest to northeast) and the other being a descending node trajectory (northwest to southeast). While a spacecraft theoretically could be launched from KSC on either trajectory, launches toward the south are never done since the ascent path would take a rocket over Cuba.

Data Transfer

On the U.S. side of the ISS, the command-and-control computer boxes in each module, known as MDMs (Multiplexer De-Multiplexer), use 80386 processor chips, similar to what home computers used in the late 1980s. This may seem like an outdated technology, given that several generations of more sophisticated and powerful processors have been made over the past two decades. Even smartphones today have faster computing power, but NASA had some good reasons for doing this. First, an older technology is a well-understood technology, which means problems are less likely to develop if older systems are used. Second, microscopically compared to a current-generation processor chip, an 80386 chip set is big and not as densely packed with circuit pathways. So if a stray cosmic ray were to hit the chip directly, it would be less likely to hit a critical area and may pass completely through without doing damage at all. As for the computing power needed to run systems on the ISS, the processors are up to the task since they don't need to run anything that requires a lot of calculations, unlike graphics found in modern computer games.

The Russian side of the station still used its own computers based on *Mir*-type architecture. Three Russian computers were used for the station's primary attitude control, while an additional three were used to handle other systems. One of the delays in getting the RSM ready for flight involved how to network them with the systems on the U.S. side. Part of that task fell to a recently selected astronaut named T. J. Creamer.

A little less than a year before the scheduled launch of the RSM, Creamer visited Russia to hash out details for a network. He had three plans for consideration: a primary plan, a backup plan, and a backup for the backup. Upon meeting with his Russian colleagues, he proposed his ideas. The first idea involved running a coaxial cable between the Russian and U.S. segments across the open hatchways. That idea was rejected since the Russians didn't want to run any exposed cables between the hatches in the aftermath of the Progress collision with *Mir* and the criticism of that practice. The second plan would involve using a wireless network router. At this time (early in the year 2000), there was no accepted standard for router architecture. There were a couple of off-the-shelf wireless systems that could be used, but the Russians were concerned that the radio frequency interference would violate the EM (electromagnetic) and RF (radio frequency) constraints required for the scientific payloads planned for the station. So idea number two was rejected. The third idea involved line-of-sight transmission of the network signals with an infrared light system across the open hatchways. The use of infrared signals instead of radio signals is something commonly seen today with radio-controlled toys; in some applications, it is a nice alternative to using radio frequencies. But the Russians rejected that idea because it could set off the fire detectors in the station, since infrared energy is also heat. So in one meeting, all three of Creamer's ideas were rejected.

After a night at the hotel coming up with alternatives, T. J. Creamer went back the next day and asked the Russians if they had any unused data cable on the outside of the RSM. It turns out there was. After looking at the specifications for the cables, Creamer and the Russians hashed out a plan to use the cables on the outside with some adapters at each end to make them compatible with networking the American and Russian computers together.

Since the cables were already built into the RSM, the cost to add the capability was reduced down to just creating the adaptors. Since the cables also ran outside of the hatchways, they wouldn't get in the way in case a hatch needed to be closed in an emergency. So one issue of many was overcome.

More Outfitting

About five months after STS-88's mission, the space shuttle *Discovery* was the next one to visit the station on STS-96. This time, the shuttle was packed with supplies in a Spacehab module to outfit the station's interior in preparation for the first crew. *Discovery*'s cargo bay also included a new design of pallet rack known as the Integrated Cargo Carrier. It contained equipment that the crew would attach to the ISS during a pair of space walks. Among the equipment were parts for a Russian Strela crane similar to what *Mir* utilized and a smaller type of U.S. crane known as the ORU Transfer Device. The Strela would be an important piece of equipment for use in EVAs conducted from the Russian side of the station using Orlan suits.

After *Discovery*'s visit, the RSM was still not ready for launch, and it would be over a year before the next shuttle would visit. Part of the reason for this was mission scheduling, as the next three shuttle missions were dedicated to the launching of the Chandra X-Ray Observatory, a servicing mission to the Hubble Telescope, and a space radar topography mapping mission. Problems were also encountered during the launch of STS-93 as a short circuit shut down two of the space shuttle *Columbia*'s main engine controllers at liftoff, forcing the vehicle to switch to onboard backups. There was also a slight fuel leak in one of the shuttle's main engines. *Columbia* made it into orbit and was able to deploy the Chandra observatory without any problems, but when the orbiter returned from its mission, an inspection revealed that the short was due to chaffed wires in *Columbia*'s engine bay caused by worn insulation. The rest of the shuttle fleet was grounded for inspections, resulting in a six-month delay before the Hubble and radar mapping missions could fly.

The space shuttle *Atlantis* flew the next mission to the ISS on STS-101 in May 2000. Some EVA construction work was done to outfit the station's Strela crane with another segment and fit additional EVA handholds, but the primary task was to give the Zarya module a tune-up, as some of its onboard batteries needed replacement. Air filters and fire extinguishers were replaced with fresher units, and some work was done to reconfigure how air circulated inside the station. Since the station's internal electronics need air circulation to help with equipment cooling, it couldn't be depressurized, so the filter replacements were necessary. *Atlantis* spent five days docked with the station before returning home. At a glance, everything seemed fine with the orbiter, but inspection of a wing panel after landing revealed that there was internal damage due to a small heat leak during reentry. The damage was logged and repaired in preparation for the next mission. The ramifications of this incident weren't fully realized at the time.

The RSM (Zvezda) Flies

On 12 July 2000 the Russian Service Module was ready to launch from Baikonur. A lot more was riding on this launch than on any other, as Energia had no backup module and no insurance against loss due to their financial troubles. They would be out a lot of money should disaster befall the module either at launch or on orbit. To hedge their bets, NASA developed an Interim Control Module (ICM) based on the Transtage of a Titan rocket to act as a power module capable of reboosting the station complex until a new RSM was ready for flight. If the RSM failed, the ICM would be launched.

Thankfully, the ICM wasn't needed. The Proton rocket performed its job as advertised, placing the module into the proper orbit. About two weeks later, on 26 July 2000, Zvezda docked with Zarya's aft port without any problems. The next day, primary control functions for the ISS were transferred from Zarya to Zvezda. A couple of more shuttle missions would have to fly first for additional construction work, but the wait for the first ISS crew to occupy the new station would not be much longer.

Outwardly the Zvezda RSM looked about the same as *Mir*, but only two solar arrays were fitted on either side of the module. These arrays would only be temporary, as the plan was to deactivate and fold them up once the massive set of arrays on the station's main truss were permanently fitted. Like *Mir*, the Zvezda had a multiple–docking port adaptor on the front of the station, but it had only three ports, with two mounted vertically, compared to *Mir*'s five ports. The upper port was originally intended for use by a Russian-designed solar array tower, but budget problems led to cancellation of the Russian array. The bottom docking port could accommodate both Soyuz and Progress craft, and a similar port at the front of the Zarya

43. The American-built Unity node attached to the Zarya FGB and Zvezda RSM, both based on Chelomei's Almaz designs. Courtesy NASA.

module could do the same as well. The station also had an aft-mounted Soyuz or Progress docking port located between the module's engines, just like its predecessors. The RSM added 43 feet (13.1 meters) to the station's length, giving the complex a length of just under 130 feet long. The new complex was now about 30 percent longer than *Skylab* and growing.

Major Construction Commences

By mid-1999 pretty much all the U.S. and European hardware intended for the ISS had been delivered to KSC in preparation for ISS construction. With the RSM docked to the station, the last major stumbling block had been lifted. Hardware could finally be sent up as a busy schedule of shuttle missions began launching on a regular basis at a rate that hadn't been seen since the early days of the shuttle program.

The space shuttle *Atlantis* visited the ISS in September of 2000 on STS-106. The main purpose of this flight was to deliver the last of the supplies to the station in preparation for the first crew and to check out Zvezda's systems. Astronaut Edward Lu and cosmonaut Yuri Malenchenko conducted a six-hour space walk to hook up power, communications, and T.

J. Creamer's computer-network data cables between Zvezda and the rest of the complex. They also installed a magnetometer to help the station know its exact orientation in order to minimize propellant use in orbit.

A month later, the space shuttle *Discovery* on STS-92 arrived at the station, carrying two new pieces of hardware in the form of a third PMA and the Z1 truss. For ISS assembly, NASA uses specific nomenclature and a coordinate system to help describe the location of modules to one another on the station. So using the orientation of the Zvezda RSM, looking from its aft end forward, with Zarya and Unity docked to the front, modules placed on top would be designated by the letter *Z* for "zenith." Modules docked underneath would be *N* for "nadir." Borrowing naval ship terms, modules docked to the left side would be *P* for "port," and ones on the right would be designated *S* for "starboard." So the Z1 truss would be the first module docked to the top, or the *zenith*, of the station, and it was mated to the zenith port on the Unity node. The third PMA was docked opposite of it on Unity's nadir port.

The Z1 truss formed the early backbone of the station, as it contained electrical hookups for the first large solar array and four control moment gyroscopes for orientation of the complex. It contained a couple of additional unpressurized CBMs as well to help temporarily stow components, such as PMAs and equipment racks, for later construction. Communications antennae were also mounted in the truss, allowing for data transmission of early scientific experiments and television-downlink capability. Over the course of six docked days and four space walks, the truss was positioned and hooked into the station's systems. Additional time was spent transferring supplies in preparation for the arrival of Expedition 1 less than a month later. The ISS was ready to host its first crew.

The Fate of *Mir*

Meanwhile, in another part of the sky, *Mir* continued to orbit. After the last Shuttle-Mir mission, the station's next residents, Gennady Padalka and Sergei Avdeyev, spent much of their mission inventorying the station's supplies and stowing equipment that was no longer used. They also performed an EVA to adjust some of the power cables between the disabled Spektr's solar arrays and the rest of *Mir*.

French CNES astronaut Jean-Pierre Haigneré took part in the last long-duration mission to the station as part of a program known as Mir-Altair. Haigneré launched into orbit on 20 February 1999 with Viktor Afanasyev and Slovak guest research cosmonaut Ivan Bella (Czechoslovakia divided into the independent Czech and Slovak Republics in 1993) on *Soyuz TM-29*. Avdeyev stayed on board with Afanasyev and Haigneré while Padalka went home with Bella after a week of joint operations, on *Soyuz TM-28*. During the mission, the Russians conducted space walks to retrieve the last of the experiment cassettes placed outside for space exposure. On 28 August 1999 the three men closed down the station and returned *Mir* to autonomous operations, something it hadn't done in about a decade. They returned home a few hours later on *Soyuz TM-29*.

If things had gone according to NASA's wishes, *Mir* would have been deorbited shortly after the *TM-29* crew returned home, leaving Roscosmos and Energia to concentrate their interests fully on the ISS. But it didn't quite happen that way. A commercial space venture known as MirCorp was formed by investment entrepreneurs from the United States and Europe. They put together enough funding to launch a crew to *Mir* to recondition it and make it ready for use as a possible destination for space tourists and the winning contestant of a planned reality television show called *Destination Mir*.

Soyuz TM-30 cosmonauts Sergei Zalyotin and Aleksandr Kaleri would take part in what was billed as the first privately funded mission to space. They launched for *Mir* on 4 April 2000 and docked with the station successfully a couple of days later. Over the next month, the cosmonauts did their best to refurbish what they could on the station, using supplies ferried up by two Progress modules. These particular Progress craft were newly developed Progress MIs, which differed from the previous Progress M variants by sacrificing a water tank and some bulk storage capacity for more fuel. They would allow the craft to make longer reboost or deorbit burns as needed.

A space walk was conducted in mid-May to inspect and repair some of *Mir*'s external damage, plus to take panoramic images of the complex. While the station was in okay shape, it was not great, as a burned wire on one of the Kvant 1 solar arrays had disabled it. Power was in short supply as degradation of the older solar arrays in combination with the damaged arrays of Spektr meant there wasn't enough power to fully activate the Priroda module. *Mir*'s useful life was coming to an end.

It had been hoped that enough funding would be in place to send another crew to *Mir* or to extend the *Soyuz TM-30* mission, but the crew instead came home on 16 June after spending only seventy-two days in space. Mir-Corp had signed up its first space tourist for the flight, American billionaire Dennis Tito, much to the protests of NASA and members of Congress. The plan was for Tito to visit *Mir*, but finally Energia concluded that it could no longer support *Mir* as MirCorp's funding sources dried up and the expenses needed to continue operating the station were becoming too much of a burden. Political pressure also began to mount as NASA wanted to see the Russians focus all their attention on the ISS.

A Progress MI module, *MI-5*, was launched on 23 January 2001 and docked with *Mir* a few days later. The Russians had a manned Soyuz waiting for launch as a backup to help guide the Progress in with a TORU system from *Mir* if the Kurs malfunctioned or if *Mir's* attitude-control computer needed replacement, but the Soyuz wasn't needed. The Progress would deorbit the station, but the maneuvers wouldn't be conducted until *Mir's* orbit had decayed to below 220 kilometers in altitude in order to conserve fuel on the Progress. The natural decay of *Mir's* orbit took longer than expected. But on 20 March the first of several deorbit burns was conducted to lower *Mir's* orbit.

On the late night of 23 March 2001, the final three deorbit burns took place over the course of five hours. The first two lowered the station's apogee and perigee, while the third one set up *Mir's* final reentry point over the South Pacific Ocean to the east of Australia and New Zealand. On *Mir's* final orbit, onboard television signals picked up by Russian tracking stations showed *Mir's* final passage over Central Russia during orbital daytime. Controllers in the TSUP paused for a bit and watched the television feed for a while. *Mir* would not see another sunrise.

On the ground, coverage of *Mir's* reentry in some ways mirrored that of *Skylab*. New Zealand issued warnings to ships about possible debris. Japan, which was directly under the orbital path of the station but north of its reentry point, also urged its citizens to remain indoors during the last forty minutes of *Mir's* descent. Private individuals and media organizations visited South Pacific islands and chartered passenger aircraft to fly in the vicinity of *Mir's* reentry path to both observe and film the fireball trails. The American restaurant chain Taco Bell put up a large target just off the eastern coast of Australia and promised every person in the continental United States a free taco if a piece of *Mir* debris hit it. It was just an advertising gimmick, though, as there was no chance of a debris strike since *Mir*'s reentry path was hundreds of miles to the east of the Taco Bell target.

The station's atmospheric reentry began at 05:44 GMT. Breakup began with the solar arrays, and eventually atmospheric stress became so great that the station broke apart along its five major modules. By 06:00 GMT the remaining debris from the station that didn't burn up impacted safely in the ocean. Video footage of the breakup was captured by reporters on the island of Fiji. A little over fifteen years since its core module was launched and after a total of 86,331 orbits around Earth, *Mir* was no more.

Expedition 1

On 31 October 2000 the first ISS crew was ready to lift off from Baikonur aboard *Soyuz TM-31*. As with other important firsts, the launch would take place from the same R-7 rocket launchpad that had been used by Sputnik and Yuri Gagarin's Vostok flight. The commander of ISS Expedition 1 was American Bill Shepherd. Accompanying him would be Russian cosmonauts Sergei Krikalev and Yuri Gidzenko, who were both *Mir* veterans. Compared to the body of experience both Russian cosmonauts had, Bill Shepherd's accumulated time in orbit among three shuttle missions was only just over two weeks. The Russian managers on the ground were a bit concerned about that, but the crew themselves managed to integrate with no problems. The three men were a team.

Bill Shepherd's path to spaceflight was somewhat unusual. Shepherd was a graduate of the Naval Academy and earned a degree in aerospace engineering, but rather than taking on pilot training or some other field related to aircraft, Bill instead decided to take on the intense training required to become a Navy SEAL. The SEALs are the U.S. Navy's special forces arm, and the training required has a very high dropout rate compared to most of the special forces of other U.S. military branches. Even if a SEAL trainee can handle the mental pressures, sometimes their bodies can give out physically, and it might take more than one attempt to get through all levels of training successfully. Bill Shepherd succeeded and joined the elite SEAL ranks in the early 1970s, while also earning an ocean engineering degree from MIT in 1978.

Shepherd joined NASA's astronaut ranks in 1984 after three tries. His

44. The crew of ISS Expedition 1: Krikalev, Shepherd, and Gidzenko. Courtesy NASA.

SEAL training became unexpectedly useful as he was involved in salvage operations off the coast of Florida to recover debris from the destruction of the space shuttle *Challenger* in 1986. His first mission into space was on STS-27 in 1988, which deployed a classified satellite payload for the DOD. On that mission, Shepherd gained a reputation as a fun-loving practical joker, according to the book *Riding Rockets*, written by fellow STS-27 astronaut Mike Mullane. On that flight, Shepherd let a piece of breakfast sausage loose in the cabin and had the other crewmembers believing it was a piece of fecal material from an apparent malfunction of the shuttle's toilet before realizing they'd been had.

Shepherd gained a reputation as a capable astronaut during the next decade as he was assigned management jobs in NASA's space station program in 1993 to try, as he described it, "to bring order to what had been eight years of chaos." As Shepherd described the ISS situation in those days, the problems weren't with the astronauts, cosmonauts, or engineers in both programs but rather with the higher echelons of management and at the political level. Due to his hard work on the ISS, he was the choice among some members of NASA's management to become the first astronaut to visit *Mir*, but Norm Thagard was chosen instead. Still, Shepherd became a natural choice when he was selected by NASA to become the first commander of the ISS.

During his training in Star City, Shepherd started a new tradition by creating a bar in the basement of one of the American-owned duplexes. Known today as Shep's Bar, it provided a nice place for training-station crewmembers of all nationalities to unwind in a casual yet somewhat private setting. More than just a place to drink, the bar has a jukebox, a piano, a billiard table, and several benches. The bar is still used to this day, even though Bill Shepherd himself has long since retired from NASA.

By the time Expedition 1 launched, the crew had been training together for about five years. After launch and a little over two days in orbit spent catching up with the station, the crew docked *Soyuz TM-31* successfully. The crew entered the station, and on the early morning of 2 November 2000, the ISS was occupied. From that point on, except for brief periods when crewmembers would transfer a Soyuz craft to a different docking port, the ISS would be manned continuously.

The first month was very busy, and the crew spent it unpacking and activating equipment. Shepherd commented during a space-to-ground interview that the biggest challenge for him was packing thirty hours of work into an eighteen-hour day. The crew communicated with one another speaking half-English and half-Russian, sometimes with both languages in a sentence in a form Shepherd called "Rusglish." Krikalev was a fluent English speaker, and Gidzenko was not far behind, while Shepherd did his part to keep his own Russian-language skills current. All communications with controllers at the TSUP were done in Russian. Each crewmember had their assigned tasks in orbit and sometimes didn't get to work in the same areas of the station, but each day at mealtimes, they all tried to eat together in order to share stories and keep each other updated as to the status of things.

Not long after arriving at the ISS, the crew came to a decision regarding the name of the station. Naming the ISS had been a point of political contention for several years. In NASA circles, it was known as "Freedom" and later "Alpha," while the Russians had their own ideas for what to call it. The name International Space Station is rather cumbersome to say in Russian, so Shepherd used his privilege as Expedition 1 commander to give the station the radio call sign "Alpha," which is the name of the first letter in the Greek alphabet. While it wouldn't officially become the name of the ISS, it was at least a start, and the station was known as "Alpha" during Expedition 1's stay.

Visitors

"Alpha" received its first Progress module on 18 November. The cargo vehicle did a normal approach, but the onboard docking system failed, so Gidzenko had to dock the craft using the station's TORU system. While TORU had been a source of controversy on *Mir* due to its use in something it wasn't originally intended for, it is a very capable system when used properly. One advantage the ISS TORU system has over its *Mir* counterpart is that it has a built-in simulator, meaning it can be used to help keep docking skills current.

After a month in orbit, the space shuttle *Endeavour* paid a visit to the station on mission STS-97. The primary payload for this flight was the P6 truss, containing the first set of solar arrays for the ISS. The small arrays on the Russian segment had been barely adequate to support the needs of the first crew. With these new arrays, the station's power generation capability would be increased fivefold and support the next phase of station expansion. Also delivered on this mission was a set of folded-up heat radiators. Once unfurled, these radiators would allow waste heat from the station's electronics to be expelled in order for the interior of the ISS to be maintained at a comfortable temperature.

The P6 solar array truss contained two large solar wings, which each measured 112 feet long (34 meters) by 39 feet wide (12 meters). Once the P6 truss was installed, the array wings were carefully unfurled like a giant set of venetian blinds. Eventually, a total of four sets of solar array trusses would be delivered, but only the first set was installed on this flight on a temporary location at the top of the Z1 truss. Once the station's main truss had been completed, the P6 arrays would be folded up and relocated to their permanent spot.

The two craft docked with one another on 2 December, with the shuttle using the PMA on Unity's nadir port. PMA-2 on the front of Unity was temporarily relocated to a CBM on one side of the Z1 truss to make room for a new module. The hatches between the ISS and the shuttle weren't opened until 8 December, after three space walks to hook up the truss had been completed. Once the hatches were opened, the shuttle crew spent two additional days hauling supplies back and forth between the shuttle and the ISS before *Endeavour* undocked and headed for home.

It would be two months before the space shuttle *Atlantis* paid the ISS its next visit on mission STS-98. The primary payload for this flight would be the heart of the U.S. segment of the station, the Destiny laboratory module. Measuring 28 feet in length (8.5 meters), Destiny is about a third longer than a pressurized Spacelab module and is slightly larger in internal diameter as well. Destiny would become the primary module for housing experiments on the U.S. side, but it would also become the station's main command and control center for U.S. astronaut crews. It contained a sleeping berth in addition to its other equipment.

Atlantis arrived and docked with the ISS on 9 February 2001. Before the Destiny module could be attached, astronauts Tom Jones and Robert Curbeam conducted a space walk to remove covers from the module and prepare its exterior. They then moved on to other tasks while mission specialist Marsha Ivins used the shuttle's RMS to carefully move the module from inside the payload bay to the front port of the Unity module. The Destiny module was one of the larger structures to be hauled up by a shuttle to date and dwarfed the skinny-looking RMS arm, but Ivins succeeded in delicately berthing the module at its permanent location. At the conclusion of the first space walk and during the second one, Jones and Curbeam hooked up data and cooling lines. They also made preparations to move PMA-2 from its temporary location. Ivins next used the RMS to relocate the PMA from its Z1 location to the front of the Destiny module. For the next few years, PMA-2 on Destiny would be the primary docking port for visiting shuttles.

Crew Change

On 8 March 2001 the space shuttle *Discovery* lifted off from KSC on mission STS-102. Among the equipment intended for the ISS was a Multi-Purpose Logistics Module (MPLM). Three MPLMs were built by the Italian Space Agency (ASI) in Europe. The modules each measure about 21 feet long (6.4 meters) and 15 feet in diameter (4.6 meters). A shuttle equipped with an MPLM can dock it with a CBM; once the hatches are opened, equipment can be transferred easily between it and the ISS. Compared to the much smaller Progress craft, an MPLM has thirty-one cubic meters of internal volume and can haul up to nine metric tons into orbit and bring almost that much home. The ability to bring large loads of equipment back to Earth would come in handy to help limit ISS clutter.

The three MPLMs were named Leonardo, Raffaello, and Donatello after famous Italian artists from the Renaissance. They also happened to be the names of three of the Teenage Mutant Ninja Turtles, characters from the popular comic book and movie series. So a NASA patch logo for the MPLM program was developed, featuring a ninja turtle wearing an orange shuttle pressure suit in honor of the MPLMs.

For this mission, the Leonardo MPLM was loaded primarily with equipment racks for the Destiny module. By using an MPLM, a heavy module could be flown empty and outfitted with heavy equipment later on. Plus, last-minute changes could be made to the equipment manifests before outfitting a module in orbit. While missions to haul MPLMs to and from the ISS weren't as glamorous as assembly ones, these flights were among the most important of the ISS program.

During almost nine days of docked operations, the combined crews conducted two assembly space walks and transferred equipment back and forth between the ISS and the MLPM. But the shuttle would also be conducting a crew transfer as well, bringing up the Expedition 2 crew of Russian cosmonaut Yuri Usachev as commander and astronauts James Voss and Susan Helms as flight engineers. Like Expedition 1, this would be another minority-commander situation, but this time, it was a Russian in charge of two Americans. Shepherd, Gidzenko, and Krikalev returned to Earth on *Discovery* after spending over four months in orbit. Their mission to activate the ISS and set it up for continuous occupancy was hailed as a success. After a slow beginning, the ISS was off to a good start.

Expedition 2

During Expedition 2's time in orbit, things continued to run smoothly. The crew continued the tasks set up by their predecessors and took time to conduct some of their own research as well. They didn't have to wait long for their first visitors, as *Endeavour* visited on STS-100 in early April. Its main cargo was the Canadian-built Canadarm 2 RMS and the Raffaello MPLM. At a glance, the Space Station RMS (SSRMS) isn't all that different from its shuttle cousin. It is designed to allow for payloads to be grappled by the station in a similar fashion, but what makes the ssrms unique is that it has a grapple fixture at both ends. Through the use of special Power Data Grapple Fixtures fitted to various points along the iss, the ssrms can move itself from point to point like an inchworm to conduct various construction tasks as needed. With this new capability, assembly work would continue to take place independent of a shuttle, and the second arm would be useful in attaching modules where a shuttle arm couldn't reach. Before *Endeavour* departed, the ssrms was installed, activated, and checked out fully. The ssrms control station was located in the Destiny laboratory.

A Tourist Visits

Just prior to *Endeavour*'s undocking, *Soyuz TM-32* lifted off from Baikonur. The mission was part of a standard swap of the Soyuz lifeboat aboard the iss for a fresh one, and the two-person crew of Talgat Musabayev and Yuri Baturin were sent to handle the mission since the shuttle was being used for early iss crew rotations. The third seat was occupied by American Dennis Tito. Tito was not a career astronaut but rather the first commercial passenger who paid entirely for his own trip into space. Tito has a bachelor of science degree in aeronautics and astronautics and spent part of his professional career working for nasa's Jet Propulsion Laboratory in the 1960s before changing his focus to investment management, where he made his fortune. He was originally contracted to fly to *Mir* before MirCorp's funding dried up, but then he became the first commercial passenger signed up by the company Space Adventures, based in Virginia. After *Mir* was deorbited, the Russians chose to send him to the iss.

The decision did not go over well with nasa, as administrator Dan Goldin was dead set against it. While there was technically nothing nasa could do to stop Dennis Tito from flying to the iss since he was a guest of the Russians, nasa would not have a hand in his training at all. Members of Congress also expressed their disapproval in Tito's flight, calling into question his patriotism as an American. *Soyuz TM-32* docked with the iss on 30 April 2001, one day after *Endeavour* departed. The crew on board gave their colleagues a warm welcome, although the short-duration crewmembers were encouraged to keep Tito only in the Zvezda module during his time on board. Tito's experience helped to pave the way for other space tourists with large enough pocketbooks to follow in his footsteps, and he reportedly enjoyed his experience of almost eight days in orbit immensely.

Assembly Continues

In June the Expedition 2 crew made preparations for the arrival of the Pirs airlock module on the Russian side. Yuri Usachev and James Voss donned a pair of Orlan suits and conducted an IVA inside the depressurized Zvezda module to attach a docking cone to the nadir port on the front of it. Pirs (a Russian term meaning "Pier") would not only allow Soyuz and Progress spacecraft to dock there, but it would also act as an airlock for cosmonauts using Orlan suits without the need to depressurize the RSM. Fuel transfer lines in Pirs would also allow Progress ships to top off both Zvezda's and Zarya's thruster fuel tanks.

Construction work continued with STS-104 arriving at the station in mid-July. The main payload in the cargo bay of *Atlantis* was the new Quest airlock. Like the SSRMS delivered on the previous shuttle mission, the Quest would allow ISS crewmembers to perform construction tasks on the station without the need for a shuttle. While technically EVAs could be performed using Orlan suits from the Russian side of the station, NASA's EMU would allow more comprehensive assembly tasks to be conducted.

The end of Expedition 2 came on 12 August 2001 with the arrival of STS-105 at the station. Aboard the space shuttle *Discovery* was Expedition 3 commander Frank Culbertson Jr. and Russian flight engineers Mikhail Turin and Vladimir Dezhurov. In addition to the new crewmembers, the Leonardo MPLM in the payload bay was stacked to the gills with science racks and equipment, as work was underway to begin an intense phase of scientific activity during the next few months. Frank Culbertson was no stranger to spaceflight, as he had been a veteran of two shuttle missions, serving as commander for one of them. For the past several years, he had also served as the manager of NASA's side of the Shuttle-Mir Program from 1995 until its conclusion. After eight days of docked operations and two EVAs, Usachev, Voss, and Helms returned home with the shuttle crew.

A Dark Day in September

By Tuesday, 11 September 2001, the Expedition 3 crew had settled into a regular routine. They were three days away from receiving the Pirs docking module, and all was proceeding well with no major problems. The Expedition 3 mission patch was rather unique, as it featured a book of space history turning the page from what had come before with *Mir* to the next, blank page featuring the future, with the ISS orbiting overhead. To people who witnessed the events of that day on the ground, it almost seemed like a page had turned in a book, albeit in a very dramatic fashion.

That morning, four commercial airline flights originating on the East Coast of the United States were hijacked by Middle Eastern terrorists. Two of the planes were crashed into the twin towers of the World Trade Center in New York City, and one was crashed into a wing of the Pentagon in Washington DC. The fourth aircraft crashed into a field near Shanksville, Pennsylvania, when the passengers on board tried to take back the flight before it could bring further destruction down on another target. Passengers on all the flights, in addition to many on the ground, were killed when the first three planes impacted their targets. The crash and fire damage done to the World Trade Center caused the two towers to collapse, killing hundreds more people still trapped inside and many first responders from New York City's fire and police departments.

Compared to previous space stations with their ground-based tracking networks, the ISS had a pretty good hookup to the ground as communications through NASA's TDRS network afforded almost-continuous two-way communications coverage. Crewmembers on board also had access to a new phone system, delivered with the station's Destiny laboratory, that could be used to call home. Regular emails were also sent to the station's laptop computers, but the crew did not have the Internet in those days.

Despite the improved communications, the Expedition 3 crew still didn't entirely know what was going on, but even from 250 miles up, they could see something bad was happening in New York, as they could clearly see a smoke-and-debris cloud enveloping Lower Manhattan when one of the station's orbits took it over the East Coast of the United States. The crew began filming a video of what they were seeing when the World Trade Center north tower finally fell at 10:28 EDT. While everyone on the crew was shocked by what they saw, Frank Culbertson had a unique perspective, given that he was the only American in space at the time. When the ISS orbited over New York City the next day, he used part of the station's communications loop to express his feelings of outrage as to what happened,

45. Frank Culbertson's photo of the New York–New Jersey area on 11 September 2001. Winds blew the smoke plume south. Courtesy NASA.

but he also took the time to tell New Yorkers to not give up hope and that their city still looked very beautiful from space.

While the attacks on 11 September touched everyone who witnessed them that day regardless of whether or not they were personally affected by it, Frank Culbertson also lost a friend. One of his Naval Academy classmates, Charles "Chic" Burlingame, was one of the pilots of American Airlines' flight 77, which was sent crashing into the Pentagon by the hijackers. Frank's two crewmates did their best to support their friend through this period, and the crew got back to the work at hand, since they could do nothing otherwise. Ultimately, though, all three men knew that the world they had left would not quite be the same one they would return to in December.

Pirs Arrives

On 14 September, a modified Progress carrying the Pirs airlock and docking compartment arrived at the ISS. The Pirs was very similar to the shuttle docking adaptor used on *Mir*, but this module would only accommodate Russian spacecraft. Once the new module was docked and activated, it wasn't long before the crew conducted a total of four space walks from it during the next three months, using the Orlan suits to complete its installation to the Zvezda module. Three of the EVAs were conducted by Turin and Dezhurov, with Culbertson taking part in the fourth space walk with Dezhurov.

Soyuz TM-33 arrived in October for a standard Soyuz ferry swap. On board were cosmonauts Viktor Afanasyev and Konstantin Kozeyev and French CNES astronaut Claudie Haigneré for a brief visit. Formerly known as Claudie André-Deshays, she had married fellow French astronaut Jean-Pierre Haigneré in the years since her visit to *Mir*. After eight days in orbit, the ferry crew returned home on *Soyuz TM-32*.

Expedition 4

Docking day for mission STS-108 arrived on 7 December 2001 when the space shuttle *Endeavour* paid the ISS a visit. Not much in the way of assembly took place on this flight, as most of it was dedicated to MPLM hardware transfers. But *Endeavour*'s personnel also included the Expedition 4 crew of cosmonaut commander Yuri Onufriyenko along with astronauts Carl Walz and Dan Bursch. *Endeavour*'s time at the station was extended by a day to allow the shuttle crew to repair a problem with the station's treadmill and replace a failed compressor in one of Zvezda's air conditioners. At the end of the docked period, the three newcomers took over occupation of the station while Culbertson's crew went home. The results of several experiments that Expedition 3 had performed also returned home. The ISS was starting to deliver scientific results.

During the start of 2002, the Expedition 4 crew performed a total of three construction EVAs. Two were performed on the Russian side from the Pirs airlock, using Orlan suits, while Walz and Bursch also took part in one space walk from the Quest airlock, using NASA EMUs in order to give the module a full checkout. It was the first time a staged EVA had been performed using EMUs without a shuttle present.

On 10 April 2002 the space shuttle *Atlantis* arrived on mission STS-110, carrying the first part of the station's main solar array truss, the S0 segment. It was attached to a special port located on top of the Destiny laboratory; over the course of four EVAs, the shuttle crew installed it, relocated the SSRMS to a fixture on the truss, and installed additional EVA handrails. It was the most ambitious construction mission to date, requiring two EVA teams, with two space walks each, to alternate the tasks. Team one was made up of astronauts Steve Smith and Rex Walheim. Team two consisted of astronauts Jerry Ross and Lee Morin. Both teams completed all four space walks without any major problems.

No crew swap was scheduled for this shuttle visit. About a week later, *Soyuz TM-33* arrived for a ferry swap. In command of the Soyuz was former Expedition 1 crewmember Yuri Gidzenko. Joining him were Italian ESA astronaut Roberto Vittori and the second paying space tourist to fly, Mark Shuttleworth. Shuttleworth was the first citizen of South Africa and the first resident of the African continent to fly into space. Possibly due to the smooth nature of Dennis Tito's visit the year before and given that Shuttleworth wasn't a citizen of the United States, no major government protest was lodged against this visit.

After the ferry-swap crew returned home aboard *Soyuz TM-32* in early May, Expedition 4 continued with its normal operations until STS-111 arrived in early June with the Expedition 5 crew. Cargo in *Endeavour's* bay included a Mobile Base System (MBS) for the SSRMS. In combination with a mobile transporter cart that was installed with the S0 truss, the MBS would allow the SSRMS to travel the length of the station's main truss once it was completed. Among the shuttle crew was French CNES astronaut Philippe Perrin. After this flight, CNES decided to disband its astronaut program, so remaining members of the CNES astronaut group were transferred to the ESA.

Expedition 5

The three members of Expedition 5 included Russian commander Valery Korzun, Russian flight engineer Sergei Treshchev, and NASA astronaut Peggy Whitson. With this crew, a change was made to how the crews were selected, as no longer would it be a minority-commander situation. Instead, the commander would typically share nationality with one of the crewmates. Whitson was the project scientist in the Shuttle-Mir Program before she joined the astronaut ranks in 1996, and she was the first dedicated NASA scientist-astronaut to fly to the ISS. During Whitson's stay in orbit, she was given the title of the first science officer aboard the ISS, as she conducted a total of twenty-one different experiments in life sciences and microgravity disciplines while also observing results from various commercial payloads that were on board during her stay.

On 16 August, Peggy Whitson also conducted the first space walk by an American female using an Orlan suit as she and Valery Korzun installed new debris shields on the Zvezda module. The four-hour space walk was not easy, though, as given Peggy's relatively small build, she could not completely cross her arms in front of her. Yet they completed the task with no problems. Korzun and Treshchev conducted a more ambitious space walk ten days later. They retrieved some Japanese experiment racks, installed EVA tether points, and added antennae for the station's ham radio.

STS-112 arrived at the ISS in October, bringing with it the S1 segment of the station's main solar array truss. The truss segment was massive, as it took up almost the entire cargo bay of *Atlantis*. Three space walks were conducted to help join the S1 and S0 truss segments together and fully integrate their systems. All three space walks on this mission were conducted by *Mir* space walk veteran Dave Wolf and rookie NASA astronaut Piers Sellers. Both men accomplished the tasks assigned with no problems. Sellers was born in the United Kingdom and had a meteorology background. He and his wife moved to the United States in 1982 so that he could take a job at NASA's Goddard Spaceflight Center, and he became a naturalized U.S. citizen in 1991. Sellers was selected to become an astronaut in 1996.

With the arrival of the Expedition 6 crew on mission STS-113 in late November 2002, the ISS program had a string of successes to look back upon. The station had grown to over twice its original size and had greatly expanded its capabilities. As Expedition 6 crewmembers Ken Bowersox, Nikolai Budarin, and rookie astronaut Don Pettit left the space shuttle *Endeavour* to take up residence in the ISS for a five-month mission, hopes were high both on the ground and in orbit that 2003 would be a great year.

12. *Columbia*

1 February 2003, 08:05 CST—Hemphill, Texas.

Ask someone to point out the town of Hemphill on a map, and they likely will not be able to. Located on the extreme eastern side of Texas in Sabine County, Hemphill is about the farthest east that one can get in Texas before ending up in Louisiana. The town of about 1,100 residents sits on the western bank of the Sabine River basin, near the Toledo Bend Reservoir. Thanks to the region's dense forests, a large portion of the town's residents work for the U.S. Forest Service, while others have jobs in support of camping and fishing activities. There is also the usual mix of farmers and livestock owners.

Hemphill's residents are typically early risers. So on that Saturday morning, almost everyone heard a strange sound. For many, it sounded like a dull rumble that increased in volume, peaked, and then faded out to silence about fifteen seconds later. It was loud enough that people taking showers heard it through solid walls and over running water. It shook homes and businesses all over the region. Nobody was sure what they heard, but it sounded as though something bad had happened very close by. Some residents outdoors looked up to see debris streaking overhead and falling out of the sky, but no one knew what had caused it. Almost right after the noise faded and debris began hitting the ground, the town's 911 dispatch center and local sheriff's office began receiving calls reporting everything from a plane crash on one side of the county to a possible train derailment on the other.

STS-107's Crew and Mission

The crew of mission STS-107 aboard the space shuttle *Columbia* was very well known in the astronaut office, as they had trained together for quite some time. The repeatedly slipping launch date kept them together long after other crews had flown. Mission commander Rick Husband, a former test pilot and colonel in the air force, had flown as the pilot on STS-96, one of the first supply missions to the ISS. He was one of three veteran astronauts on the crew. Mike Anderson was a veteran of STS-89, the final shuttle mission to *Mir*, which brought Andy Thomas home.

The third veteran was Kalpana Chawla, a veteran of STS-87. Chawla was born in India but moved to the United States to get her degree in aerospace engineering in the 1980s, and she became a naturalized U.S. citizen in 1990. She was also an aerobatic private pilot. When Chawla joined the astronaut corps in 1995, she was given the nickname KC as it was easier to pronounce than her Indian name. At work, she usually had a smile on her face and the most infectious, upbeat attitude of anyone there.

There were also four rookies aboard *Columbia* for this mission. William McCool, *Columbia's* pilot, was a naval aviator with attack-jet experience who graduated second in his academy class. He was also blessed with perpetual boyish good looks. Laurel Clark and Dave Brown were both flight surgeons from the U.S. Navy. Clark was married to NASA flight surgeon Jonathan Clark. Flight surgeon Dave Brown had also served time in A-6 and F/A-18 attack-jet squadrons before joining NASA.

The fourth rookie was Israeli payload specialist Ilan Ramon. Ilan was a child of Holocaust survivors from World War II. He was a former F-16 fighter pilot and a colonel in the Israeli Defense Forces. Two decades prior, he was the youngest member of a group of pilots in the Israeli Defense Forces who were handpicked to conduct a bombing mission against the Osirak nuclear reactor, deep within Iraq's borders. The purpose of the mission was to deny Iraq the ability to create its own nuclear weapon. The strike was successful, and all aircraft returned safely. As Israel's first astronaut, he was a celebrity in his home country. Even though he was designated a payload specialist, the length of training time caused by mission delays meant that he was almost as well trained as a mission specialist. His wife and youngest children lived with him in Houston during his training.

The STS-107 mission carried a Spacehab double module outfitted as a science laboratory. The module was built by a private company known as Spacehab and was the brainchild of the company's founder, Bob Citron. Like the pressurized Spacelab module, the Spacehab was designed to fit within the shuttle's cargo bay, and it could fly as a single, short unit or as a longer unit known as the double module.

46. The STS-107 crew inside the Spacehab lab module. *Clockwise from top*: McCool, Anderson, Ramon, Clark, Husband, Chawla, and Brown. Courtesy NASA.

But the Spacehab differed from the Spacelab module in a few key ways due to advances in technology. It was a bit bigger internally and could be outfitted either as a logistics module or as a dedicated laboratory, unlike Spacelab's sole use as a scientific payload. The Spacelab module, when used by itself, also required loading into the rear of the shuttle's payload bay due to center-of-gravity issues, limiting what other cargo could be carried on the same mission. Built from lighter materials, the Spacehab had no such limitations.

The Spacehab had first flown on STS-57 as a single-module science lab, and later it had flown in both single- and double-length configurations as a logistics module on flights to *Mir* and to the ISS. The use of Spacehab for this dual purpose led to lower mission costs, and it led to the eventual phasing out of the Spacelab module, which flew its last mission in 1998 aboard *Columbia* on STS-90. STS-107 was the first flight of a Spacehab as a double-length science module. STS-107 was also the last planned shuttle mission dedicated to science gathering. After this flight, the ISS would become NASA's primary orbital science laboratory, with shuttle missions flying only in support of it.

The mission lasted over two weeks on orbit; to accommodate that, *Columbia* was equipped with an Extended Duration Orbiter (EDO) pallet rack at the rear of the payload bay. The EDO rack contained additional tanks of cryogenic liquid oxygen and hydrogen for the shuttle's fuel cells, allowing for extended operations compared to the eight or nine days of a normal shuttle mission. The space shuttle *Endeavour* had originally been outfitted with an EDO kit as well, but the system was removed to make it lighter for ISS support flights. *Columbia* kept its EDO kit since it was already too heavy for most ISS cargoes, although NASA had plans to refit the orbiter for ISS support flights once STS-107 concluded.

For this flight, around-the-clock science was the order of the day, so the crew were divided into red and blue teams. A total of thirty-two scientific payloads flew on STS-107. Crewmembers would act as subjects for biomedical data collection, which is primarily why two flight surgeons were on board. But experiments were being conducted in other scientific disciplines as well, making this a jack-of-all-trades flight. Scientific payloads involving rats, insects, and primitive organisms such as flatworms were being flown on behalf of several schools on Earth to study the effects of microgravity. A new water-recycling system was also being tested for possible use on the ISS. The properties of flames in zero gravity and how to detect them was also being studied in a continuation of research conducted on previous shuttle flights.

Outside the Spacehab, additional experiments were mounted in the payload bay on a bridge rack known as the "Hitchhiker." They were part of the Fast Reaction Experiments Enabling Science, Technology, Applications, and Research (FREESTAR) package. Among them was a payload known as the Mediterranean Israeli Dust Experiment (MEIDEX). It was a camera and instrument system designed to study the composition and density of dust particles over the African, Mediterranean, and Israeli desert regions and to measure their effects on weather. Ramon was on board primarily to handle the MEIDEX experiment, in addition to other duties.

At the conclusion of almost sixteen days on orbit, *Columbia*'s scientific mission was considered a huge success. Over 80 percent of the scientific data had already been electronically transmitted to the principle investigators, along with numerous photographs shot by the crew's digital and video cameras. Scientists were already digesting the data while waiting for the experiments themselves to return. The crew were proud of what they had accomplished and were ready to return home.

"Get Ready"

1 February 2003, Kennedy Space Center, Florida, near the Shuttle Landing Facility.

On the ground at KSC in Florida just after 09:00 EST, the families of the STS-107 crew were waiting around the VIP area for the two sonic booms that would announce the arrival of *Columbia* and its final approach to the Shuttle Landing Facility. Pretty much everyone was in a relaxed mood, anticipating the return home of seven astronauts to be reunited with their families.

For shuttle missions, the NASA astronaut office typically assigned several members of the astronaut corps to become escorts for the families. They would help act as guides to the spouses and children while at KSC during launch and landing. The escorts would also keep the families insulated from the media. But the escorts also had an important secondary role. They acted as assistants to the families when a disaster occurred. That role had come into play only once before during the loss of the space shuttle *Challenger* on STS-51L.

For STS-107, astronauts Steve Lindsey, Scott Parazynski, Terry Virts, and Clay Anderson were assigned as escorts and were all on hand for *Columbia*'s liftoff two weeks prior. Only Lindsey, Virts, and Anderson were available to make the trip to Florida for *Columbia*'s landing. Anderson and Virts were rookies who hadn't flown in space yet.

Since 11 September 2001, security at KSC had been tightened for launches and landings to guard against possible terrorism. Security for this mission was tighter than normal due to the ongoing tensions between Israel and its Arab neighbors. Given Ramon's assignment on STS-107's crew, the ongoing war against terrorism in Afghanistan, and all signs pointing to war between the United States and Iraq in a few months time, NASA security, the FBI, and other federal agencies had a larger presence for this mission than normal.

While members of the astronaut office typically know one another, it is a big group, and some relationships are closer than others. In Clay Anderson's case, he had several friends on this flight. One particular STS-107 astronaut with whom he had worked before was Mike Anderson. Obviously the two men weren't related to one another since Mike was of African American descent, while Clay had a much lighter complexion and a noticeably receding hairline. Clay was born in Ashland, Nebraska, and had grown up there. He became the first person born in Nebraska to be selected as an astronaut. One of Mike Anderson's first duty assignments in the U.S. Air Force right out of flight school was flying jets at Offutt AFB in Bellevue, Nebraska. He had also taken classes at Creighton University in nearby Omaha. So both men were very aware of each other's Nebraska connection.

In January 2000 both men were the first to fly a NASA T-38 after concerns about the infamous Y2K computer bug. As an air force pilot, Mike flew front seat while Clay, being a civilian, flew in the rear seat. The flight went without incident. While Y2K concerns seemed more hype than reality, the flight made for a good story shared between them as two astronauts, two Andersons, and two Nebraskans (one by birth, one by association).

Back at KSC, about ten minutes after 09:00 EST, Clay Anderson and Steve Lindsey were chatting with STS-107 family members. While watching the countdown of the mission clock to the estimated time of landing, both men could hear CAPCOM astronaut Charlie Hobaugh back in Houston on the mission control feed saying "*Columbia* Houston, UHF comm check" over and over again. Both men wondered why mission control was calling for a communications check on UHF. UHF stands for "ultrahigh frequency." It is a radio frequency band used by aircraft, but the shuttle is equipped with a UHF backup radio in case of a failure with the normal radio systems. Both astronauts thought it was odd for the call to be made, since UHF isn't normally used, but they didn't think much of it otherwise. A minute or two later, Anderson figured something was up when he watched the face of a female security person listening to a walkie-talkie go white as she heard something on her radio. Lindsey stepped away for a bit, apparently to be informed of what had happened. When Lindsey came back, his only words to Clay and the other escorts were, "Get ready."

According to Clay Anderson, "Once [Lindsey] said 'Get ready,' I got a sinking feeling in the pit of my stomach, and I knew something was wrong. We hadn't seen [*Columbia*] come in; we hadn't heard the sonic booms. We heard comm that was not of the ordinary." After that, the escort astronauts gathered their designated families together and walked them to the parked vehicles used to drive them over earlier that morning. While the families didn't know exactly what, they figured that something had happened, as they were being taken from the stands without a shuttle on the runway. Clay made sure to shut the radio off in the vehicle he was driving, because he guessed correctly that news agencies had already picked up on the story; he didn't want the families to hear anything before an official notification had been made.

When the group arrived at the crew quarters building after a short drive, everyone was taken into a conference room. At that point, nobody had any firm answers yet, and the escorts were just as much in the dark as the families. Anderson walked down to the management wing of the building and met with astronaut Bob Cabana, who, at the time, was part of the Flight Crew Operations Directorate. While even Cabana didn't know exactly what had happened, he had heard reports, was in contact with JSC, and likely had also seen the video of *Columbia* breaking apart in the sky over Texas on the news. Cabana made the decision that the families had to be told. *Columbia*'s crew had perished on reentry. The families were all told at once. Their reactions were sudden, as kids screamed and adults began to cry. A lot of emotions were released once the announcement was made. It wasn't even 10:00 EST, and already it had been a long day.

After Clay Anderson called his wife to tell her what had happened, he went to work, as the escorts still had a job to do. They went back to the hotels to gather the families' packed bags and the rest of their belongings. The group then drove back to KSC and loaded the luggage onto a NASA transport for the long flight home. The plane arrived at Ellington Field near JSC in the midafternoon, and the families were escorted home by JSC personnel and local police. Once the families of the STS-107 crew were back home, Clay Anderson had to go back to JSC for an all-hands astronaut meeting and got pulled over by a police officer for speeding on his drive back, but he was let off with a warning. The mood in the communities surrounding JSC was very somber that evening and was shared by all.

1 February 2003, International Space Station Expedition 6, orbiting 250 miles up.

Meanwhile, in Earth orbit, ISS crewmembers Ken Bowersox, Nikolai Budarin, and rookie astronaut Don Pettit were off duty. Typically, the weekends were free for them to pursue what activities they wanted to do, with no assigned mission tasks. They were over halfway through their planned four months of assigned duties, and things were going pretty well for an ISS that had been hosting crews for a couple of years. Even with its relatively short time of occupation, the station was already starting to amass a bit of clutter as supplies were stockpiled in the FGB and Unity nodes. But it was operating well, with excess energy capacity from the new solar arrays.

There was a planning conference scheduled for that day to update the next week's scheduled activities, and it didn't start on time. The crew, instead, was told to standby. The Expedition 6 crew knew that *Columbia* was heading home that day; a few days earlier, the crews had communicated with one another through the TDRS satellite network. Don Pettit had also been maintaining a chess game in orbit with Willie McCool, and they exchanged moves with one another via email.

Eventually, JSC director Jefferson Howell got on the communications loop to tell the crew what had happened. Howell had only been the director of JSC since 2002, replacing the somewhat controversial and mercurial George Abbey, who retired in 2001. But Howell was a retired Marine Corps general, and people respected him. For him to communicate directly to the ISS crewmembers rather than have it go through the assigned CAP-COM astronaut meant that this announcement was very important. Howell kept it short and to the point: "I have some bad news; we lost the vehicle." That was all he had to say. The crew was next brought up to speed quickly and heard the sobering details.

Investigation and Aftermath

For everyone in the world who had access to a television set, what had happened in the skies over Texas that Saturday morning was pretty clear. *Columbia* had tumbled out of control and almost completely broke up, with its destruction looking eerily similar to *Mir*'s deorbit two years prior. The debris trail extended about two thousand miles, from the western-most piece of debris found just east of the Texas–New Mexico border in the state's panhandle to a valve that was recovered near Baton Rouge, Louisiana. Most of the debris landed in central and east Texas in a nearly three-hundred-mile path extending from south of the Dallas–Fort Worth area to Hemphill. It was a small miracle that nobody on the ground was injured or killed by the falling debris. Some of the pieces were rather large, and the rumbling that people heard was pieces of debris traveling at supersonic velocities as they fell to Earth. One of *Columbia*'s main engine bells was recovered at a golf course when groundskeepers discovered a new water hazard that they hadn't installed. The bell hit with such force that it buried itself deeply enough for groundwater to fill the small crater it left.

Video footage shot by enthusiasts along *Columbia*'s reentry path showed that the orbiter was starting to shed debris as it flew over California. In Dallas, Texas, news cameras, which were pointed to the sky to capture the reentry, filmed *Columbia*'s final breakup, as did a U.S. Army Apache helicopter on training maneuvers near Fort Hood, Texas. A still photograph of the breakup taken by Dr. Scott Lieberman, a cardiologist and amateur photographer from Tyler, Texas, became one of the most published photos of the event in newspapers and magazines around the world.

Groups of volunteers and members of the National Guard were mobilized into debris search teams all over central Texas, as well as New Mexico and Louisiana. Hemphill, Texas, became a major staging area for the recovery since a couple of local residents had discovered what were believed (and later confirmed) to be human remains.

The recovery activities became quite a strain on the town's resources as the area's population swelled almost overnight with federal agencies and news media, but people got on with their jobs in a professional manner. Ultimately, remains of all seven astronauts and portions of *Columbia*'s nose and crew cabin were found within Sabine County's borders. Local funeral home directors loaned their time and services to help with recovery and transportation of the astronauts' remains. It was all done privately and with the utmost respect.

In that part of Texas, the volunteers for the search teams had a very difficult task in searching for debris, as they had to walk through some of the most densely packed woods in the country, standing shoulder to shoulder with one another so that no piece of debris, no matter how small, would be missed. When a piece was found, a handheld GPS device would mark the location, and photographs were taken before the piece was removed for analysis.

The work to recover the remains of *Columbia* was painstaking and took several weeks. When the spring rains came, it got cold, wet, and very muddy. But day after day, the volunteers went out and searched. In the end, more than seven hundred thousand acres were searched over a four-state area. It also had a human cost, though, as Francis "Buzz" Meir, a contract helicopter pilot with the Forest Service, and Charles Krenek, a manager of one of the search teams, were killed when their Bell 407 helicopter went down on 27 March 2003 during an aerial search for debris in a wooded area. Three other search team members on board survived the crash with serious injuries. In Texas the two men are memorialized alongside the seven astronauts.

Once in a while, new pieces of debris are still being discovered. About five years after the event, a resident of Hemphill, while making preparations to sell his home, got ready to patch a leaking roof. He noticed that a hole had been drilled through the roof and into the side of some storage boxes in his attic. When he opened one of the affected boxes, it contained a screw from *Columbia*. As before, NASA was contacted, and they sent representatives to recover the debris. A sphere from *Columbia*'s fuel system was also recently recovered when a lake was drained.

The residents of Hemphill to this day are proud of the work they did. To their credit, unlike some other places where debris was found, no residents of Hemphill or Sabine County made any attempt to keep or sell debris from *Columbia*. Today, the Patricia Huffman Smith NASA "Remembering Columbia" Museum (named for the late wife of the person who donated the land to build the museum) in Hemphill stands as a memorial that the general public can visit. Thanks to exhibits from NASA and items donated by the STS-107 crew's families, the museum helps to tell the story of the space shuttle *Columbia*, STS-107, the recovery, and how things changed in that region of Texas on 1 February 2003.

Debris from the shuttle was collected in a hangar at KSC. As with an aircraft crash, the recovered pieces were laid out on a grid on the floor and placed in their proper locations on the vehicle. Some debris was burned beyond almost all recognition, while other pieces were almost completely intact with no visible signs of damage. One piece of equipment recovered in good shape was a data recorder fitted to *Columbia* for its test missions. Because the recorder was buried deep inside the shuttle, engineers did not remove it after the testing program had concluded. The recorder and its data tapes helped provide clues as to the sequence of *Columbia*'s breakup after all contact was lost with mission control. A videotape from a camcorder used by mission specialist Laurel Clark was also recovered in good shape. Clark had been recording *Columbia*'s reentry from the flight deck; while the portion of the tape that might have recorded the breakup had been too badly damaged, the rest of the tape revealed that up to ten minutes before the shuttle disintegrated, the crew was in good spirits and there were no signs of impending disaster.

Footage captured during launch revealed that the orbiter was struck by a large chunk of foam that came off an aerodynamic ramp on the shuttle's ET. The foam was used to insulate the cryogenic fuel in the tank and keep heavy ice from forming on it. While the foam is flexible, contraction of the ET during fueling can still produce cracks. Pieces can come loose on ascent either from the cracks or from trapped air bubbles, causing foam sections to break off as outside air pressure decreases.

Prior to *Columbia*'s reentry, a few NASA engineers had expressed concerns in private emails that the foam might have damaged *Columbia*'s tiles on the left wing. But mission managers dismissed these concerns as shuttles had returned from orbit with tile damage before with no problems. Some engineers asked to have the DOD use its reconnaissance satellites to take images of *Columbia*'s wing, but management also rejected those plans. Managerial members of the shuttle program felt that if there were damage, it would just cause problems during turnaround of a shuttle for its next flight and would not be a safety risk on this one. They were very wrong in making that decision.

Enhancement of the launch footage revealed that the chunk of foam that hit the wing was about twenty-one to twenty-seven inches long by over twelve to eighteen inches wide and struck at a velocity from four hundred to nearly six hundred miles per hour due to the shuttle still accelerating from the thrust of its SRBs. While managers felt that tile damage wouldn't be a concern, no attention was paid to the reinforced carbon-carbon (RCC) wing panels, which were designed to handle the highest heat loads of reentry. The RCC panels are very stiff. Although many consider them able to withstand a lot of abuse, the panels are also very thin compared to the tiles, and the areas behind them are hollow.

To test the theory that ET foam could damage RCC, engineers with NASA and other agencies took some RCC wing panels to the Southwest Research Institute in San Antonio, Texas, and fired foam blocks at them using angles and speeds suggested by the data analysis. During a full-scale test using launch video data, the foam block blew nearly a ten-inch hole in the middle of an RCC panel, and an audible "Holy shit!" could be heard on video footage of the test. People were shocked by what they saw. Regardless of what some had believed before, everyone now knew that a foam strike was what had doomed *Columbia*.

During reentry, heat plasma is normally kept away from the shuttle's surface by its entry attitude and shape. A shuttle enters the atmosphere with its nose pitched up at forty degrees so that the bottom bears the brunt of the reentry. This creates a bow wake until the shuttle slows down enough that its nose can be lowered safely for a glide to landing. The shuttle's heat-resistant tiles, blankets, and RCC panels keep the rest of the structure from overheating due to the air friction, but the tiles and RCC can't resist the direct heat of plasma. The loss of a single tile is not enough to cause the plasma wave to be disturbed, but a large hole, especially in a critical area such as the leading edge of a wing, can cause the plasma to approach closer to the surface of the shuttle's exterior and perhaps touch it. Due to the hole in the RCC panel, hot plasma entered *Columbia*'s wing.

Over the next several minutes, the plasma plume torched through wires, causing sensor dropouts, and melted the wing's internal structure. The plume also opened up the hole in the wing wider. The wing's shape itself gradually began to distort, shedding tiles along the way and exposing more of the orbiter's aluminum structure to reentry heating. *Columbia*'s computers maintained control of the wing's elevons and, near the end, activated thrusters to try to compensate for the increasing drag on that side. But eventually, *Columbia*'s hydraulic lines in the left wing were breached, rendering the elevons useless.

Columbia started to lose control at 07:59:37 CST when it lost hydraulic pressure, and nine seconds later it pitched up into a flat spin to the right. Seeing the indication on his monitor, pilot McCool flipped switches on a cockpit panel in an attempt to restart the shuttle's auxiliary power units to try to get hydraulic pressure back, but it was too late. At 08:00:02 portions of the left wing and OMS pod broke away. The fuselage of the orbiter stayed together until aero forces ripped it apart, with the front fuselage and crew cabin separating from the rest of the orbiter about sixteen seconds later.

Later, tests revealed that the crew cabin started to depressurize when it separated from the fuselage. The crew would have had less than ten seconds of consciousness before passing out from lack of oxygen. Analysis of the retrieved suit remains revealed that none of the crewmembers had closed their helmet visors or activated their emergency oxygen packs. Even if they had done so in a now-darkened cabin, they likely would have been killed by the blunt force trauma sustained as the cabin continued to tumble. Even if someone had survived that, they still would have died when the crew cabin started coming apart a few seconds after 08:01.

The *Columbia Accident Investigation Board Report* concluded that NASA management was at fault for not paying attention to the warning signs of a possible problem and not taking steps to try to prevent foam loss on the ET, instead accepting damage caused by foam shedding as a normal part of shuttle operations. Put simply, NASA was playing a case of russian roulette. As with *Challenger*, there was a bullet in the chamber. It was a different bullet this time than *Challenger*'s SRB O-ring seal, but the results were the same.

It would be two and a half years before a shuttle was ready to fly again as NASA spent the stand-down period designing cameras to detect debris from the shuttle on ascent and new equipment to inspect its bottom once it reached orbit. Work was also done to minimize foam loss from the ET. New materials and repair procedures were developed in case heat shield damage was detected. Plans were also put in place to make sure a backup shuttle was ready to launch within a few weeks in case damage to a shuttle was too extensive and a crew had to use the ISS as a safe harbor. Even then, the ISS as a safe harbor wasn't an available option for STS-107, since *Columbia* was in a completely different orbit from that of the ISS.

But a bigger change was in store for NASA's long-term future. The administration of President George W. Bush (son of the elder George H. W. Bush) and Congress decided that shuttle flights would end by 2010. NASA would focus its efforts on design of a new capsule-based spacecraft called Orion as part of the Constellation program. Constellation would take crewmembers to the ISS, on missions back to the moon, and on to the planet Mars. The shuttle program was expensive to operate, with some inherent risks related to its configuration, while capsules on top of rockets were deemed to be safer. Future spacecraft would carry either crew or cargo but not necessarily both.

The shuttle would still continue to operate for the next few years, as its heavy-lift capabilities were needed to finish construction of the ISS. In 2006 NASA projected that it would need about twenty more flights to finish the station. While the work to complete the ISS was underway, NASA and its contractors started work on Constellation. Hopefully, if NASA had the budget it needed, the delay between the last shuttle flight and the first within the Constellation program would not be a long one.

Expedition 6

All these decisions about NASA's long-term future were still far off in the days and weeks after STS-107 was lost. Aside from the loss of friends to each ISS crewmember, *Columbia*'s loss would also have a big impact on life aboard the station, both in the short-term and for the foreseeable future. While the ISS was new and in better shape than *Mir* was during the Shuttle-Mir Program, logistical resupply from the shuttle was no less important. Three people on orbit consume a certain amount of supplies while generating trash and waste. The station was in an unfinished state as well, with some of its equipment operating in temporary configurations only intended for a short period of time. The shuttle was also vital for periodic reboosts of the station's orbit. While a Progress craft could reboost the station, a shuttle had bigger engines and carried a larger fuel reserve.

During the stand-down period, ISS program managers and engineers had to develop new procedures to help maintain the station in a semimothballed state until it could be supported by a shuttle once again. Naturally, this meant that STS-114 would not be coming up in three months to take the crew home. Instead, the current crew would have to return to Earth on the Soyuz. There would be a fair amount of work ahead for the Expedition 6 crew before they could return home. The next crew would also require some changes and retraining before flight, since they would have to launch on a Soyuz.

Staged EVAs

Thanks to delivery of the Quest and Pirs airlock modules the previous year, the ISS was no longer completely dependent on shuttles for assembly tasks. While EVAs conducted from the Russian side carried out important work, EVAs conducted from Quest were capable of doing more-complicated tasks. The shuttle-era EMUs allowed for more mobility. Russian suits operate at higher pressures, meaning less of a need for prebreathing to flush nitrogen from the blood. American suits operating at a lower internal pressure of

3.7 psi allow for easier movement and less fatigue. During early ISS space walks, EVA crewmembers would conduct brisk exercise while prebreathing pure oxygen to flush nitrogen from the blood quicker, but this would leave them slightly winded. When the Quest airlock arrived, EVA crews gained the option of camping out overnight in the airlock at higher oxygen levels at a slightly reduced pressure to flush nitrogen. So on EVA day, the amount of preparation time needed before heading outside was reduced.

In addition to the normal EVAs conducted by shuttle crews, station crews conducted what NASA calls "staged" EVAs. While the tasks were usually simpler than a shuttle assembly mission, staged EVAs were no less important, as they might involve tasks for getting ahead, such as relocating equipment already present, fitting small pieces of hardware, or cleaning up a previous work site. EVAs are not scheduled on a whim, as the tasks have to be critical for continued station operation and the risk has to be low before a space walk is authorized.

Expedition 6 crewmembers Ken Bowersox and Don Pettit conducted two staged EVAs. One was conducted in January, and the second one was conducted in April, after *Columbia*'s loss. The first one primarily involved removal of launch locks from a heat radiator on the main truss so that it could be deployed properly and removal of debris from a PMA on Unity. The second EVA involved reconfiguring electrical cables on the truss and fitting backup power cables to one of the gyros. They also replaced a power-control module and fitted a light fixture on to one of the truss's Crew Environment Translation Aid carts. These carts act in a manner similar to a railcar and would allow the SSRMS or other equipment to be moved to the ends of the truss just short of the main solar arrays once the truss was fully completed.

Don Pettit and *Saturday Morning Science*

Nikolai Budarin was a typical Russian civilian cosmonaut. He spent two years in the Soviet Army in the 1970s before getting his degrees in aircraft and electrical engineering and joining NPO Energia in various postings. Prior to Expedition 6 he had also been part of two expeditions to *Mir* late in its lifespan, and his engineering background came in rather handy for its maintenance.

Ken Bowersox, on the other hand, was the stereotypical American astronaut of the shuttle program. He was an Eagle Scout, a graduate of the U.S.

Naval Academy with a degree in aerospace engineering, a naval aviator, and a graduate of the U.S. Air Force Test Pilot School at Edwards before joining the astronaut corps in 1987 as a shuttle pilot. He was also a navy captain. He flew two shuttle missions as a pilot and two more as a mission commander.

Compared to those two men, Donald Pettit was a unique individual. Selected as an astronaut in 1996, Pettit had been assigned as a backup for Expedition 6, but he was moved up to prime crewmember when original Expedition 6 science officer Don Thomas was medically disqualified after an issue cropped up during a preflight physical. Pettit was a smart individual with kind of a thin build. Like Bowersox, he was also an Eagle Scout, and he had a love of the outdoors, having grown up in Silverton, Oregon. He had a degree in chemical engineering and a doctoral degree from the University of Arizona before working for the Los Alamos National Laboratory. He had the smarts, but he was still a rookie astronaut on his first assignment.

Don Pettit also had a unique curiosity for how things work. Back on Earth, he set up a laboratory in his garage to do scientific experiments on his days off. A self-described "übernerd," as it were, Don Pettit is almost like a cross between Don Herbert's classic television persona Mr. Wizard and fictional television scientist MacGyver, with some Bill Nye the Science Guy thrown in for good measure. While he didn't have the looks of Richard Dean Anderson's character MacGyver, Pettit was just as creative. In one case, Pettit took three of the crew's personal CD players and used them for an experiment in gyroscopic properties. One CD player with a CD spinning in it just tumbles. Two taped together at ninety-degree angles from one another are more stable, but the assembly still tumbles in one direction. Tape a third CD player to the others at another ninety-degree offset, and the resulting creation is rock-solid stable. Pettit used his new creation as a flashlight holder while doing maintenance inside access panels in the Destiny laboratory.

Pettit was also an avid photographer. Most members of the astronaut office were well trained in photography, since the skill is practically a requirement for many missions, but Pettit tried to push the envelope of what was possible with Earth photography aboard the ISS. One of his activities in orbit was to take pictures of cities on Earth at night. But even with the best camera settings, it can be very tough to get a good exposure, due to how fast the planet rotates underneath the station. A slow exposure with a handheld camera tends to produce a blurry image with streaks.

To compensate for these problems, Pettit used a leftover piece of equipment, an IMAX camera mount. IMAX cameras were flown up during two ISS missions on associated shuttle flights to film footage for the IMAX film *Space Station*, narrated by Tom Cruise. The IMAX cameras themselves were not on board for Expedition 6, since the film can't be stored in orbit for a long time as radiation exposure eventually clouds it. But the rig used to mount a camera to one of the view ports on the ISS for filming outside was still on board.

Pettit modified the rig to support a digital still camera with a telephoto lens. To help compensate for the station's movement, he modified a pivot mechanism on the mount and used a cordless screwdriver with a jackscrew to act as a movement tracker. When a city came into view, Pettit would use the screwdriver to track the camera on the target and compensate for the station's motion while it took a long photo exposure of a city at night. The results were magnificent, as bright details from the city lights could be seen clearly.

As fun as the night photography was for Pettit, one of his proudest accomplishments was a series of videos he filmed on Expedition 6 that he called *Saturday Morning Science*. Rather than packing his personal flight kit mostly with trinkets and collectibles to fly for family members, Pettit packed it instead with some simple tools, such as small picks, syringes, tubing, stiff wire, and the like. In combination with other items found on the ISS, Pettit used the items he brought along to conduct simple scientific experiments on his days off. Using the wire to make hoops or various diameters that he could steadily brace in front of a camera lens, Pettit would fill these hoops with water droplets of various sizes. The water's surface tension would keep the droplets from floating free, and Pettit could perform all sorts of experiments, such as injecting them with food coloring, putting air bubbles into them, or seeing how droplets just behave in zero gravity all by themselves.

A very interesting experiment, which Pettit called "Bubble War," was the result of sticking an Alka-Seltzer tablet in a water droplet. The tablet fizzed and produced dozens of tiny carbon dioxide bubbles inside the droplet. Eventually, the small bubbles would touch one another, rupture, and the gas pockets would join to produce larger bubbles. The process would continue until all that was left was one large bubble at the center of the water droplet. With a syringe, Pettit could pull that bubble completely out, leaving just the droplet once again. After his flight, Pettit used these videos as part of his public-speaking engagements. Pettit wouldn't tell the audience what was going to happen in the videos. It was up to the viewers to hypothesize the result before he would let the video run and show the results.

Pettit also found that the surface tension of water could be used to anchor a can of food to a table during mealtime in the Zvezda module. It's similar to picking up a glass of ice water from a smooth coaster on a table. Sometimes the surface tension of the condensed water keeps the coaster stuck to the glass. In orbit Pettit would take a can and moisten the back of it with a drop of water, and it would stick to a nice flat surface, which would keep the can from drifting away. For Pettit, eating in zero gravity became an art form as he could coax food out of a can with only a set of chopsticks, using surface tension to keep the food on the sticks until it got to his mouth.

These experiments, performed with just simple tools and a curious mind, make a great example for why humans should be flying in space. When Don Pettit conceived the ideas for these experiments, he didn't know what the results were going to be, but he performed the first few and used the results to help come up with a set of ideas for what to try next. A person can do that much easier than a machine and can do it potentially faster than a space probe that might require days of reprogramming on the ground and transmission of new instructions.

Handover

On 28 April 2003 a visiting spacecraft finally arrived at the ISS in the form of *Soyuz TMA-2*. On board was the two person Expedition 7 crew of cosmonaut Yuri Malenchenko and NASA astronaut Ed Lu. With their arrival, Expedition 6 would be free to return home on *Soyuz TMA-1*. Expedition 7 was to have been a three-person crew, but it was reduced to two in order to stretch out the station's supplies and to allow for support only from Soyuz and Progress craft. Permanent occupancy on the ISS would continue while the shuttle was grounded, but the scientific workload would be cut back, as Expedition 7 would only conduct fifteen experiments from the Destiny laboratory during their occupancy period.

While Expedition 6 had trained on the Soyuz systems as part of their pre-mission exams, their original ride home was supposed to be the shuttle. However, every resident ISS crewmember is fitted for a seat liner and a Sokol pressure suit. Even if a shuttle is used for the trip up, the seat liners and Sokol suits are carried since the Soyuz acts as the lifeboat. These items are transferred into the Soyuz once the shuttle docks, and ISS crewmembers run a final check of their Sokol suits before the shuttle leaves to make sure everything will work. The fitting of the suits and liners takes place in Russia at the Zvezda facility, which has been making pressure suits going back to before the days of Yuri Gagarin. The suit liners are custom-fitted to each crewmember as a person sits in a seat mock-up and a fast-drying plaster is poured around them to get a casting of their backside. Then the crewmember being fitted gets out of the seat, and technicians sculpt the mold to remove pressure points. Once a mold is taken and a liner is made from it, the crewmember then dons a Sokol suit and sits in the liner for additional fit checks.

The need for fit checking and snug suit liners would become of great importance on *Soyuz TMA-1*'s return to Earth. *TMA* stood for "Transport Modified Anthropometric," in Russian. It featured a few minor changes from the older TM series Soyuz. The internal size of the TMA's descent module was expanded slightly to allow taller crewmembers to fit inside, so astronauts like Scott Parazynski who are slightly too tall to fit into a Soyuz TM can potentially fly on a Soyuz TMA spacecraft. Improvements were also made to the TMA's Argon computer system and displays. It was unusual, compared to previous practices, to send a new design of spacecraft, even if it was a modification, into space manned on its very first flight, but Energia and Roscosmos were still cash-strapped and couldn't afford a dedicated test flight. The systems alterations were very minor compared to previous Soyuz upgrades anyway.

The *TMA-1* had docked properly months before, and all seemed well when it undocked from the station on 3 May 2003. In addition to the crew, some experiment results were being brought home, and Pettit had several digital videocassettes packed in a bag on his lap. Many of the cassettes contained his *Saturday Morning Science* footage. While the ISS had the ability to downlink high-resolution photographs to the ground, high-resolution video still took up too much memory on available computers back in 2003. Don Pettit had to haul his own videos home.

Reentry began as normal with a controlled descent, but a fault developed with the computer system, causing it to switch to a backup reentry mode. Instead of coming home at a relatively gentle 3 to 5 g's, *Soyuz TMA-1* experienced a ballistic descent of over 15 g's. One might think that compared to six months of zero gravity, this crushing type of reentry would be unbearable. But according to Don Pettit, the custom suit liner and seated positions used for this type of reentry made it at least tolerable, and the crew didn't black out. They landed safely, albeit three hundred miles short of their planned landing point.

Upon landing, the crew were on their own for over an hour, as there were no recovery forces in their area and they didn't land near any communities. Because one of the antennae was ripped off in the descent and two others did not deploy, they had no communication with recovery forces until Budarin made use of an emergency satellite phone. Getting out of the craft required a lot of effort; even with regular exercise to prepare them for gravity, their bodies weren't used to 1 g. But eventually, the crew were recovered safely and returned home with no injuries. Of the three crew-members of Expedition 6, Budarin and Bowersox wouldn't fly another mission, but Don Pettit would fly again a couple of more times.

Expedition 7

During the long months that Lu and Malenchenko were in orbit, a couple of space firsts were achieved. Yuri Malenchenko became the first person to get married in space. He married his girlfriend over the radio while Ed Lu played wedding music on an electronic keyboard. Both men had flown in space together and conducted an ISS assembly space walk on STS-106 in 2000. Being of Chinese American descent, Ed Lu was in a unique position to congratulate China on being the third country to launch a citizen into orbit on a rocket and spacecraft of their own when Chinese taikonaut Yang Liwei rocketed into space aboard *Shenzhou 5* (a Chinese term meaning "Magic Vessel"). The flight took place on 15 October 2003 and lasted twenty-one hours before Liwei returned safely.

Of the cosmonauts who have been part of the ISS program, Yuri Malenchenko is a hard worker and a very experienced cosmonaut, having taken part in a space walk using American equipment on a previous shuttle flight. Malenchenko's piloting skills were successfully tested during a *Mir* stay as he used the TORU to successfully dock a Progress when the automatic system failed. He flew fighters for many years as a pilot in the Soviet Air Force before becoming a cosmonaut in 1987. In official photographs taken of Malenchenko, he never smiles on camera, primarily due to his Russian cultural upbringing, but many who have worked with him say that he is known to crack big sincere smiles when the cameras aren't rolling.

47. The ISS as it looked from late 2002 to 2006. Visible are the main truss, the P6 solar array temporarily attached at Z1, and the Destiny Laboratory module. Courtesy NASA.

Expedition 8

On 18 October 2003, *Soyuz TMA-3* lifted off from Baikonur with three crewmembers on board. *Mir* veterans Michael Foale and Aleksandr "Sasha" Kaleri would crew the ISS as part of Expedition 8, with Foale in command. Joining them was Spaniard Pedro Duque, an ESA astronaut and veteran of STS-95. Duque would use his brief time in orbit during the crew handover to conduct experiments using ESA equipment in the station's Destiny laboratory as part of a Spanish ESA program called Mission Cervantes in honor of the sixteenth-century Spanish writer Miguel de Cervantes. During his eight days in orbit, Duque devoted about forty hours of his time to conducting his experiments.

In his diary, Duque helped to debunk the myth of the space pen and pencil. A popular Internet story stated that NASA had spent plenty of money to engineer a special pen to work in the zero gravity of outer space, while cosmonauts just used a pencil instead. Duque found out during training that the Russians had always used standard ballpoint pens on their missions, even though many of his colleagues said a pen would not write properly in orbit. So to help prove that a special pen was not required, he took a common ballpoint pen into orbit to write in his diary, and it wrote just fine with no problems.

Duque returned home with Lu and Malenchenko aboard *Soyuz TMA-2*. The descent module performed a controlled reentry with no repeat of *TMA-1*'s problems. Foale and Kaleri were left to continue the caretaker role aboard the ISS. For both men, this mission was much less eventful than their previous visits to *Mir*. Both men conducted a space walk on 26 February 2004 from the Pirs airlock, using Orlan suits to do some maintenance work on the Russian side before a malfunction in Kaleri's suit forced the EVA to be cut short. It was the first time a space walk had been conducted on the ISS using only two people with no one inside to mind the store while they were away.

The Return of Shuttle

During the next three ISS expeditions, things went rather well in orbit. Expedition 9 crewmembers Gennady Padalka and Michael Fincke took over from Expedition 8 in April 2004. Joining them for the handover was Dutch ESA astronaut André Kuipers for a week-long set of experiments. Foale, Kaleri, and Kuipers returned home aboard *Soyuz TMA-3*. Salizhan Sharipov and Leroy Chiao then took over for Padalka and Fincke, on Expedition 10 from 16 October 2004 to 24 April 2005. Joining Expedition 10 for the trip up was military cosmonaut Yuri Shargin, who apparently performed some classified tasks while in orbit. The trip home for Padalka, Fincke, and Shargin was a little eventful as they had to perform a manual undocking due to a faulty battery on the Soyuz. But they returned home safely.

Sharipov and Chiao conducted two space walks from the Pirs compartment with Orlan suits during their stay, in order to prepare the station for the arrival of the first ESA Automated Transfer Vehicle (ATV). The ATV was designed to be a heavy-lift vehicle, capable of hauling up cargo that was too big for Progress when a shuttle wasn't available. A total of six ATVs were originally planned for flights to the ISS, with the launches taking place from the ESA facility in French Guiana and using the Ariane 5 rocket. While the first ATV flight was planned for 2006, one wouldn't finally visit the station until 2008. For docking, the ATV would use the aft Soyuz port on the station's Zvezda service module.

Expedition 11 aboard *Soyuz TMA-6* came next. Sergei Krikalev was in command, with NASA astronaut John Phillips as flight engineer. Italian ESA astronaut Roberto Vittori occupied the third seat of the Soyuz and came home with the Expedition 10 crew of Sharipov and Chiao aboard *TMA-5*. The reentry of the Soyuz was perfect, and the craft touched down right next to Russian recovery forces.

The Expedition 11 crew would be on board when the shuttle program finally got off the ground again. The shuttle *Discovery* would conduct a newly revised STS-114 mission. In addition to its ISS resupply duties and EVA assembly tasks, the crew would also test out several new shuttle repair techniques using different tile-filler materials and a patch material designed to fill small holes in the RCC panels.

The shuttle would also carry a new device into orbit, the Orbiter Boom Sensor System (OBSS). The OBSS was built using a fifty-foot segment of an unused shuttle RMS, and it was fitted with special cameras and a laser range imager at one end, a grapple fixture at the other, and a second grapple fixture midway along its length. Using the OBSS, the shuttle's RMS could inspect the shuttle's RCC and tile panels for damage in spots that couldn't normally be reached by just the RMS. The cameras and sensors built into the OBSS could scan the tiles and generate a three-dimensional picture if damage was found, to determine if a dent in a tile was deep enough to cause concern.

Additional cameras were fitted to the shuttle's SRBs and ET, and these would film the shuttle's ride into orbit. The SRB camera footage wouldn't be available until the boosters were recovered after splashdown, but the ET camera would deliver real-time video footage back to mission control until just after the shuttle separated from it. Still cameras in the shuttle's fuel line attachment wells would take pictures of the tank during jettison, and astronauts on board would take additional photographs with handheld cameras.

As a final inspection procedure, before beginning its final approach to the station from a few hundred meters away, the shuttle would conduct a rendezvous pitch maneuver, a type of slow backflip. When the bottom was visible from the ISS, crewmembers aboard the station would shoot pictures with their onboard digital cameras, using 400 mm and 800 mm telephoto lenses. The images would be transmitted back to Houston for detailed analysis to clear the shuttle's heat shield for return. All this preparation was designed to see if a debris strike caused any damage to the orbiter.

On 26 July 2005 all was finally ready. The mission was commanded by Eileen Collins on her fourth spaceflight. She was the first female shuttle commander of the program, and this was her second mission in this role. Jim Kelly would serve as pilot. Space walks on the mission would be conducted by Stephen Robinson and JAXA (Japanese Aerospace Exploration Agency) astronaut Soichi Noguchi. *Mir* veteran Andy Thomas as well as Wendy Lawrence and rookie Charlie Camarda would help with EVA operations from inside the shuttle and ISS, while also helping with the transfer of station supplies.

The space shuttle *Discovery* lifted off from pad 39B at exactly 10:39 EDT. On the way up, it scored roadkill as the nose of the ET managed to hit a turkey buzzard on ascent. The bird did no damage to the shuttle as it was a relatively slow-speed collision, but the unconscious bird was vaporized by the shuttle's exhaust plumes as it fell to Earth. The rest of the climb into orbit seemed uneventful as the new cameras did their job. Inspections the next day revealed no damage to the critical areas of the shuttle's heat shield, but analysis of the ET camera footage revealed that a large chunk of foam broke off from the tank about 127 seconds after liftoff. While the foam chunk didn't strike the orbiter, it was big enough to potentially do serious damage. NASA administrator Michael Griffin announced that future shuttle missions would be delayed until the problem of foam loss was fully corrected.

Discovery arrived at the ISS two days after launch and performed the first rendezvous pitch maneuver of the program. The photographs shot by Krikalev and Phillips during the backflip were breathtaking and proved to be quite a highlight for space enthusiasts and the general public. After docking and hatch opening, the combined crews got to work, performing the ISS resupply tasks with the Raffaello MPLM, including delivery of fresh water supplied by the shuttle's fuel cells. Robinson and Noguchi got to work on their repair technique tests during their first two EVAs, while NASA analyzed the photographs of *Discovery*'s bottom. On the second space walk, the pair replaced a failed control gyro on the station.

A couple of days after analyzing the photographs, NASA announced that the third space walk would include an attempt by Robinson to remove a pair of protruding gap fillers from *Discovery*'s tiles on the underside of the shuttle's nose. Gap filler is a thin, stiff felt material that is fitted between the shuttle's tiles. Each tile has a small gap between them to account for thermal expansion and contraction as the shuttle's aluminum structure heats and cools. The fillers keep the tiles from touching one another and chafing their edges. Two of these pieces of gap filler had loosened slightly during the flight, and there was concern that they might disrupt the plasma flow around the shuttle's nose on reentry, potentially causing hot spots on the tiles.

Engineers and astronauts on the ground spent two more days trying different techniques to either remove the gap filler or cut them flush and then instructing the spacewalkers on the techniques. The third space walk took place on 3 August 2005. Before trying the repair, the two spacewalkers installed an amateur radio satellite called PCSat2 on the outside of the station, in addition to some exposure cassettes. For the gap filler removal, the OBSS was fitted with a foot restraint, and the RMS maneuvered Robinson in the restraint to the bottom of *Discovery*'s nose. Robinson would use his gloved hands in the first attempt to remove the fillers. It turned out that using his hands worked just fine, as both pieces of gap filler popped free with only a tug. Nothing else was needed.

Given that it was the first and what would ultimately be the only time an astronaut got to go underneath a shuttle on a space walk, it was a rather tense time for engineers. But the RMS performed well, even with a relatively heavy mass (Robinson and the OBSS) on the end of it. The OBSS turned out to be a stable work platform. Three days later, *Discovery* left the ISS and ultimately returned home on 9 August. The mission was hailed as a success. But nobody knew exactly when the shuttle would be cleared for flight again.

Expedition 11 finished out its tasks and eventually returned to Earth on 11 October 2005. Deorbit and reentry were normal. During this mission, Sergei Krikalev added to his time from previous flights to set a record for total time in space of 803 days, nine hours, and thirty-nine minutes, surpassing the previous record of 747 days set by Sergei Avdeyev.

Expeditions 12 and 13 would fly during the next year. Astronaut William McArthur and cosmonaut Valery Tokarev occupied the ISS from October 2005 to April 2006. Pavel Vinogradov and Jeff Williams took over for them starting in April. They were joined by Brazil's first astronaut Marcos Pontes, who returned home with McArthur and Tokarev after the crew handover. Pontes was flying thanks to Brazil's ISS hardware contribution.

While engineers modified the ET design, the next major shuttle delay came from mother nature as Hurricane Katrina clobbered the Gulf Coast of the United States in late August. ETs were manufactured at the Michoud assembly facility near New Orleans, and several buildings at the site and one ET itself were damaged by the winds. Thankfully, Michoud was spared damage from flooding. The levees protecting New Orleans failed, causing parts of the city to flood. Thanks to the repairs being completed faster than anticipated, Michoud was back in business by mid-October. It was a credit to the workers at the factory, as several of them were rendered homeless by the storm. It took many months (and in some cases years) for communities in the rest of New Orleans to get back to some semblance normalcy.

STS-121

By 30 June 2006 all was ready as *Discovery* sat on pad 39B for a launch date of 1 July for the second of two "Return to Flight" missions. The six-person crew was made up of shuttle commander Steve Lindsey; pilot Mark Kelly; and mission specialists Mike Fossum, Lisa Nowak, Stephanie Wilson, and Piers Sellers. Joining them for the ride into orbit was German ESA astronaut Thomas Reiter, who would be joining Expedition 13 once *Discovery* reached the ISS.

There were two launch scrubs due to weather on 1 and 2 July. A cracked piece of foam that broke off of one of the ET struts threatened to postpone STS-121's third launch attempt, but NASA managers decided to go ahead since the foam piece was not large and they didn't find any other problems with the tank. *Discovery* was ready to go on 4 July. It would be the Fourth of July fireworks show to remember, as thousands flocked to the Florida coast near KSC to watch the display.

At 14:37 EDT, the space shuttle *Discovery* rocketed into a clear, blue Florida sky. The weather was perfect, and all who watched were treated to quite a show. While some tiny pieces of foam were observed coming off the tank, they were too small to do any damage. The new cameras didn't detect any debris strikes, and OBSS inspections of the tiles revealed no problems with the shuttle's heat shield that required additional inspection or repair work. With the foam issues having been dealt with satisfactorily, normal shuttle flight operations could now resume.

Discovery approached the ISS, and after the now-standard rendezvous pitch maneuver, it docked safely. Upon the ship's docking, the Expedition 13 crew rang the station's bell two times, and Jeff Williams announced with the now-standard greeting, "*Discovery* arriving." That began the start of nine days of docked operations, as transfers were made from the Leonardo MPLM to the ISS from a CBM on the Unity node. Michael Fossum and Piers Sellers conducted three space walks on this flight, with their tasks split between doing some testing work with the OBSS, testing additional shuttle tile repair techniques, and changing out some minor equipment on the U.S. side of the ISS. At the end of nine days, Leonardo was placed back inside *Discovery*'s cargo bay, and the shuttle returned home with no problems. With these two shuttle test missions out of the way, the path was now clear for ISS assembly to resume.

13. Construction Resumes

With the flight of STS-121 and the success of the thermal protection system inspection and repair techniques tried out on two shuttle flights, ISS construction missions resumed in late 2006. When plans for retirement of the shuttle were announced, NASA indicated that they would fly no more than twenty missions, with nineteen of them being dedicated to ISS assembly and resupply. With STS-114 and STS-121 completed, that left seventeen more missions for the ISS. The remaining flight would be a repair mission to the Hubble Telescope.

STS-115 came next as *Atlantis* delivered the P3 and P4 solar array truss segments when it docked on 11 September 2006. This was one of the heaviest payloads flown during the shuttle program, weighing about sixteen metric tons (one ton being equivalent to one thousand kilograms, as opposed to an English-measured ton, which weighs slightly less at two thousand pounds). The crew was reduced from seven to six crewmembers to save weight in onboard consumables. During the mission, Canadian Space Agency astronaut Steve MacLean became the first Canadian citizen to operate the SSRMS when the new segments were transferred from the shuttle's cargo bay to their location on the station's truss. The solar array delivery and attachment phases went well with no problems. MacLean also became the second Canadian to perform a space walk, which he did with NASA astronaut Heidi Stefanyshyn-Piper.

Heidi Stefanyshyn-Piper, being one of the few women in the astronaut corps to train for EVAS, was very qualified for her work, as she was a salvage diver in the U.S. Navy before joining NASA. Her name is a bit long and cumbersome since she kept her maiden name, Stefanyshyn, to honor her father's Ukrainian heritage. Heidi was born in the United States in St. Paul, Minnesota, but she is a fluent Ukrainian speaker thanks to her family's upbringing. Heidi Stefanyshyn-Piper became only the eighth woman to conduct a space walk.

STS-116 came next in December 2006. As part of that flight, *Discovery* delivered the P5 truss segment, which was then bolted outboard the P3 and P4 segments with the SSRMS. The arm was used exclusively for the attachment task since there were no plans for any astronauts to conduct an EVA that far away from the station's airlock. The P5 segment would serve as the attachment point for the P6 solar array truss when it finally got relocated from its central Unity Z1 location on a later mission.

The space walks conducted this time in addition to procedures inside were primarily to reconfigure the station's electrical system from the Z1 to the main truss, so half the P6 solar panels were unhooked from the power grid and the P4 arrays became the main power supply. Controllers then commanded the port-side arrays on the center-mounted P6 unit to retract in order to give room for the P4 arrays to track the sun properly.

There were some problems during retraction, though, as the arrays kept getting snagged and kinked during the operation. They were only intended to be unfurled in their temporary location for a few months to a year at most, but over three years had passed since they were erected. The array could not remain half-erected, as trying to move unsecured panels with an RMS later could cause all sorts of problems. They needed to be completely folded.

It took a space walk with astronauts Bob Curbeam and Christer Fuglesang using some special tools to help "fluff" the arrays like a set of blinds during retraction to get them to fold in properly. The balky panels finally did so, giving the P4 array a full path of travel. The starboard P6 panels, still generating some power, would be retracted on a later mission. Christer Fuglesang, an ESA representative, was Sweden's first astronaut, and he conducted three of the four space walks on STS-116 with EVA leader Bob Curbeam, a veteran of STS-98's Destiny laboratory delivery.

On board, Sunita Williams traded places with Thomas Reiter to join Expedition 14 crewmembers Michael López-Alegría and Mikhail Tyurin, after she performed one of the STS-116 space walks. Expedition 13 crewmembers Pavel Vinogradov and Jeff Williams had returned home in September aboard *Soyuz TMA-8* along with spaceflight participant Anousheh Ansari, who had arrived with the Expedition 14 crew aboard *Soyuz TMA-9*. This type of staggered crew arrangement, with a single astronaut launching and landing on shuttle missions overlapping two ISS Expeditions, would be the norm for the next couple of years. When López-Alegría and Tyurin returned to Earth, Sunita Williams would join Expedition 15's Russian crew of Oleg Kotov and Fyodor Yurchikhin when they arrived in April 2007.

Sunita Williams's time on orbit was a pleasant one as she bonded rather well with her Expedition 14 and 15 crewmates. Her ethnic background is half-Indian on her father's side and half-Slovenian on her mother's side. She became only the second astronaut of Indian descent after Kalpana Chawla to join NASA and fly in space. Prior to joining NASA, Williams was a naval helicopter aviator. She flew missions in support of Desert Storm and relief to communities in Florida in the aftermath of Hurricane Andrew in 1992.

During her time on the ISS, Williams conducted three space walks with Michael López-Alegría—or Mike L. A., as he is known in Houston. López-Alegría also conducted one Orlan-suited space walk with Mikhail Tyurin prior to STS-116's arrival. During this period of ISS operations, construction tasks, more than scientific activities, were the order of business, and the staged EVAs got a bit more ambitious as the ISS grew in size. Among the tasks completed were disconnecting coolant lines from an Early Ammonia Servicer unit that would be jettisoned on a later flight and relocating power connection lines on the trusses. Cables for the ISS's Station to Shuttle Power Transfer System were also hooked up.

On all previous flights to that point, the shuttle remained powered up in order to supply the ISS with added energy for its own systems. Thanks to the Station to Shuttle Power Transfer System, starting with STS-117, the ISS would be generating a surplus of power, and the shuttle's systems could be shut down to extend their mission life. By doing this, a shuttle could remain docked 50 percent longer on future missions in order to conduct more extensive assembly and resupply tasks.

Extravehicular Activity, American Style

While the Russian Orlan was a very capable suit design, the experience NASA gained on over three decades of EVAs made the EMU a space-going equivalent of a swiss army knife with a plethora of tools at its disposal. During the shuttle program, when a newcomer was accepted into the ranks of the NASA mission specialists, they could opt to focus primarily on robotic arm operations with the RMS or train for EVAs. Not every astronaut necessarily has the

48. The ISS EMU contains everything an on-orbit handyman needs. Courtesy NASA.

build to fit in the suit, but the number of different sizes available for lower torsos, arms, and gloves allow the EMU to fit a large variety of body types.

Even then, many consider one of the physical qualities that best helps an astronaut with EVAs is having large arms, or a "gorilla build." Astronauts such as Dave "Ox" Van Hoften, Dave Wolf, and Scott Parazynski seem better-sized for the suit than some of their colleagues. Since Russian cosmonauts are typically required to perform EVAs in the Orlan, most also have a physical build well suited for the EMU.

Getting into a suit requires a lot of preparation, starting with a prebreathe phase to flush nitrogen from an astronaut's blood supply. During the shuttle days, astronauts had to prebreathe pure oxygen for about two hours before going outside. During the early ISS assembly program, to help cut down prebreathe time, astronauts would do vigorous exercise on the station's velo-ergometer while breathing pure oxygen to flush nitrogen at a faster rate. In both cases, if these tasks weren't done, nitrogen bubbles could have formed in human tissues due to the reduced pressure of the suit, causing the bends or worse. However, the exercise routine also could be exhausting to crewmembers performing EVAs potentially lasting up to eight hours at a stretch.

With the Quest airlock in place, astronauts could now camp out in a pressurized environment sealed from the rest of the station the night before and flush out the nitrogen in their bodies by breathing pure oxygen at a slightly reduced pressure. The next morning, they would be ready for their EVA tasks and could suit up as normal. Typically one or two other crewmembers would be on hand to help suit up the astronauts going outside.

Donning the suits starts with the waste-management devices. For the males, it is a urine-collection device. It attaches to a male astronaut's anatomical region with a condom; when he urinates, the fluid is collected in a bag thanks to a one-way valve. Up to 950 cc of urine can be held in the bag. Female spacewalkers wear the Disposable Absorption Containment Trunk (DACT). It looks like a set of tight-fitting boxer shorts and acts like a diaper that can hold up to 900 cc of moisture. The Disposable Absorption Containment Trunk is disposed of after use, unlike the urine-collection device, which has reusable elements. Since women spacewalkers are few in number, the Disposable Absorption Containment Trunks work just fine, and plenty are kept aboard the ISS.

Donning continues with the liquid cooling and ventilation garment.

Sweat inside a suit can be dangerous, as it could fog the helmet visor if the humidity level gets too high. To prevent this, cooling water is circulated around the garment through tubes built into it. The Orlan uses a similar garment. Astronauts regulate the water temperature with controls on the chest pack of the EMU. Once the liquid cooling and ventilation garment has been put on, the upper torso is next, and an astronaut puts it on like a large stiff shirt by snuggling up through the waist. The pants and boots come next. Then come the familiar "Snoopy" communications headset (so named for its black-and-white color), the gloves, and finally the helmet.

Unlike the Apollo-era suits, the EMU's backpack containing the oxygen, batteries, cooling water, and carbon dioxide scrubber bed is integral to the torso, as is the chest pack with the controls. The EMU's power systems are recharged via a docking rack inside the Quest airlock. The astronauts typically are docked with the racks during cabin depressurization and repressurization to save the batteries for the EVA itself.

The EMU helmet contains a pair of flip-down visors for sun shielding and lights to illuminate the work area during night passes. For the ISS program, NASA also added a lipstick-sized television camera above the visor to provide support astronauts inside and controllers on the ground with a first-person view of what the wearer sees. Inside the helmet, near the astronaut's face, is a drink bag with a straw to help prevent dehydration. An optional food stick can also be fitted in the helmet, but very few astronauts use one, to prevent food particles from floating free.

Most EVA crewmembers typically have at least three pairs of EVA gloves on hand due to wear and tear caused by the work outside. For ISS crewmembers on long-duration missions, new gloves can be flown on cargo supply craft as needed. Because an astronaut needs a little bit of tactile sensation to do their work, the gloves are thinner than other parts of the suit. Even though steps were taken during design of the ISS to minimize sharp edges and corners that could possibly damage a glove, years in Earth orbit can cause surfaces to develop sharper edges and surface pits due to exposure to atomic oxygen and micrometeoroid debris impacts. While moving between work sites, a gloved hand might not feel a tug or a snag. So during an EVA, regular glove checks are performed. If a glove shows abnormal wear to a certain point, an EVA can be terminated early, regardless of how important the tasks are. During EVAs, crew safety comes first.

49. Visible in this photograph of the EMU is the SAFER unit. Courtesy NASA.

Strapped below the EMU backpack is the SAFER unit. Invented by a team at Lockheed Martin, which included *Skylab* astronaut Joe Kerwin, SAFER stands for "Simplified Air for EVA Rescue." It is a self-contained jet pack. While each spacewalking astronaut uses a tether to clip themselves to the station, there are times when they have to unhook while transiting from one place to another. If an astronaut should lose grip of the station and begin drifting away, the pack can be activated with a joystick controller, which can be unstowed very quickly from the pack's right side.

Unlike the Manned Maneuvering Unit, the SAFER unit is strictly for emergency use. To date, the only times it has ever been activated were during its test missions. A prototype version of SAFER was first flown on STS-64. A production version was tested on STS-86 by Scott Parazynski, but it failed to work when a pyrotechnic charge designed to open the path for fuel to the thruster ports didn't fire. The pack was a dud. Thankfully, the test was conducted with Parazynski's feet firmly attached to the shuttle's RMS, so there was no risk of him floating away. The bugs were finally worked out, and the pack was successfully flight-tested on STS-92.

The Orlan suits aboard the ISS are designed to use SAFER as well. Modifications were also done to the Orlan helmets to allow them to use the EMU helmet cameras. Both the EMU and the Orlan can use each others' airlocks, but this is only done in emergencies.

Once outside, the astronauts clip on a special tool "belt." Rather than being a belt worn around the waist, however, the large metal bracket to which the tools are attached connects at three points, one on the chest pack and two on the torso. The tools themselves are suspended from the belt in a position slightly below and in front of the chest pack, where they can be seen by the astronaut wearing them.

The tools at an EMU astronaut's disposal would make actor Tim Allen of the television show *Home Improvement* jealous. The Russians like to joke about the EMU by saying "too many tools," compared to what the Orlan has at its disposal. Having a lot of tools may seem like overkill, but it comes in handy.

The primary device is the pistol-grip tool. At a glance, it looks similar to a cordless drill or a NASCAR impact wrench, and it is powered by a rechargeable battery pack. Most of the ISS assembly tasks involved either removal of launch locks from modules or bolting items together. With this power tool, an astronaut can set how much torque to apply and how many revolutions to turn a screw.

The fastener attachments on the ISS were designed so that if a screw were loosened to remove a launch lock, the launch lock would stay attached in its bracket and not float away free. For jobs that require a more precise touch or the tightening of bolts in spots too cramped for the pistol-grip tool, astronauts can use a handheld ratchet. This tool is similar in design to a standard half-inch ratchet, but it is designed to work with gloved hands. Both the ratchet and the pistol-grip tool have rings on their ends so that they can be secured to the tool belt's tethers.

Another tool used from time to time is a caulking gun. The device was originally designed for application of tile filler material during repairs to the shuttle's thermal protection system, but it can be loaded with other materials as well, such as grease for bearings. To collect loose items in orbit, each astronaut was also equipped with one or more cloth tool bags with drawstrings. Since an item could potentially become debris that was free to float anywhere if it were let go of, mission procedures called for removed items like launch locks or material samples to be placed inside the bags.

EVA Training

Most astronauts who have joined NASA during the shuttle program have some EVA training since every shuttle mission included at least two qualified spacewalkers and a set of EMUs for emergency use. But for those astronauts who focus primarily in EVAs or are assigned to ISS missions, they received more-specialized training.

In the 1990s Scott Parazynski and engineers on the ground created the EVA Skills Development Program. This program taught spacewalkers everything from how to communicate during an EVA in a common language to the proper procedures for movement, tool usage, and handling certain types of equipment. The training was beneficial since everyone got used to working on the same page. It also helped to minimize the development of bad habits and to maximize productivity.

To help with hands-on EVA training, JSC has two main facilities at its disposal. The first is the Sonny Carter Training Facility, Neutral Buoyancy Laboratory (NBL), located near JSC. It was a replacement for NASA's smaller NBLS since a larger facility was needed to work with the massive assemblies planned for the ISS. It is named for astronaut Sonny Carter, who was tragically killed in a commercial airliner crash in 1991. In the NBL, astronauts wear similar equipment to what they use on orbit, and safety divers accompany them to their work sites. While maneuvering their tools, the trainees use lightweight mock-ups, which they hand off to the safety divers for the real (and heavier) pieces of equipment to do their tasks. While an astronaut wearing a properly weighted suit is neutrally buoyant, their bodies are not weightless in the suit, so the training runs can be uncomfortable sometimes, depending on their orientation. The shuttle EMU's arms also drape out to the sides in a position more natural for zero gravity, but the arm position can sometimes cause discomfort in neutral-buoyancy training. It isn't uncommon for astronauts to develop arm and shoulder pain during a long training run if they are on their sides or inverted. The training can also be brutal to the hands, given the thickness of the gloves.

Astronauts training for a specific space walk could spend two or three training sessions a week, each lasting many hours in the pool. The experience gained during these runs outweighs the physical discomfort, though. Training runs are monitored closely in a control center at the facility. Should an astronaut or a diver develop a problem, they can be pulled out quickly by the safety crews. The facility has a hyperbaric chamber on standby for emergency use and properly trained medical support personnel at its disposal.

The second training aid is JSC's virtual reality laboratory, where astronauts can conduct dry runs of their space walks while wearing normal clothes. While sitting in swivel chairs, they don a torso unit with an EMU chest pack mock-up and a set of special gloves and goggles that allow them to see a three-dimensional, computer-generated world. To an outside observer, training astronauts look like they are grasping for things in midair, but the astronauts in the virtual reality environment see something very close to what they would see on orbit. While the NBL helps to teach the mechanics and tasks of specific EVAs, the virtual reality lab acts as a way to help polish the mission procedures further without the time, personnel, and expense required for NBL training sessions. But even as good as virtual reality is, it likely will never completely replace NBL training.

Even with all the skills and training procedures, preparation doesn't end there, as the astronauts study the equipment they are assigned to work on. So if a problem develops with a system not powering up properly, an astronaut doing an EVA can help troubleshoot the problem more quickly since he or she is typically on-site or close by. This came in handy when the SSRMS was hooked up during an EVA on STS-100, as it didn't power up at all when the systems were tested. The spacewalkers outside knew that they had done their tasks properly, so they moved to an area that had been hooked up on a previous mission and tested the electrical connections there. Sure enough, after unhooking and rehooking a couple of the connectors, the problem was fixed, and the SSRMS powered up normally.

Hail Damage

Sunita Williams set many records during her time in orbit both as part of her official duties and during other activities. This included surpassing Shannon Lucid's spaceflight duration record from *Mir*. Williams also ran the 2007 Boston Marathon. She was listed as an official entrant; when the starter gun went off on Earth, Williams ran the 26.2-mile course, with the station's treadmill adjusting its resistance at the proper distance points. She completed the event in four hours and twenty-four minutes, with her crewmates acting as support with drink bottles and encouragement.

At KSC, STS-117 was scheduled to launch in mid-March, when a severe hail storm damaged the foam on the shuttle's ET. So *Atlantis* was rolled back into the vehicle assembly building, and engineers got to work patching the dents. This resulted in a three-month delay for the mission. Sunita Williams was scheduled to exchange places with Clay Anderson on STS-118. Keeping the original mission order would have run the risk of putting Williams too close to her cumulative radiation exposure limit. Exceeding the limit would have jeopardized her chances of doing another ISS mission later in her career, so Clay Anderson was moved up to the STS-117 crew.

Anderson was ready, as he had been involved in ISS training for several years. Originally, Anderson had trained with cosmonaut Aleksandr Lazutkin as part of the same crew. Unfortunately, Lazutkin suffered a heart attack and had to leave the cosmonaut program after having lived through the *Mir* fire, the Progress collision, and the other problems caused by *Mir*'s advanced age. Anderson was very sad to see his colleague retire before he got a chance to fly a second trip into space, one that promised to be less eventful than Lazutkin's previous space flight. The loss of Lazutkin also meant that Anderson had to be assigned to a different crew on a later mission assignment.

Language Training

With the ability to read and speak Russian being a requirement among astronauts assigned to ISS duties, NASA made some changes in how its employees learned the language since the missions to *Mir*. During the later stages of the Shuttle-Mir Program, NASA began setting up a language training center at JSC so that their people could learn what they needed to without having to attend classes elsewhere. Language studies in English for visiting Russian and Japanese personnel were also set up, since both countries were sending people to Houston for astronaut training and ISS support work.

One of the NASA people involved with development of the language center was Susan Anderson, who, at the time, was a member of the Astronaut Training Office. Her husband, Clay Anderson, worked as an engineer at JSC before his selection as an astronaut in 1998. Due to the potential conflict of interest with having a member of the training office married to an astronaut, Susan transferred to NASA's education department since she had been a high school teacher before joining the agency.

Susan Anderson explained, "For the person who is just traveling over there for a business meeting and [who] doesn't need to be in intensive negotiation, the [Defense Language Institute] method is probably fine because a lot of our folks are very intense and they can learn something very quickly. But for the level that [people] needed to perform at on a regular basis, for our upper management involved with the Shuttle-Mir Program, with [astronauts] living on board the *Mir*, with training over in Star City and so forth, it was quite necessary for us to have a more extensive program."

Anderson further explains how, for ISS astronauts, language training typically continued once they were in Russia: "When we would send our folks over to Russia for training, they were also providing Russian-language training. So we learned how to integrate the two programs, because we have Russian-language instructors in the United States and you have Russian-language instructors in Russia, they might teach a little bit differently. And so we developed an integrated relationship. They don't teach in the same methodology, but they keep one another informed of the progress of the students." This type of integration worked well, especially if a student might need some additional education in a specific aspect of the Russian language. While the system wasn't fully set up by the time

Expedition 1 flew to the ISS, the crews after Expedition 3 got the full benefit of the new program.

Never Give Up, Never Surrender

Clay Anderson's path to orbit was a study in perseverance. He had known from a young age that he wanted to be an astronaut, having seen *Apollo 8*'s mission to the moon as a child growing up in Ashland, Nebraska. His mother was a schoolteacher, and his father worked for the Nebraska Department of Roads as a highway engineer. Both parents taught Clay and his siblings to be well rounded in their education and experiences, as all three of the Anderson children studied well in school, learned to play musical instruments, and were involved in sports. Clay Anderson played basketball in college and became an NCAA Division 1 basketball referee for a time. He earned a bachelor of science degree in physics from Hastings College and a master's degree in aerospace engineering from Iowa State University. Prior to his master's degree, Anderson also was accepted for an internship working at JSC for two summers. Upon graduation Clay Anderson became an engineer at JSC, doing work in the Mission Planning and Analysis Division before moving on to assignments in the Mission Operations Directorate, which included being leader of the team that helped plan the launch trajectories for the Galileo space probe to Jupiter.

All during this period, Anderson regularly submitted updates to his astronaut application every two years and was rejected every time. One might figure that after four, five, or even six applications, one might give up and decide that it was not in the cards. Clay Anderson applied fifteen times before he was granted an interview in 1996. While he didn't get selected that year, he resubmitted in 1998, as he knew his chances were better having made it to the interview stage previously. It turns out his instincts were correct, as the Nebraska native joined the 1998 class of ASCANS (astronaut candidates).

Each astronaut class has a name. Some are mundane, while others are humorous. But all have a story. The tradition of naming astronaut classes typically has the previous class selecting the name of the incoming class in much the same way that established pilots in the military give a newcomer his nickname or call sign. For instance, the 1996 astronaut class was known as the Sardines since they were the biggest astronaut class to date.

With that many astronauts already in the program, it looked like flights for newcomers might be few and far between. So flightless birds became the theme of the 1998 class. The Sardines were going to name the newcomers the Dodos since the mascot was a flightless bird and an extinct one at that. But members of the 1998 class with foreknowledge of how the naming process went decided to mount a preemptive strike. The group voted to call themselves the Penguins; in order to drive home the point, the night before the 1998 class officially reported to JSC, members who were already JSC employees decorated the classroom with all sorts of penguin memorabilia before anyone from the 1996 class caught on. So the 1998 class officially became known as the Penguins.

Clay Anderson's assignment to the astronaut corps got him involved in ISS training after a few years of working on electrical system designs. This involved language training and regular visits to Russia, in addition to the work at JSC. Anderson was away from his Houston home for six months of each year during the three and a half years he spent training before his ISS flight, but both he and Susan Anderson managed to stay in touch the whole time through regular emails, phone calls, and other forms of contact. The pair also managed to raise two healthy and active kids in the process, a son named Cole and a daughter named Sutton.

Bumping up a crewmember from one shuttle mission to an earlier one had never been done before, and it required a bit of reshuffling both at home and at work. But the STS-117 support staff made the process as smooth as possible for a relatively last-minute switch. Once STS-117 launched, Anderson made himself useful by acting as an assistant to the flight crew, doing jobs in whatever capacity was needed, since even with assigned tasks, things can be rather busy when a shuttle reaches orbit.

One important change brought about by the crew switch involved scheduled EVAs for Anderson. On STS-118 his first two EVAs were to be conducted with Rick Mastracchio. After *Endeavour* left, he was scheduled to perform a staged EVA with Fyodor Yurchikhin to remove and release the Early Ammonia Servicer unit. With the rescheduled flights, Anderson and Yurchikhin would have to perform their staged EVA prior to STS-118's arrival. This would be the first time two EVA rookies, including one Russian, would be performing such a task with another Russian inside operating the arm completely solo.

This revised tasking caused some friction between the Expedition 15 crew and NASA management. Before Expedition 15 launched into orbit, a last-minute test was arranged for Anderson and Yurchikhin in the virtual reality lab at JSC while more than the typical number of managers and observers watched. The test was different from the one originally planned, but the two men did their job well and drove home the point that at least they knew what they were doing and could execute their tasks properly, even if NASA management might not have had the same confidence. No changes were made to the EVAS.

ISS Computer Failure

The space shuttle *Atlantis* was finally ready to fly on 8 June 2007 on mission STS-117. In its cargo bay was the S3 and S4 truss segment with the station's third set of solar arrays. The mission launched as scheduled at 19:38 EDT into a clear, blue Florida sky. *Atlantis* docked with the ISS forty-seven hours and fifty-eight minutes later. The transfer of the new set of solar arrays went like clockwork as the station and shuttle crews utilized the SSRMS to transfer the new truss segment to its location on the port side of the truss. With the new arrays unfurled, the ISS was now double its 2003 width and in a symmetrical configuration, meaning orbital reboosts could once again take place without center-of-mass concerns.

After the new arrays were hooked up to the power grid, a problem developed. On day seven of STS-117's mission, the three primary computers on the Russian side of the station unexpectedly powered down, knocking out the station's environmental control systems and orientation thrusters. Thankfully, the ISS didn't lose total control, as its control-moment gyros were still operating and the shuttle was available to stabilize it as necessary. The computers were successfully restarted a few hours later, but they triggered a false fire alarm that woke the crew during a sleep period.

Initially, Russian managers guessed that perhaps the new arrays had sent a power spike through the systems or that perhaps the mass of the ISS allowed it to build up an excessive static charge, but inspection of the computers by Kotov and Yurchikhin over the next few days, plus engineering analysis on the ground, revealed the real culprit. Wires in a power monitor box, a form of surge protection for the ISS, had corroded; thanks to a short circuit, it had sent a false power-down command to the computers. The problem was corrected when the faulty power box was bypassed. The corrosion was caused by excess condensation in the Zvezda module. Bypassing the faulty box and hooking directly to the power feed made the computers potentially susceptible to a possible power surge, the same as plugging in a home computer to a wall outlet instead of a power strip. A new monitor box was sent up on the next cargo vehicle. Life aboard the ISS returned to normal for the three Expedition 15 crewmembers after the STS-117 crew wrapped up their tasks and headed for home with Sunita Williams on board.

Three Musketeers in Orbit

Fyodor Yurchikhin, Oleg Kotov, and Clay Anderson got along well together during their time in orbit. It was rather lucky, given that Clay had spent more time training with Expedition 16 crewmembers Peggy Whitson and Yuri Malenchenko on the ground. But Anderson and Kotov had at least known one another as they conducted Soyuz survival training together. The Russian-language training for Anderson paid off, as he could speak pretty good Russian to his crewmates. Oleg Kotov could also speak very good English, and Fyodor Yurchikhin was not far behind. The crewmembers called themselves the "three *mushkatiery en orbitya*" ("three musketeers in orbit").

When the time came for Anderson and Yurchikhin to perform their space walk, they did it professionally with no problems. Anderson was particularly proud to wear the red stripes on his suit, meaning that he was designated the lead spacewalker on the EVA, even though Yurchikhin was the Expedition 15 commander. After completing other tasks, which included installation of a television camera stanchion and service work to an S-band antenna, the pair unbolted the refrigerator-sized Early Ammonia Servicer tank and a second piece of related equipment. With Kotov driving Anderson on the end of the SSRMS to a launching point, the astronaut let go of both assemblies in a path opposite of the station's direction of travel. These new satellites were dubbed Nebraska 1 and Nebraska 2, and each spent a few months in orbit before they finally reentered and burned up in the atmosphere.

STS-118 visited the station in August, carrying the S5 truss, the External Stowage Platform 3, and replacement control-moment gyros. Internal supplies were also carried inside the final Spacehab module to fly on a shuttle. This was the first flight of the space shuttle *Endeavour* since the *Columbia* accident, and it was also the first time the Station to Shuttle Power Transfer System was tried out, allowing the shuttle to power down its fuel cells in orbit for a while. Four EVAs were conducted on this flight, and Clay Anderson took part in two of them as part of his original STS-118 assignments. But the first EVA had to be cut short when Rick Mastracchio's glove showed a possible tear to its outer covering. The pair had completed their assigned tasks already and were doing some get-ahead tasks when the EVA was abruptly concluded.

Astronaut Barbara Morgan was part of the STS-118 crew. In 1985 she was selected as Christa McAuliffe's backup for the Teacher in Space program. After McAuliffe was lost on *Challenger*, Morgan returned to her teaching job in Idaho. In 1998 Morgan was selected to become a mission specialist astronaut, and she became one just prior to NASA's creation of the title educator-astronaut, for other teachers wishing to become astronauts. As part of STS-118, she took part in a classroom lesson for schools on Earth, which demonstrated the unique aspects of weightlessness and Isaac Newton's physics laws.

STS-118's mission was cut short by a day due to a threat to the Houston area by Hurricane Dean. The storm was in the Gulf of Mexico when Canadian Dave Williams and Clay Anderson were conducting the fourth EVA. The pair got a very good look at the massive hurricane as it passed below them. Upon seeing it, Anderson commented, "Holy smoke!" Dave Williams had a look and replied, "Man, that's impressive," to which Anderson responded, "They're only impressive when they're not coming toward you." The space walk was concluded two hours early with all the assigned tasks completed. The hatches between the two vehicles were closed that night.

Early the next day, *Endeavour* undocked and headed for home. After the two crews parted company, Anderson appeared overcome with emotion since he was good friends with the STS-118 crew. He was in regular contact with his family at home and knew about the evacuation preparations for the hurricane. NASA started preparations in case of a direct strike, and primary control of the ISS was transferred from JSC to the TSUP in Russia so that workers in Houston could assist with their own families' storm preparations. The precautions weren't needed, though, as Hurricane Dean ended up going south and making landfall on the Yucatán Peninsula instead of Texas. *Endeavour* returned to Earth safely, and ISS operations continued without interruptions.

Ballistic Reentry, Part 1

Expedition 15 gave way to Expedition 16 when *Soyuz TMA-11* docked with the station on 12 October 2007. This mission was a first, as astronaut Peggy

Whitson was assigned as the station's first female commander. To commemorate the event, Russian managers gave her a pair of traditional Kazakh riding whips as a sign of respect just prior to launch in order to honor her command role. For the launch, and later for the return, flight engineer Yuri Malenchenko was the commander of the Soyuz craft. Joining them on this mission was Malaysian spaceflight participant Sheikh Muszaphar Shukor, flying a short-duration mission to the ISS as Malaysia's first astronaut.

After eight days of joint operations, Fyodor Yurchikhin, Oleg Kotov, and Sheikh Muszaphar Shukor boarded *Soyuz TMA-10* to return home. Undocking and retrofire went well, but the reentry didn't go as planned, as *TMA-10* underwent a ballistic reentry instead of a controlled one and landed about 340 kilometers short. The crew landed safely, but concerns were raised about the Soyuz TMA design, as this was the second ballistic reentry to occur within a five-year period. A postflight investigation said that the problem was due to a faulty cable. The *TMA-11* spacecraft was given a clean bill of health by the Russians.

Peggy Whitson had come quite a long way since her days as a project scientist representing NASA during the Shuttle-Mir Program. She conducted many experiments as Expedition 5's science officer, and while construction was the order of the day for Expedition 16, her management skills came in handy. Her scientist husband continued research in the work she had begun when they both started working for JSC in the 1990s. Part of their research dealt with the use of a calcium citrate dietary supplement to help prevent the formation of kidney stones in astronauts due to the leaching of minerals from bones during a long stay in orbit.

Expedition 16 had a total of six astronaut members assigned to it. Clay Anderson's increment would only last a couple of weeks before his replacement, Dan Tani, would arrive aboard STS-120. French ESA astronaut Léopold Eyharts from STS-122 would then take over for another month before being replaced by STS-123's Garrett Reisman, whose duties would dovetail into Expedition 17 before his scheduled return home on STS-124.

Finding Yourself on NEEMO

Many ISS astronauts have taken part in a program called NEEMO, which stands for NASA Extreme Environment Mission Operations. The NEEMO missions take place aboard an underwater habitat known as *Aquarius*, which is located just off the Florida Keys. The lab is funded by the National Oceanic and Atmospheric Administration; scientists use it to study sea habitats in the region, since very long dives can be conducted from there.

Occupants of *Aquarius* stay on board and don't return to the surface until the end of their missions. By living in the habitat for several days, astronauts (or rather aquanauts as they are called during NEEMO) learn teamwork and skills to operate in hostile environments, since the ocean floor can be just as deadly as space. It also acts as a form of screening for possible ISS astronaut candidates by putting them in a stressful environment. Peggy Whitson commanded NEEMO 5 with Clay Anderson and Garrett Reisman as two of her crewmates. Dan Tani took part in NEEMO 2 with Sunita Williams.

Pambo

STS-120 lifted off on 23 October 2007. In *Discovery's* payload bay was ISS Node 2, a module known as Harmony. As with Unity, the new module had CBMs all around it, and its placement on the ISS would allow for the Columbus laboratory module from Europe and the Kibo laboratory from Japan to be attached at later dates. Harmony would add 20 percent to the station's internal volume. Originally, the ISS was supposed to have a dedicated habitation module to act as the living quarters for three crewmembers among its other uses, but budget cuts in 2001 forced a revision. New sleep stations and other amenities would be provided in Node 2 instead. The node's addition was necessary before the ISS crew could be increased from three to six crewmembers.

The mission would also move the P6 solar arrays from their central location to the end of the P5 truss, where they would once again be unfurled. The final set of S6 solar arrays were scheduled to arrive on STS-119 in a few months. With the final arrays installed, the ISS would be fully mission-capable from an electrical power standpoint.

In command of STS-120 was Pam Melroy. A veteran pilot of ISS assembly flights STS-92 and STS-112, Pam Melroy was the third female shuttle pilot selected by NASA and only the second female shuttle commander. Astronaut Susan Kilrain, a naval aviator, was the second female shuttle pilot; even though she had two missions under her belt, Kilrain retired from the astronaut corps in 2002.

Melroy was a U.S. Air Force veteran and flew KC-10 tankers after flight school. Her squadron flew in support of the F-111 bombing raids on Libya in 1986 as part of Operation El Dorado Canyon, but unfortunately Melroy was unable to participate since she was conducting air force survival, escape, and evasion training at the time. She would later fly support missions for the invasion of Panama in 1989 and the first Iraq war in 1991. Melroy attended the Air Force Test Pilots School at Edwards AFB and was involved in early testing of the C-17 cargo plane before her selection as an astronaut.

Around the astronaut office, Melroy was known as Pambo in reference to the Sylvester Stallone *Rambo* films. Gifted with an interesting sense of humor, Melroy was also sometimes known as Tank Girl, since one of her favorite movies was the 1995 postapocalyptic film of the same name. The main character of the film is a seemingly ditsy blonde girl who drives a tank. A copy of *Tank Girl* was placed in the ISS movie library by Melroy during one of her visits to the station as a house-warming gift of sorts.

In addition to the crew and the mission cargo, the space shuttle *Discovery* had a *Star Wars* lightsaber movie prop in a storage locker, flying as part of the mission's Official Flight Kit, which contained memorabilia items selected for a trip into space. NASA has pretty strict guidelines as to what can and can't be flown in Official Flight Kits and the astronauts' own Personal Flight Kits. The lightsaber generated a bit of media attention on the ground, and it gave Clay Anderson an idea for some wakeup music.

During Sunita Williams's stay, she had begun a tradition of transmitting a wakeup call from the ISS to the ground and selecting music to play for mission control. When Anderson arrived, he decided to continue that tradition, calling it "Radio Station K-ISS." He would periodically select music that was appropriate for that day in similar fashion to the wakeup music usually transmitted to shuttle flights. So the morning after *Discovery* lifted off, Anderson played the song "Whip It" by Devo in honor of the two female commanders Peggy Whitson and Pam Melroy, since one had been given whips and the other was packing a lightsaber. Peggy Whitson laughed of course, but Anderson joked that he was glad she didn't have her riding whips in orbit or she might have used them on him.

STS-120's *Apollo 13* Save

Docking day came as expected, and the combined crew got right to work. The Harmony node was berthed on a CBM on the port side of Unity for temporary stowage. After the shuttle returned home, PMA-2 on the front of Destiny would be relocated to the front of the node, and the pair would then be berthed on the front of Destiny for future shuttle dockings. Astronauts Scott Parazynski and Doug "Wheels" Wheelock performed a space walk to prepare the module's exterior and set up an S-band antenna on it, as well as set up an external stowage platform. The Harmony node was opened for the first time on 27 October, and the crew said hello to their new living quarters.

Parazynski and Tani conducted a second space walk to prepare the P6 for its relocation a couple of days later. They also checked out the starboard truss's Solar Alpha Rotary Joint (SARJ), as it had apparently been acting up for about a month with vibrations as it moved. The SARJ units allow the truss assemblies to rotate the arrays and track the sun. Each array also has a joint allowing side-to-side movement as well. Tani's inspection revealed some metal shavings, indicating that the SARJ's bearings were wearing excessively; he used a piece of tape to collect some samples for ground analysis. Controllers made plans for Parazynski and Wheelock to inspect the P4 and P5 trusses on the third space walk to see if there was a similar problem. The P6 array was then moved off the Z1 truss and parked on the SSRMS before movement to its final location.

During the next EVA, the P6 was finally relocated, and the proper electrical connections were made. The arrays were then unfurled again. The first pair opened up fine, but a big problem developed on the second one. Like a set of window blinds, the arrays also contain guide and support wires to make sure they unfurl properly. A guide wire on one of the arrays frayed and got caught on a support wire. The snag began tearing a hole in one of the solar panels and kinked up the hinges in others, producing an ugly-looking bulge in the delicate array. Things did not look good at all.

The movement was put to a stop while everyone analyzed the problem. The array could not be kept partially unfurled, and it couldn't be fully retracted again either. In its position, it was unstable as station reboost firings could cause the structure to oscillate wildly and perhaps snap from the stress. Engineers had to come up with a fix and do it fast. To complicate matters, the ISS had not been designed to support EVAs to the arrays. There were no handholds that far out and also no provisions for a space walker to climb up the arrays themselves.

Fortunately, the ISS had all the materials it needed to perform a repair.

Engineers came up with a procedure to cut away the snagged wire. Several "cuff links" were made with thick, nylon-coated wire wound together like a rope with a loop at each end. Inside these end loops were a set of flat blocks. When the array was unfurled, the cuff links would help guide the array to flatten in its proper position, giving a firm platform for the panel hinges to pivot against, at least in theory.

Don Pettit, on the ground, helped come up with the procedures for making the cuff links and a couple of tools that would be used for this repair. He helped shoot a video that was uplinked to the ISS along with written instructions. It was an interesting parallel to *Apollo 13*, when the crew of that mission had to assemble adapters for their carbon dioxide scrubber cartridges to work in the LM using available materials and instructions radioed up.

Both shuttle and ISS crews worked together as a combined team to get ready for the repair. There was a lot riding on this; everyone knew that if the repair wasn't successful, the arrays might have to be jettisoned. That would cut the power-generating capability of the ISS by a significant percentage and potentially cripple the station's mission. There was no backup set of arrays on the ground and no budget to build a replacement, let alone have a replacement available before the shuttle's planned retirement. For the two commanders, Whitson and Melroy, the waiting that night before the repair EVA was agonizing as they ran the possible scenarios in their minds concerning what might happen.

On 3 November the repair EVA was conducted. Scott Parazynski set up a foot restraint on the OBSS and strapped himself on it. To get to the work site, the SSRMS grasped the OBSS at its midpoint grapple fixture, and the astronaut positioned on its end would be moved to the damaged area of the array. Since the SSRMS had never been operated in quite this manner, Doug Wheelock acted as the arm operator's eyes at the base of the arrays while Parazynski directed the movements from his end.

The midpoint grapple fixture on the OBSS had to be used since the SSRMS wasn't set up to attach at the base of the boom like the shuttle RMS. It couldn't use the cameras and sensors on the OBSS either. The mission also had a time constraint, since the electronics in the OBSS could only remain unhooked from shuttle for about ninety minutes before its onboard batteries would run out of power, potentially causing the sensors to freeze from lack of heating. While no problems had been detected with *Discovery*'s heat shield, a final OBSS inspection is typically done after undocking to check for last minute problems, just in case. The OBSS sensors were still needed.

Scott Parazynski picks up the story:

This was an unprecedented work site. We had never planned to go [that] far away from the safety and comfort of the airlock, and we had never planned to go near a solar array that was only partially deployed like this, which is very unstable. And finally, we had never planned to go near a solar array that had sustained damage because one thing that could conceivably happen is [electrical] arcing from this energized solar array into the spacesuit, which of course is just a balloon filled with 100 percent oxygen. Thinking things through clearly and carefully, it was felt . . . correctly so . . . that the likelihood of arcing to my suit was relatively low, provided I didn't come in direct contact with [the array]. Anything that I was going to contact the array with would be isolated.

As Parazynski was riding to the array's damage site, shuttle crewmate Paolo Nespoli was reading to him a page of cautionary notes and warnings for what Scott was to observe when around the array as brief reminders. As Parazynski described it, "I quipped, 'Well that's a lot of stuff. Is there anything that I *can* touch?' [Nespoli] said, 'Wait, I'm only halfway done.' It was really an impressive list of things [I] could not touch."

Everything Scott used to touch the solar array was specially insulated. He had a set of cutters for the guide wire with insulated handles and an L-shaped, hockey stick–like tool to allow him to either push the panels away or pull them toward him. Getting to a guide hole, Scott would use the hockey stick to pull a panel toward him with one hand. Next, he would align the cuff link block at one end lengthwise with the wire and push it through the guide hole with his other hand while making sure not to let his hands contact the array itself. Threading a cuff link into a guide hole on an array is sort of like trying to stick a piece of thread through a hole in a sewing needle. Once the block was completely through the hole and on the other side, Scott would try to catch the block on the back side of the array to fold it out. To do this, he would use the hockey stick to push the array away from him while putting tension on the cufflink to make sure it was fitted securely.

Parazynski admits this was a job best suited for three hands, but he made it work. He attached a total of five cuff links to the arrays, having

50. Scott Parazynski on the OBSS arm extension waves after attaching five cuff links to the damaged P6 solar array. Courtesy NASA.

to thread blocks through holes ten times. He made it look easy with skills honed as one of NASA's most experienced spacewalkers. With the job done, the array was carefully unfurled the rest of the way while Parazynski and Wheelock watched. The cuff links worked, and the array finally snapped into its proper position with no unsightly sags or gaps. The best part was that the array was still generating electricity at 100 percent capacity. None of the panels themselves were damaged.

With the repair completed, the STS-120 crew wrapped up their tasks while Anderson and Tani finished their crew swap. *Discovery* undocked a few days later. Pam Melroy, using her commander's prerogative, managed to convince NASA to sleep shift the crew to perform a daylight reentry from a descending node track over the continental United States as opposed to an ascending node descent over Mexico twelve hours later in the dark. *Discovery* performed the first reentry over the United States since *Columbia*'s breakup. The reentry path took the shuttle over central Nebraska, and a sonic boom could be heard near Clay Anderson's hometown of Ashland. *Discovery* arrived home safely, and Anderson was reunited with his family, including his mother, Alice, who had been undergoing treatment for cancer. Unfortunately, she died a few weeks later in mid-November before her astronaut son returned to Nebraska for his official welcome-home ceremony.

Dan Tani's Heritage

Dan Tani was born to parents of Japanese ancestry in Ridley Park, Pennsylvania, in 1961. His parents originally lived in California but were rounded up during World War II as part of the forced internment of Japanese American citizens by the U.S. government. Tani's parents were interned at a camp in Topaz, Utah. Tani's father was an insurance salesman before the war, but one of his clients worked for the Evangelical and Reformed Church. They helped to find a job for Mr. Tani and helped to sponsor the family's release from the camp during the war. Dan Tani's dad became a minister for the church. Over the years, it merged with other churches to eventually become known as the United Church of Christ.

Dan Tani's early childhood in the 1960s included watching spaceflights on television, as he has fond memories of watching Neil Armstrong and Buzz Aldrin doing their walk on the moon. Model rockets were a big thing back then, and Tani was heavily involved with the hobby in his childhood. Going to high school, the space program in combination with his rocketry hobby shaped his goals as he wanted to become an engineer, but he didn't necessarily want to become an astronaut.

After getting his bachelor's degree, Tani went to work for Hughes Aerospace on various satellite projects. One of them included helping to build hardware to repair the malfunctioning Leasat 3 satellite, which had been stranded in a useless orbit after it failed to deploy properly on STS-51D. The satellite was both repaired and redeployed on mission STS-511. Even with this project's success, Tani looked for ways to get out of aerospace engineering and pursue a different career path.

A few years later, Tani's former boss at Hughes invited him to visit a new aerospace company called Orbital Sciences Corporation for a job interview. Tani wasn't really interested as he had found a great job, working for a think tank near Cambridge, Massachusetts, studying how people make decisions. But he went for the interview anyway. He liked what he saw and was hired. Orbital was a new company. Unlike a large firm like Hughes, Orbital employed a small handful of engineers who got the chance to do more exciting things. Orbital's first major success was the air-launched Pegasus rocket, which began sending small satellites into orbit in 1990. After working for Orbital for a few years, Dan Tani turned his sights upward, eventually becoming an astronaut in 1996.

After conducting a space walk on STS-108, Tani transitioned into the ISS program, where circumstances ended up with him on the ISS as part of Expedition 16. Initially, his time on the station was only supposed to be three months, as STS-122 was supposed to bring him home in December. But problems with hardware for the next shuttle mission forced a postponement, and ultimately Tani spent four months in orbit. Tani and Peggy Whitson conducted four EVAs on the ISS preparing the Harmony node and PMA-2 for their relocation to the front of Destiny and then finishing out exterior work after the module was moved in November. They also inspected the balky SARJ to try to discover the source of its problems. In January, during their fourth EVA, they replaced an electric motor used to drive the starboard solar arrays. During these four EVAs, the pair conducted the one hundredth ISS space walk, and the time spent outside allowed Peggy Whitson to set a record as the female spacewalker with the most EVA time.

Not everything was pleasant, though. On 18 December 2007, as the crew was in their pre-sleep period, they got a call from the ground to prepare the private communications line for Tani. Mission control never called in during pre-sleep unless it was important, and the use of a private line usually meant something bad. Tani called his wife, Jane, and was relieved that she was okay, but she was heading to JSC to get the word herself as to what was up. Tani had left instructions that if bad news was being delivered, it would be done by his wife or his flight surgeon. Once his wife arrived at JSC, Dan Tani received the sad news. His mother, Rose, had been killed in a car crash.

No astronaut had ever lost a family member while on a mission before. The shuttle was grounded and wouldn't fly until February, and Dan Tani wouldn't be able to attend the funeral for his mother. Tani credits his support team on the ground and the flight controllers with helping him cope. He still had a job to do, but they gave him plenty of leeway to do what he needed. To Whitson, the loss of Rose Tani was easily the lowest point of the mission. But looking back, Dan Tani felt that it could have been worse.

Video conferences were arranged with family and friends. Clay Anderson also got to show his support in a private video conference since he had lost his own mother in November. Even though the situation was bad, it didn't impact the science, support, or assembly schedules for the ISS. If Rose had died when things were busier two weeks before, the packed schedule at that point could have been impacted. As a song from the Beatles says, Dan Tani got by "with a little help from his friends."

Finally, February came, and *Atlantis* docked on STS-122 to take Tani home. Looking back on his experience as an astronaut and being a third-generation Japanese American, Tani sums up his success in life and his heritage in this way: "What makes my story great, I think . . . my parents were, for all effects, imprisoned by the U.S. government for several years of their life. [With] my father, they took radios and cameras away from him, and yet just one generation later, I'm given the opportunity to represent the United States and entrusted with a space station and entrusted with many radios and many cameras." The Japanese American community flourished after World War II's internment to become important contributors to the American experience in culture, art, and science. Dan Tani is very proud to be a part of it, and it has been an honor of his to represent them in space.

Columbus

When STS-122 arrived, its main payload was the Columbus laboratory built by the ESA. Like the MPLMs, the structure for Columbus was built at a factory in Rome, Italy, by Alcatel Alenia Space. Alcatel had also been involved in construction of the Unity and Harmony nodes, and all the modules shared similar structures. Even the Japanese Kibo laboratory module's basic structure used the same design. The Columbus would allow the ESA to carry out its own dedicated scientific activities. The module was also directly related to the old Spacelab pressurized module, but it featured many improvements to conduct science inside and on racks mounted outside.

Expedition 16 crewmember Léopold Eyharts and STS-122 ESA astronaut Hans Schlegel of Germany were heavily involved in outfitting the module once it arrived on orbit. Schlegel was also scheduled to perform the first space walk of the mission with astronaut Rex Walheim. But a minor medical issue sidelined Schlegel, and Stan Love performed the space walk with Walheim instead. Schlegel performed the second space walk of the mission just fine as Columbus was firmly attached to the ISS and opened for business. Walheim and Love performed STS-122's third space walk as well.

STS-124 came next as the space shuttle *Endeavour* transported up the first

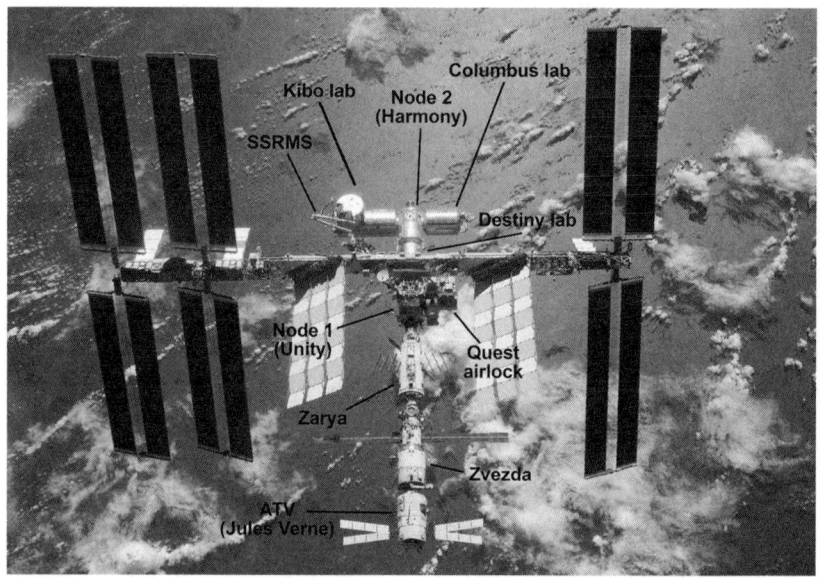

51. The ISS as viewed from *Discovery* on STS-124 after undocking. Courtesy NASA, photograph labels by the author.

segment of the Japanese laboratory, Kibo. The Kibo complex is made up of a short Experiment Logistics Module segment, the main module segment, and an external equipment rack with its own robotic arm. The first module flown was the Experiment Logistics Module. This mission also delivered the Special Purpose Dexterous Manipulator, also known as Dextre. Dextre is a multiarmed robot that can be positioned at the end of the SSRMS to perform complex assembly and support tasks, either preprogrammed or with an astronaut operating it from inside the station using telepresence. It was built by Canada as their final major piece of hardware for the ISS.

Garrett Reisman took over for Eyharts after only a month in orbit for the Frenchman. Five EVAS were conducted on this mission to hook up the new modules and Dextre. A final task involved temporarily stowing the shuttle's OBSS on the ISS, since there wasn't enough weight available on the next shuttle to carry it into orbit aboard STS-125. After the OBSS was stowed, a power cable was hooked up to it to maintain its systems for its two months of scheduled hibernation. Additional inspection work was also conducted on the balky starboard array's SARJ. *Endeavour* stayed docked with the ISS for eleven days before departing.

Jules Verne

On 9 March 2008 the first ESA-built ATV was launched into orbit from ESA's spaceport in French Guyana aboard an Ariane 5 rocket. The ATV is an automated spacecraft designed to haul cargo to the ISS. It is much bigger than a Progress craft, and it can handle three times the payload. The ATV is designed to dock with the aft Russian docking port on the Zvezda module, so it can replenish the station's fuel and water supply using the standard Progress hookups. The engines on the ATV can be used for reboosting the station's orbit as well.

The first ATV was named after the great science-fiction author Jules Verne. After nearly a month in orbit undergoing a thorough checkout, *Jules Verne* achieved a successful docking. It stayed docked for many months, acting as a storage room after delivering its supplies. It was loaded up with trash before it undocked to burn up in September.

Ballistic Reentry, Part 2

Less than a week after the first ATV arrived, it was time for Expedition 16 to conclude and hand over the reigns to Sergey Volkov and Oleg Kononenko. Garrett Reisman would remain on board to handle the Expedition 17 transition. Both Volkov and Kononenko were flying their first missions into space. Sergey Volkov was the first second-generation space traveler to fly, as the son of *Salyut 7* and *Mir* cosmonaut Aleksandr Volkov; at age thirty-six, he also became the youngest commander of the ISS to date. Joining them on *Soyuz TMA-12* was South Korean spaceflight participant Yi So-yeon. She was flying as part of a Korean astronaut program and had originally been the backup until prime crewmember Ko San was scrubbed due to violating Star City's rules for taking training manuals out of the classroom. Yi's mission involved eighteen microgravity experiments, ranging from taking three-dimensional pictures of her face for the purpose of measuring how fluid in the human body redistributes itself in zero gravity to studying the behavior of fruit flies in orbit. After ten days in space, Yi returned to Earth with Whitson and Malenchenko.

Soyuz TMA-11 undocked just fine and conducted a normal deorbit burn, but a very serious problem developed as the propulsion and descent modules didn't separate from one another cleanly. Some of the explosive bolts holding the pair together failed to fire. Due to the spacecraft's center of mass, *TMA-11* began a ballistic reentry front first, like *Soyuz 5* almost four decades earlier. Both Whitson and Malenchenko could feel that they were in trouble as the craft seemed to rock back and forth during reentry. To Peggy, it felt like a pitching motion, while Yuri perceived it rocking from side to side, but their inner ear perceptions couldn't be entirely trusted, since neither crewmember had felt gravity in many months.

Finally, the struts broke apart due to the pressure buildup and the descent module righted itself as it was designed to do, continuing a ballistic descent. *Soyuz TMA-11* would land about three hundred kilometers short of its final destination. The front of the module was heavily charred, and the communications antenna had burned off. Photographs of *TMA-11*'s landing show that the hot Soyuz also managed to start a smoldering grass fire, which scorched a fair amount of land before it was extinguished. Malenchenko climbed outside the craft and used a satellite phone in the craft's survival pack to contact the recovery forces. Even though forces were pre-staged in preparation for a possible ballistic reentry, it still took the helicopters forty-five minutes to arrive and render assistance.

At the conclusion of the flight, Yi So-yeon spent a few days recovering at a hospital in Korea due to pain in her back from her spine lengthening a little in orbit. Otherwise, everyone recovered fully from their ordeal. A postflight investigation eventually focused on the cause of the problem, a suspect batch of pyrobolts in the struts that connected the descent and propulsion modules together. It would take time to decide if corrective action for *TMA-12* was needed. The head of Roscosmos at the time, Anatoly Perminov, created some additional controversy when he also speculated that the incident in part might be connected to a Russian nautical superstition about the number of women on the flight outnumbering the men on board causing bad luck. He said that he would take steps to avoid having that many women on future flights.

In 2009 Peggy Whitson took over the job of NASA's chief of the Astronaut Office, joining the ranks of other famous names that held the position, such as Deke Slayton, John Young, and Hoot Gibson. This is the highest-ranking management position among active astronauts, and Whitson was the first woman selected. Her selection marked a fundamental shift of the role traditionally being held by a pilot-astronaut to one occupied by a nonpilot with a scientific background and paralleled the fact that the ISS was becoming NASA's primary space program with the space shuttle's retirement date edging ever closer.

14. Final Construction

"How Do You Go to the Bathroom in Space?"

One of the biggest challenges to overcome in long-duration space missions is trying to create a closed loop system to minimize the amount of resupply needed. Even though the ISS wasn't flying to a distant planet, the need to make things as self-sufficient as possible was a critical one with the shuttle's retirement looming. For a closed loop system, recycling liquid human waste such as urine and sweat becomes a high priority.

The Russian toilet on the ISS was an evolution of the designs first flown on the stations *Salyut* and *Mir*. It works by using a vacuum to help draw off the fecal material into the bowl or liquid urine through a hose. Urinating into the system is relatively easy to do, but learning how to use the bowl properly can take skill to master. Part of the crew transition period in orbit involves instruction in the proper technique to ensure that the feces goes down the hole and doesn't bounce out to float free in the cabin.

The feces gets collected in individual bags in a storage container at the base of the toilet, and it is compacted down to save space. The urine collects into a sealed tank. In order to reuse the tank, its contents are offloaded into the liquid storage tanks of a Progress before undocking. Since some of the instruments on the ISS can be affected by contamination from water ice crystals, overboard urine dumps are not carried out. Due to the acidic quality of urine, there is a use life for the urine tanks, and they are periodically changed out as fresh ones are brought up on Progress vehicles. Used fecal storage containers are loaded into the Progress for disposal.

For three crewmembers on a long-term stay or six crewmembers on a short-term one, this system works just fine. But with increasing the ISS crew size from three to six people, water recycling became more critical. Once the shuttle fleet had retired, NASA would lose the capability of using the shuttle's fuel cells to generate water; since water weighs about six pounds per gallon, devoting weight in a Progress or an ATV to launch water means less weight for other cargo.

Waste management is handled by the Environmental Control Life Support System departments at the NASA centers. In preparation for six-person occupancy, engineers at Marshall developed a new urine-recycling rack that would be hooked up to a second Waste and Hygiene Compartment (or lavatory), planned for launch on STS-126. This newer toilet is the same design as the one in the Russian segment and uses many of the same parts to make replacement easier, but rather than off-loading urine to the storage tank, the urine would be sent to the recycler instead. The tank system is kept as a backup in case a problem develops in the rack. On orbit, the urine-recycling racks are the responsibility of JSC's Environmental Control Life Support System department in Houston.

In a distillation process on Earth, heat is typically used to boil off water; in space, high heat can be dangerous. Therefore, water is boiled off from urine with reduced pressure instead. A rotary centrifuge is used to help the separation process, making sure that the heavy parts of the urine are split off from the lighter-weight water. Additional filters help to catch any residual waste products. The ulraconcentrated urine is stored in a tank, which the engineers refer to as brine storage. Once the tank is full, it is replaced with a fresh tank and disposed of on a Progress. One tank can carry a month's worth of concentrated urine from six crewmembers.

The separated water from the urine-recycling rack is sent to the water-recycling racks, where it is combined with water recovered from the station's internal humidity (including sweat). From there, it can be used as a technical-grade water, or a biocide (something that kills bacteria) such as iodine can be added to make it safe for drinking. Many people might get squeamish at the thought of drinking water from urine, but the station's recycling process isn't all that different from what happens in nature. The water can also be used to generate oxygen for the life-support systems by using electrolysis. Gaseous hydrogen from the electrolysis process is vented overboard.

"Somebody Call a Plumber?"

Before the new recycler could launch on STS-126, a minor problem developed with the Russian toilet on orbit. The separator system wasn't working properly and needed two flush cycles to empty when only one should have been needed. Engineers on the ground traced the problem, and replacement parts were loaded at the last minute aboard *Discovery* just before it launched on STS-124. The problem received coverage in the media and on the Internet, generating a bit of humor in the process. *Discovery* launched into orbit on 31 May 2008. Its seven-person crew, commanded by spaceflight veteran Mark Kelly, had a busy mission in store, as their primary task was delivery of the pressurized Kibo laboratory module. JAXA astronaut Akihiko Hoshide would oversee the outfitting and activation of the module. *Discovery* docked with the station on 2 June 2008; upon hatch opening, the first words from Mark Kelly to ISS crewmember Garrett Reisman were, "Somebody call a plumber?" After the malfunctioning separator parts were changed out, the toilet returned to normal function.

Kibo

The main payload aboard STS-124 was the pressurized Kibo laboratory module. It was the second part of the Japanese Experiment Module (JEM) package built by JAXA. In the Kibo laboratory, astronauts can conduct research directly funded by Japan in various disciplines from materials science to biomedical research. Given Japan's large population, biomedical research into disease control and the effects of aging are considered very important to their program.

JAXA was formed by the merging of three former Japanese space agencies: the National Space Development Agency, the Institute of Space and Astronautical Science, and the National Aerospace Laboratory. The National Space Development Agency was the largest of the three agencies when JAXA was formed in 2003. So to Japan's ISS and astronaut programs, it was mostly a name change with a new patch logo.

The Kibo module was docked with the Harmony node's port side on 3 June 2008; after an EVA lasting over five hours conducted by Mike Fossum and Ron Garan, it was opened up the next day. Once the early checks had been completed, the combined station and shuttle crews entered the module. Kibo is the largest dedicated ISS laboratory module with a length of just under thirty-seven feet and a diameter of fourteen feet. Its empty interior without a lot of science racks installed was very spacious.

The combined ISS and shuttle crews took a few minutes to enjoy some zero-g tumbling exercises inside Kibo. On the downlinked video to mission control, the size of the empty module made the astronauts almost look like goldfish swimming in an aquarium. Garrett Reisman also managed to get himself stuck in the middle of the empty space to see if he could maneuver his body to get to a wall. It took a few moments and some interesting body contortions, but he was successful. His activity was acknowledged by ISS CAPCOM Chris Cassidy saying, "Well done Garrett, well done. Golf clap."

Kibo's interior wouldn't remain empty for long. Two days later, the Experiment Logistic Module that had been sent up previously was berthed to Kibo's zenith CBM; once the approval was given, the crews began outfitting the lab with science racks and covers for the floor and ceiling areas. Kibo can hold up to twenty-three science racks. Four were launched into orbit with it, and eight more were added from the logistics module. Others would arrive on later flights. The OBSS was also retrieved by *Discovery* and survived its time in orbit in good shape.

Kibo had its own dedicated robotic arm installed and checked out by astronauts Karen Nyberg and Akihiko Hoshide from the lab's newly installed work station. The arm acts as an interface to transfer experiments from Kibo's scientific airlock to external racks in front of the module. The racks would be installed on STS-126 and allow for the placement of experiments outside without the need for EVAS.

Reisman was replaced on Expedition 17's crew by astronaut Greg Chamitoff, who arrived on STS-124. Two Orlan space walks were conducted by Volkov and Kononenko in July; among their various tasks, they removed a suspect pyrobolt from the Soyuz for analysis on the ground. Special care was taken to insulate the bolt inside a container during *TMA-12*'s return, to lessen the chance of it detonating inside the Soyuz during reentry.

Second-Generation American in Space

In October 2008, Expedition 18 arrived at the station aboard *Soyuz TMA-13*. Commanding the crew was cosmonaut Yuri Lonchakov, with astronaut Michael Fincke serving as flight engineer. Joining them was spaceflight participant Richard Garriott. Richard Garriott was the son of Skylab and STS-9 scientist-astronaut Owen Garriott. Rather than pursuing a career in laboratory science like his father, Richard Garriott decided to focus instead on computer science and programming, as he had shown a knack for it while growing up, convincing his high school to let him take part in a self-taught programming course for credit. Richard Garriott formed the company Origin Systems in the 1980s and made a fortune designing computer games, such as the very popular *Ultima* series. But he never entirely forgot his father's heritage, as the younger Garriott was an avid collector of space memorabilia. When the right opportunity came to purchase a ride on a Soyuz as a spaceflight participant, Richard Garriott jumped at the chance.

Owen Garriott was on hand in Baikonur to see his son off as the mission launched into orbit on 12 October. Once the crew arrived at the ISS, it marked the first time that second-generation space travelers from Russia and the United States were in orbit at the same time. When Richard Garriott began his training at Star City, one of his intentions was to become the first second-generation space traveler. But with Sergei Volkov going into space several months earlier as part of Expedition 17, he was denied that chance. Richard Garriott also became the second space traveler to wear a British flag on his clothing, since he was born in England and maintains dual citizenship.

SARJ Repairs

Soyuz TMA-12 returned to Earth with Volkov, Kononenko, and Garriott on board after a nine-day transition period. The reentry and descent went just fine with no pyrobolt or ballistic reentry problems. Chamitoff remained on board as part of Expedition 18 to help with the transition. The space shuttle *Endeavour* paid the ISS a visit that November on mission STS-126. Among the cargo, they brought up two crew quarters racks for the Harmony node, the new Waste and Hygiene Compartment, a new galley (or kitchen) for the station, and other hardware to help prepare the ISS for a six-person crew.

Not much external assembly work was planned for this flight, given that much of the shuttle crew's tasks would take place inside, but four EVAs were conducted on this mission. Heidi Stefanyshyn-Piper conducted three of the space walks as the lead spacewalker. She was joined by Stephen Bowen on the first and third EVAs and by Robert Kimbrough on the second one. Bowen and Kimbrough conducted the fourth space walk. While the EVAs involved some equipment replacement, the primary focus was to finish repairs on the starboard solar array's SARJ.

Problems were detected initially in 2007 as the starboard arrays had picked up a vibration while moving. Metal shavings removed on inspection space walks determined that elements of the SARJ were wearing out prematurely. Peggy Whitson and Dan Tani replaced a drive motor during an Expedition 16 staged EVA as a precaution. After studying the problem, NASA decided to replace the trundle bearings in the SARJ, which are used to help the race ring of the joint to move properly. The repairs were very important as without the SARJ, the starboard arrays could not be moved to track the sun properly and the power supply of the ISS would be reduced. The repairs were also necessary before the launch of STS-119 in early 2009, as its cargo was the final set of starboard solar arrays.

During the four space walks, the EVA teams removed and replaced the trundle bearings. They also used grease guns to lubricate the SARJ movement tracks. A problem developed with one of the guns, though; on the first space walk, trapped air inside the gun caused grease to ooze out into Stefanyshyn-Piper's tool bag, creating a sticky, gray mess inside. While she concentrated on getting grease off her gloves, the tool bag began to float away. She considered grabbing for it, but it was too far away to grab safely and got lost in orbit. Heidi Stefanyshyn-Piper felt bad for what happened since she was the lead spacewalker on the mission, but she finished the space walk with Stephen Bowen, using the tools she needed from his bag. NASA's procedures to double up on the necessary items had come in handy once more. To other astronauts who have worked on EVAs, losing tools is a subconscious concern, and everyone acknowledges that it could have just as easily happened to them as well. The media coverage made a bigger deal out of it than it was.

The remaining space walks took place with no problems; after controllers on the ground checked everything out, the SARJ was back in operation once again. After the mission, engineers on the ground redesigned the grease guns with an additional locking mechanism on the tip to ensure that grease would not leak out until an astronaut needed to use it. The errant tool bag orbited Earth as its own satellite for a few months before it finally entered the atmosphere and burned up.

Eating in Space

Endeavour wrapped up its work and undocked on 28 November 2008, taking astronaut Greg Chamitoff home and adding astronaut Sandy Magnus to Expedition 18. Since the mission took part over the Thanksgiving holiday in the United States, the food lab at JSC had prepared a special meal for the combined crews, in the form of turkey and candied yams.

Since the early days of food preparation in NASA programs, the science of edible space meals had advanced considerably. Food preparation on the U.S. side of the ISS is the responsibility of JSC's food laboratory. Along with a special facility at Texas A&M University in College Station, Texas, they handle purchasing of commercial goods and in-house preparation of many of the meals. During the shuttle program, food packaging shifted to flying prepackaged meals in bags, instead of the cans used during Skylab.

Food preparation on the ground can vary depending on the items being sent into orbit. A large percentage of the food products are flown to the ISS in a dehydrated state to both cut down weight and give them a longer shelf life. Dehydrated products are contained inside special pouches with a septum (a one-way valve) on one end. An astronaut inserts a water-dispensing tool into the septum to inject the prescribed amount of hot or warm water needed to reconstitute the product. The septum prevents water from leaking from the port when removed from the dispenser. After a few minutes, when the water has properly mixed with the product, the astronaut can cut open the bag to eat the food. The food is designed to stick to the bag and eating utensils so that particles don't go drifting off into the cabin.

Drinks are prepared in a similar fashion, as each drink bag contains the mixing powder inside it, while instructions for how much water is needed are printed on the outside. A straw is inserted into the top of the bag after water is added. Astronauts can enjoy hot beverages such as instant coffee (with cream and sugar already added, since the bag can't be opened to mix it) or cool ones such as instant lemonade or fruit punch. One of the water dispensers is in the Zvezda module, and it can dispense two temperatures of water in the form of hot (usually very hot) and warm. A cold drink requires something else.

There is no active refrigeration on the ISS for food storage, but the galley launched on STS-126 includes a small beverage cooler capable of chilling drink bags. The cooler comes in very handy after a workout on the station's exercise equipment, since a cold drink is more refreshing than a warm one. Ice cream has been sent into orbit by the food lab periodically when empty science freezers were being sent up on shuttle missions to bring back frozen biological samples, but this was a rare luxury.

Not every space food product on the ISS is dehydrated, though. Because dehydration would damage the contents of some food, many foods are thermostabilized, meaning that only heat is used to purify them before packaging and no water is removed. Other foods are irradiated to kill pathogens before packaging. A few items, typically the bonus foods requested by specific crewmembers, are purchased off the shelf. A lot of the food is similar to what one might find for camping trips. Bread products, such as tortillas, are vacuum packed in special bags with an oxygen scavenger card (similar to what is found in bags of beef jerky) to help prevent bacterial growth. Tortillas are used because they take up less storage space than sliced bread and work just as well for making sandwiches without drying out or releasing crumbs as easily.

Finger foods, such as the ever popular "candy-coated peanuts" (NASA doesn't call them M&Ms as they don't want to be seen as endorsing a commercial product), are also available for snacking purposes. The packaging for these foods is similar to the type used for military MREs (Meals Ready to Eat). Dry, crumbly items are kept to a minimum; so in the case of hard crackers or cookies, bite-size versions are flown instead of larger ones, allowing crewmembers to consume them whole to prevent crumbs.

The taste of the food is pretty good, and astronauts get a chance to sample their menu choices at the food lab before their missions. For a standard ISS crewmember, there is no real restriction as to what they can eat unless they are taking part in a study where they have to keep track of their food consumption closely, but the food lab at JSC does closely coordinate with the medical and research departments to ensure that the crewmembers get the proper nutritional and caloric intake in their diets. Vitamin D is the only supplement added to the food, since astronauts aren't able to safely get suntans in orbit.

Each food product on the U.S. and Russian side has a bar code, and a scan of the code will reveal the date the product was packaged, what shipping container it was sent in, and its contents, to help keep track of any problems. Since some of these items are off-the-shelf commercial products, the JSC food lab keeps track of them in case of product recalls.

"What Is a Party Chicken?"

Many food items flown by the Russians to the ISS come in metal cans, which don't look all that different from the ones used on *Skylab*. Some

Russian food products, such as snack foods and bread cubes, are packaged in bags. To help prevent injury from sharp can edges and prevent metal shavings from floating free in the cabin, NASA sent up a hand-twist commercial can opener that cuts through the side of the can seal rather than through the edge of the lid. The cans can be warmed in slots built into the RSM's dinner table.

Each food label is written in both Russian and English. Since each language has some unique words, the translations have periodically provided a few conversation pieces. The Russians have a canned meat dish known simply as the "Appetizing Appetizer" (what one might call "mystery meat") in English. The JSC food lab has a thermostabilized spicy chicken dish known as "Fiesta Chicken." But when the name was translated into Russian, somebody from Roscosmos asked, "What is a party chicken?" since "fiesta" translates into "party" in Russian. A food lab scientist had to explain that "fiesta" also refers to a type of spice preparation common to the American Southwest.

Different crewmembers have different tastes, and some foods are better liked than others. Another issue is that taste perception can change in orbit. Due to fluid changes in the body, astronauts can feel as though they are stuffed up, until their bodies adjust. The sense of smell is also affected since air currents flow differently on orbit. So food that tastes good on Earth can taste a little bland in zero gravity. To help compensate for this, some of the food choices have a little more spice added to them. For instance, shrimp cocktail is one of the most popular dishes because the cocktail sauce has a kick of horseradish added to it.

Salt and pepper in liquid forms are provided for seasoning, as are small fast-food packets of ketchup and mustard. Different hot sauces, horseradish, and wasabi are also popular, and NASA periodically sends up condiment bags stuffed with such goodies on resupply flights. European foods are typically canned like the Russian ones. The Japanese have been starting to fly their own foods in unique packaging for their astronauts as well.

The NASA food lab has been improving their menu with more choices as they get feedback from ISS crews returning from orbit. A small selection of food choices that was perfectly suited for a two-week shuttle mission usually becomes boring after a months-long ISS assignment, so work continues constantly to expand the menu.

52. Expedition 20 crewmembers Kopra, De Winne, Romanenko, and Barratt enjoying dinner in the Unity node. Various American and Russian foods along with fresh produce from a Progress resupply can be seen. Courtesy NASA.

During the early days of the ISS, there were plans to fly a refrigerator and freezer in the habitation module, but when the module got axed during budget cuts, managers in charge of JSC's food lab were able to convince NASA administrators to maintain the food research funds, which were instead used to come up with more freeze-dried and thermostabilized menu choices. A benefit of that funding has been the introduction of breakfast breads and warm dessert dishes. Such comfort foods have become very popular.

To help track what astronauts eat, they each fill out a questionnaire that asks how many of each food group they consume. Additional tracking is done with bar codes on the food containers. So when a container is opened, it is scanned into the system; when the container is empty, it is scanned again to help track what was consumed and when. Each food container is about as large as a thick college textbook and contains nine meals' worth of food. The food packages themselves are also bar coded, but they typically aren't scanned unless required for a food-intake experiment. Consultations are done with the mission's supporting flight surgeons on a regular basis to help monitor any dietary issues.

More Assembly

On 17 March 2009 *Discovery* next visited the ISS on STS-119, carrying in its cargo bay the s6 truss containing the final set of solar arrays and storage batteries. Due to the proximity of the shuttle to some modules on the ISS, it took some interesting coordination. The s6 was lifted out of the bay by the shuttle's RMS and then handed off to the SSRMS while the RMS was repositioned. Then the segment was handed back to the RMS before the SSRMS was moved on the main truss to its installation location and the final handoff was made between the two arms.

The s6 segment was finally installed on the truss on the second day with astronauts Steve Swanson and Richard Arnold making the final connections during an EVA. Joe Acaba, the first astronaut of Puerto Rican descent, also performed two space walks on this flight, alternating with Swanson and Arnold as EVA partners. Both Arnold and Acaba were part of the first class of educator-astronauts selected by NASA. The educator astronauts have teaching backgrounds rather than military or science ones, but they are fully trained as mission specialists. In addition to their astronaut duties, they also act as liaisons between educators, schools, and NASA personnel, to help educate children about spaceflight.

With the addition of the fourth set of solar arrays, the ISS was now operating at full power. Sandy Magnus switched positions on Expedition 18 with JAXA astronaut Koichi Wakata, who would continue activation work in the Kibo laboratory. Wakata had been a member of the astronaut corps since 1996 and was the first Japanese astronaut to fly as a mission specialist on STS-72. Wakata was the first JAXA astronaut involved directly in ISS construction; with his transfer, he became the first Japanese ISS crewmember.

The Slight Chill

Wakata's time as part of Expedition 18 was very brief, as the day after *Discovery* departed the ISS, Expedition 19 crewmembers Gennady Padalka and astronaut Michael Barratt launched into orbit aboard *Soyuz TMA-14*. Joining them was spaceflight participant Charles Simonyi. Simonyi was only the second Hungarian to fly into space and had made his living working as a software developer for Microsoft in its early days before forming his own company. This was his second trip into space, as he had flown to the ISS on *Soyuz TMA-10* in

2007. To date, he is the only "tourist" to visit the ISS twice. After a week in orbit conducting research and communicating with schools via the station's ham radio, Simonyi returned home on *TMA-13* with Lonchakov and Fincke.

Expedition 19 got off to a slightly less-than-stellar start as word got to the press that ISS commander Padalka was not happy that he was being forced to use only the equipment in the Russian side of the station for exercise and hygiene due to a disagreement on the ground. While the relationship between space workers from the two countries had become friendlier over the past fifteen years, there was still some partisanship going at the government and management levels during negotiations. The bill for Russian launch services had gradually been going up; with shuttle retirement coming very soon, it was looking like negotiated costs to launch astronauts on the Soyuz would escalate even further.

A war between Russia and the republic of Georgia in the summer of 2008 over the disputed regions of South Ossetia and Abkhazia didn't help matters since the United States was against Russian military action. Accusations were made that once hostilities started, Russian crewmembers on the ISS had surveyed the disputed region supposedly for military purposes instead of peaceful ones. While the war between the two former Soviet republics didn't last long, it did create a slight cooling of relations between Russia and the other ISS participants. But life on the ISS still continued more or less as normal, even if the bureaucrats and politicians didn't entirely see eye to eye. However, this wouldn't be the last time that such world events would cast a shadow over the ISS.

According to astronauts who have flown to the ISS, Moscow tends to keep its operational tasks for the Russian crewmembers separate from the U.S. side (which also includes the JAXA and ESA assets). On orbit though, ISS crewmembers consider themselves a single crew, and they help out one another whenever possible, regardless of whether it is an American providing assistance as needed on the Russian side or a cosmonaut helping out on the U.S. side. Each crew is a little different, but to date, there has been no indication that an ISS commander has tried to rule with an iron hand due to some political mandate.

Six Crewmembers

Two months later, on 27 May 2009, *Soyuz TMA-15* lifted off from Baikonur with Expedition 20 crewmembers Roman Romanenko from Roscosmos, Frank De Winne from the ESA, and Robert Thirsk of the Canadian Space

Agency. Rather than taking over for the previous crew, they would instead expand Expedition 19 from three people to six to become Expedition 20. For the first time, all major agencies would be represented simultaneously by ISS crewmembers. The newcomers docked with the ISS two days later and started a new phase of operations as the station was now fully manned.

As part of one Japanese experiment, Wakata performed something that might be considered a little extreme. He tested out a new type of experimental underwear by wearing it for one month to see if the fabrics it was made from helped to control odors. The new material in the underwear did its job, and fortunately for Wakata, he did not stink while wearing them. The underwear returned with Wakata when STS-127 visited the station.

STS-127 was supposed to launch in June, but problems with a gaseous hydrogen leak at the pad forced a month-long delay as engineers replaced equipment to correct it. Even after the hardware problem was fixed, weather problems kept *Endeavour* grounded during four more launch attempts. Finally *Endeavour* got off the pad on 15 July 2009. The main payload of this flight was the external experiment rack for the Kibo laboratory, along with some additional equipment for the interior.

One experiment of note from this mission was a lidar system known as Dragoneye, built by the commercial firm SpaceX. This was a piece of equipment designed for use on SpaceX's Dragon cargo vehicle to help with rendezvous. NASA had been looking to commercial firms to take over delivery of supplies on the ISS, and SpaceX was one of the firms given a contract for services. Formed by entrepreneur Elon Musk, SpaceX was developing the Dragon spacecraft and the Falcon 9 rocket in-house, so it wouldn't need to subcontract launch services for its spacecraft to other companies.

As of 2009 SpaceX had been successful in launching the Falcon 1 rocket after three attempts and was about a year away from flying its first Falcon 9 with a boilerplate Dragon spacecraft. What made Dragon different from the Progress, ATV, and Japanese HTV vehicles was that it had a heat shield. Once the shuttle program was retired, Dragon would be the only vehicle capable of returning bulk cargo from the ISS back to Earth, compared to the tiny amount that could be crammed into a Soyuz with a returning crew. SpaceX had more ambitious plans, as the company was also designing a Dragon variant for use as a manned spacecraft capable of carrying up to seven crewmembers to the ISS.

Five EVAs were conducted during STS-127. Veteran spacewalker Dave

Wolf was the lead for three of the EVAs. U.S. Army colonel Tim Kopra was his partner for the first EVA as the pair attached the external experiment racks to the Kibo laboratory. Wolf and rookie astronaut Tom Marshburn outfitted the station with some replacement equipment in external stowage racks on the second EVA. Marshburn may have been new to spaceflight, but not to NASA, having worked as a flight surgeon at JSC before his selection in 2004. With construction of the ISS nearing an end, the primary purpose of the next few shuttle flights would be to equip the station with as many spares as possible, since once the shuttle retired, there would be no way to send the heaviest equipment up anymore.

On the fourth EVA, Dave Wolf was joined by rookie astronaut Christopher Cassidy. Cassidy was a former U.S. Navy SEAL, and his prespaceflight exploits included two six-month tours of duty in direct support of Operation Enduring Freedom only weeks after the events of 11 September 2001. The astronaut exploits of fellow SEAL veteran Bill Shepherd inspired Cassidy to join NASA. Shepherd was very supportive of Cassidy, and the two men became good friends thanks to the close-knit community of SEALs. Like Tom Marshburn, Cassidy became an astronaut with the 2004 class and performed support duties, including manning the CAPCOM console at mission control.

The third EVA had to be cut short when readings from Cassidy's suit indicated higher-than-normal levels of carbon dioxide in its air supply. Like the shuttle, the EMU uses a lithium hydroxide chemical bed to collect exhaled carbon dioxide. For some reason, only part of the chemical bed was scrubbing the air. It was getting oversaturated and becoming less effective, so the task of installing fresh storage batteries on the truss had to be cut short as a safety precaution. Two additional EVAs conducted by Marshburn and Cassidy over the next few days got the battery installation completed along with work on the Kibo's external racks. After eleven docked days with the ISS, *Endeavour* returned home with Koichi Wakata. Taking Wakata's place on the ISS was Tim Kopra.

The COLBERT

In late August the space shuttle *Discovery* paid the ISS a visit with an MPLM full of supplies during STS-128. Among the cargo was a new treadmill for the station. It was known as the Combined Operation Load Bearing External Resistance Treadmill, or COLBERT for short. This new treadmill was designed to supplement the older one. It featured improved technology for longevity and consumed less power during use. Having two treadmills on the station in addition to the bicycle ergometer meant that several crew-members could exercise simultaneously rather than having to schedule times around one another. The COLBERT was temporarily stowed in the Harmony module since its final home module wasn't scheduled to fly until STS-130.

The COLBERT was named for comedian Stephen Colbert, the host of a political-comedy television show at the time known as *The Colbert Report*. Colbert was a space fan and a friend of astronaut Garrett Reisman. During the previous year, NASA hosted a poll on its website to name the final node module slated for delivery in 2010. There was a short list of names, but voters could also write in additional choices. So as a good-natured hijack, Colbert urged his viewers to go to the website and vote to name the module Colbert. Due to public interest, "Colbert" overwhelmingly got most of the votes. But rather than naming the module Colbert, NASA decided to name the treadmill after Colbert instead and named the final node Tranquility. The whole incident generated some nice public relations for NASA, and a special patch was designed for the COLBERT, featuring Stephen Colbert in a yellow jogging suit running on the treadmill bearing his name.

Other cargo carried up on *Discovery* included a new air-revitalization system for the ISS. While the systems already in place aboard the ISS were adequate to the task of supporting six crewmembers, it is always a good idea to have excess capacity. That way, if one of the life-support systems fails or has to be taken off-line for maintenance, usage problems are less likely to crop up. Air circulation aboard a station as large as the ISS is of critical importance to help prevent carbon dioxide gas pockets from forming in the nooks and crannies of modules. A new ammonia coolant tank was also brought up, and three space walks were conducted to replace the old ammonia tank with this newer one in preparation for the arrival of the Tranquility node. Tim Kopra returned home on this mission, being replaced on Expedition 20 by astronaut Nicole Stott.

Expedition 21 Handover

Handover for the first half of the six-person crew occurred on 1 October 2009 when *Soyuz TMA-16* arrived to start Expedition 21. Jeffery Williams and Maksim Surayev would replace Padalka and Barratt. Frank De Winne took over as the first ESA astronaut to command the ISS. Williams would take over for De Winne on Expedition 22. Joining the two newcomers on the Soyuz crew was Canadian spaceflight participant Guy Laliberté, a former circus performer and founder of Cirque du Soleil. Laliberté was flying to draw attention to his charity, the One Drop Foundation. One Drop's stated goal is to fight world poverty and help ensure that people all over the world have access to clean water. After a week in orbit, Laliberté returned home with Padalka and Barratt on *Soyuz TMA-14*.

HTV Arrives

On 10 September 2009 a Japanese H-IIB rocket lifted off from the Tanegashima Space Center just south of the Japanese mainland. Its payload was *HTV-1*, a transfer vehicle that contained nearly four metric tons of supplies for the ISS. Six metric tons can be carried on an HTV, but JAXA gave the module a smaller cargo load to make room for more propellant in order to conduct a thorough check of the vehicle before it approached the station. The craft arrived at the ISS on 17 September and was snagged successfully by the SSRMS before it was berthed to the nadir CBM port of the Harmony module. The HTV remained docked for about a month and a half before it was released to burn up on reentry.

So as not to interrupt the Japanese fishing industry near Tanegashima Island, H-II rockets can only be launched during a brief period of a few weeks, meaning HTVs can only fly to the ISS once a year. But the addition of this new cargo capability has made the ISS more self-sufficient. In 2018, Japan hopes to add a cargo return capsule to the HTV so that it will be able bring back cargo as well as send it up.

Ares I-X and Constellation Cancellation

At the end of October, pad 39B at KSC hosted a new rocket, the Ares I-X, which would conduct the first test flight of the Constellation program. The Ares I rocket design was to become the new launch vehicle for NASA's Orion spacecraft, which would dock with either the ISS or mission-specific modules intended for flights to the moon or Mars. These mission modules would be lofted into orbit by the much larger Ares V rocket. Ares I would use a five-segment, shuttle-based solid-rocket motor for its first stage and a liquid-powered second stage with the Orion spacecraft sitting on top.

No agency had ever used a solid motor exclusively as the first stage of a manned launch vehicle before, and there were many unknowns. To test the configuration, NASA modified a four-segment shuttle SRB with a dummy fifth segment and upper stage to simulate the Orion configuration. While not as tall as the Saturn V, the Ares I-X was the tallest thing to fly from Launch Complex 39 in almost four decades. The ghostly white rocket on pad 39B towered over the space shuttle *Atlantis* sitting on pad 39A in preparation for STS-129 that November. After a one-day launch delay due to range weather restrictions, Ares I-X roared into the sky at 11:30 EST on 28 October 2009. The rocket performed flawlessly on ascent and gave engineers a lot of test data to digest. Everyone involved with the program that day was optimistic about the future of Ares I and the Constellation program, based on the performance of this first test flight.

Unfortunately, in early 2010 the administration of U.S. president Barack Obama decided to cancel the Constellation program. Development of the Orion spacecraft would continue, but without a launch vehicle, Orion wouldn't be flying in space anytime soon, let alone to the moon or Mars. Funding also wouldn't be as plentiful due to a financial recession that began in 2008.

Instead, the ISS program would be NASA's only major manned program for the foreseeable future, and the Russian Soyuz would be the only way crews could get to it. As expected, the price of a ride to the ISS aboard a Soyuz doubled from about $32 million to $65 million when the time came for NASA to negotiate with Roscosmos for additional Soyuz modules. More funding was awarded to SpaceX for their manned Dragon craft. But at the time the cancellation of Constellation was announced, the decision was met with a lot of criticism from former NASA engineers and astronauts who were concerned that putting the future of NASA's manned spaceflight activities in the hands of the Russians in the near term and commercial companies in the long term without a backup was inviting potential disaster. As of early 2010 no commercial spacecraft had successfully been flown into orbit.

Poisk

ISS assembly work continued with the arrival of the Poisk (a Russian name meaning "Search") module at the ISS. The Poisk, also referred to as Mini-Research Module 2, was a docking module similar to the Pirs docking compartment. During the previous summer, Padalka and Barratt conducted a couple of Orlan space walks (one inside the depressurized Zvezda) to prepare the ISS to receive the Poisk module by placing antennae and a docking cone on the Zvezda module's zenith port. The zenith port was originally supposed to host the Russian solar array tower, until budget cuts forced the tower's cancellation. The Poisk was launched on 10 November on a modified Progress tug and docked successfully with the station on 12 November. The Progress tug departed the station in early December once the module's systems were activated, exposing a new docking port for the ISS. Poisk gave the ISS a fourth Russian docking port so that three Soyuz craft could dock with it while the aft port was occupied by a Progress or an ATV. An additional space walk to install Poisk's docking antennae to prepare it for receiving its first Soyuz craft was conducted in January of 2010.

Shuttle EXPRESS

Atlantis paid the station its next visit on 18 November 2009 during STS-129. In the cargo bay were two EXPRESS logistics carrier assemblies. The EXPRESS (Expedite the Processing of Experiments to Space Station) racks were part of Brazil's contribution to ISS construction in partnership with NASA's Goddard Spaceflight Center. These racks would act as a system of outside storage where experiments and equipment could be plugged in via attachment fixtures. The experiments could then be integrated with the station's power and data lines while operating independently of one another. There are EXPRESS racks on the inside of the station as well. Equipment and experiments can be added or removed as needed, so items can be mounted to the racks either on a temporary or on a semipermanent basis. To help accommodate these racks and other exterior items, the ISS is equipped with a common attach system along various points on its structure. The common attach system acts as a combination of a data interface and a power plug, similar to USB ports for smartphones and computers.

STS-129 continued the process of stowing spare equipment on the outside of the ISS, as *Atlantis* carried up two nitrogen tank assemblies, an ammonia tank assembly, two pump modules, a part for the truss's mobile transporter, and a spare end effector (or grapple fixture) for the SSRMS. Two experiment racks were also installed, containing materials and coatings being considered for future spacecraft designs. This equipment was installed on three EVAs. Another item carried up was a UHF communications unit and command panel for the SpaceX Dragon cargo vehicle. The system would allow ISS crewmembers to monitor an approaching Dragon craft and abort the approach if a problem developed. The Dragon was not designed to hard dock with the station. Instead, a Dragon would approach close enough for the SSRMS to grapple and berth it on a CBM, like Japan's HTV craft.

At the conclusion of STS-129, Nicole Stott returned home with the shuttle crew. She became the last ISS crewmember to fly on the shuttle. For all future ISS crewmembers, a smooth return home gliding on the wings of the shuttle was a thing of the past. For a growing group of NASA astronauts, their only trip to and from space would be in a capsule.

Rocket Ride

One member of the growing American fraternity to only fly in a capsule was astronaut Timothy "T. J." Creamer. When Creamer joined the astronaut corps in 1998 after working in a support capacity at NASA for three years, little did he suspect that it would be eleven years before he took his first trip into space and that it wouldn't be on a shuttle. Born to a military family, Creamer was a highly experienced helicopter pilot and colonel in the U.S. Army, in addition to having a bachelor's degree in chemistry from Loyola College. Creamer also has a very healthy sense of humor and likes to joke that helicopters really are high-performance aircraft, since unlike most jets (except for the Harrier jump jet), they can hover and fly backward. Time logged in high-performance aircraft was a requirement for shuttle pilots.

Even though his fellow Penguin classmates from the 1998 class got to fly in space ahead of him due to their assignments and the delays brought by *Columbia*'s loss, Creamer took it all in stride and did his assignments with no complaints. Finally, he was assigned as a flight engineer for ISS Expeditions 22 and 23. He would fly aboard *Soyuz TMA-17* with Expedition 23 commander Oleg Kotov and JAXA astronaut Soichi Noguchi.

Unlike KSC and the Cape Canaveral Air Force Station, which are located close to the surrounding communities of Central Florida, Baikonur is very remote. The only people who live there on a semipermanent basis are the engineers and technicians directly supporting launch activities, plus members of the military. Most Roscosmos officials and cosmonauts typically don't arrive until launch preparations begin. This is especially true during the very cold winter months, such as when *TMA-17* was scheduled to lift off.

As with shuttle crews, Soyuz crews spend about a week or two in quarantine, keeping contact with outsiders to a minimum to prevent exposure to germs that they might then otherwise carry into orbit. The last night before a launch, they spend time staying in crew quarters on-site. On the morning of the launch, before heading off to the prelaunch festivities, new crewmembers sign the doors of the crew quarters rooms they stayed in, while veterans add mission listings to their previous signatures.

After a final medical checkup, the crewmembers don their Sokol pressure suits. The outer cover of the Sokol contains the internal pressure bladder, which would just expand like a balloon until it pops otherwise. To seal the internal bladder, the opening in the front waist area is brought together at a part known as the appendix on the suit's abdominal region. There, a technician simply double knots a rubber band around the appendix to seal the bladder. The seal is very simple, but it is also very effective.

The outer suit is next zipped up, and the crewmembers enter a conference room divided by a glass wall to see family, friends, and members of the press one final time for prelaunch good wishes. In this room, one at a time, each suited crewmember sits in a seat liner as they would on a Soyuz and hook up their air lines so that a technician can check the integrity of their pressure suit. It is all done in full view of people and cameras on the other side of the glass as a final reassurance to the invited guests that crew safety is the first priority.

Once the checks are complete and the crew say their goodbyes, their suits are each hooked up to portable air conditioner boxes to keep their body temperatures comfortable on the trip to the pad. The crew then head outside and meet with members of the launch commission. The Soyuz commander steps forward to present his crew and salutes while declaring that they have finished preparations and are ready to fly. Once the launch commission gives its approval, the crew board a bus and head out to the launchpad.

Along the way, the bus stops about two hundred yards short of the pad, and the crew get out to perform a ritual that tradition says began with Yuri Gagarin himself: they urinate on one of the bus's tires. As T. J. Creamer explains, this is a very important step: "You are about to go into a position where your legs are above your chest . . . in an elevated position for three hours with no relief tube to urinate into. Simple body fluid dynamics means you don't want to go up there with a full bladder." To do this tradition, each crewmember has to unzip the suit, undo the appendix, urinate, and then suit back up with a technician's help. Supposedly, women crew-members relieve themselves before boarding the bus, yet it is also said that they carry a small bottle of their urine, which they open and dump on the tire to honor the tradition. With this final call of nature complete, the crew reboards the bus and finishes the journey to the pad.

At the pad, the crew climbs up the gantry to an elevator that takes them to the top. Here, they are accompanied by a couple of technicians who, according to Creamer, are built well enough to play on professional American football teams. They are there to make sure that none of the Soyuz crewmembers falls, as the gantry was originally designed when the R-7 rocket was just an ICBM. The gantry wasn't intended to handle three crewmembers wearing bulky pressure suits. According to Creamer, "You become a very sensitive item at this point because you can't be replaced quickly enough." At the entrance to the elevator, one of the technicians gives each of the Soyuz crew a "good luck" kick in the rear with his knee as part of the tradition. Creamer described the kick with a smile on his face: "I wouldn't even say it was a short knee. He kicked a forty-yard field goal."

As the crew climb up and give one final wave to the crowd, they board the elevator, which Creamer says was really only designed for two people. But three travelers in pressure suits are pressed into it along with one technician for the ride to the top. It takes a bit of repositioning for everyone to fit in so that the door closes cleanly.

At the top of the gantry, the crew arrive at the Soyuz. One of the side covers on the launch shroud is removed, exposing the spherical Soyuz orbital module, where two technicians are preparing things. Entry to the Soyuz is through a side door on the orbital module, which in the early days of the program was also the egress hatch for EVAs. The purpose of at least one of the technicians is to help strap each crewmember in. Two crewmembers each climb inside and down into the descent module, where they stand on the center seat before shimmying over to their assigned seats and strapping in. A launch technician is there to help as much as possible during this difficult task. According to Creamer, this technician really earns his pay and does his job very well. Once two of the crewmembers are strapped in, the technician climbs out and the commander climbs in. Then the technician hangs upside down through the descent module's hatch and does the same job for the commander, strapping him into the center couch.

When everything looks good, the hatch to the descent module is closed and the side door on the orbital module is sealed up. The launch shroud is added next, and so begins the three-hour wait with periodic equipment checks. There isn't much space in the descent module, which has been described as everything from a bathtub with three guys crammed into it to the front seat of a Volkswagen Beetle after climbing in through the sunroof.

The spacecraft commander has a wand that he can use to press buttons on the Soyuz computer, but prelaunch procedure also involves turning a couple of valves in the Soyuz. It turned out that Creamer's arms were long enough to reach both valves even though they were in front of Oleg Kotov. So when the time came to turn them, Kotov gave one look at Creamer that said, "Please?" Among the cosmonaut ranks, Kotov is a very cordial individual. T. J. Creamer said, concerning his feelings toward his commander, "When I finally grow up, I want to be just like Oleg." The camaraderie among the *TMA-17* crew and, indeed, among the five members of the Expedition 22–23 crew on orbit was excellent and even surpassed the good camaraderie of the military units Colonel Creamer had served with in the army: "The five of us together was just an amazing and wonderful time."

Before heading out to the pad, each crewmember takes an over-the-counter pain pill to help with the discomfort, but after two hours of lying on your back in a fetal position with an hour left before liftoff, one might begin to feel like it is a form of mild torture. The quickest way to get relief is to launch, since a scrub might mean more time waiting in the capsule before technicians can get the crewmembers out. But compared to the shuttles, Soyuz craft have very rarely ever had launch scrubs.

Finally, liftoff comes, and the acceleration is smooth and constant, almost like an airliner taking off, but for a longer period. Peak acceleration load on a Soyuz ascent is typically about 3.5 g's. Cameras inside the Soyuz transmit images of the crew during ascent. Most Soyuz commanders typically have a talisman tethered to the control panel to act as a microgravity indicator. So when the engines stop as the Soyuz reaches orbit, the tether line goes slack, indicating zero gravity. The talisman is a mascot of sorts for the Soyuz commander. In Oleg's case, it was a toy black cat. For the crew, it is relief at last as the prelaunch discomfort subsides in zero-g freefall.

After an hour of work to prepare the Soyuz for orbital flight, the crew can remove their Sokol suits and put on their flight coveralls. The commander takes his suit off first since it is easier for him to move around in the orbital module, but the other two crewmembers get out of their suits soon after. The total internal volume of the Soyuz craft is about as large as a 1960s Volkswagen bus. For sleeping, two of the crew use their ISS sleeping bags in the orbital module, while the commander sleeps in the descent module so that he can handle communications with the ground as needed.

Two days later, the Soyuz rendezvouses and docks with the ISS. This is done over the Russian ground stations as the Russians typically don't use NASA's TDRS network. There are two windows on the Soyuz, plus the commander's periscope, and Creamer was treated to an awesome sight when he first saw the ISS on a fly-around inspection pass: "It was at a wonderful lighting time, and the colors [of the station] were outstanding. It pretty much burned into my mental memory [as] stunningly beautiful in terms of visual coloring and majesty in size." A reasonable comparison might be that seeing the ISS in person would be several times better than the best IMAX film image of it. In person, one can more easily comprehend the size and the complexity of a station that to that point Creamer had only seen in pictures, computer simulations, and video footage.

After the crew docks and the hatches are opened, the newcomers enter the ISS into the arms of their colleagues. When *TMA-17* arrived, it was a couple of days before Christmas, so Kotov and Noguchi entered the ISS with Santa Claus hats, while Creamer entered wearing a pair of green curly toed elf slippers. They brought a bag of "gifts" in the form of fresh food for their crewmates. Typically, a press conference with officials in Moscow and family members takes place before the combined crews get to work. Then it becomes business as usual, starting with a "tip to toe" safety briefing by the resident crew to bring the new people up to speed.

Life aboard the ISS

During his time in orbit, Creamer conducted the usual experiments and station maintenance in addition to his other tasks. Tasks are blocked out by the ground on a daily basis for what to do, but they are not necessarily scheduled in a strict fashion. The lessons learned during the Skylab and Mir programs have finally been taken to heart by NASA controllers.

Each crewmember has assigned tasks, and they might involve taking data from science experiments, monitoring equipment racks, or taking samples from surfaces, water, and air to check for pathogens. Periodically, a piece of equipment, such as a water recycler, might also need to have its flow redirected. Mission control can operate many of the systems on the ISS remotely with a simple mouse click, but some tasks require a crewmember on orbit to flip a switch and monitor the event if a problem develops.

For newcomers on the station, there is a ramp-up period where they get up to speed in how best to do things and figure out for themselves the best method for scheduling individual tasks. Since Kotov's crew arrived around Christmas, they had the benefit of additional days off to help with their transition. The new crew was well prepared when the big stuff started, as things would soon get very busy by ISS standards.

With the regular interaction of all the crewmembers, everyone was on hand to help out one another if somebody was falling behind. All the crewmembers were trying to stay ahead of the mission timeline, as they wanted to keep free time available for unscheduled activities, such as taking pictures of Earth and having coffee breaks with crewmates. The interaction between crewmembers was very good for crew morale, and having a fresh set of eyes while looking at an experiment procedure might enable one to interpret what is being asked in the procedure and to explain it in common terms if the other person isn't entirely clear about it after looking at the instructions.

T. J. Creamer had a chance to perform maintenance above and beyond the call of duty when the Advanced Biological Research System (ABRS) science rack in the Destiny laboratory malfunctioned. The rack contained several individual compartments with their own self-contained environments to grow various samples of plant microbes. The rack was the responsibility of the science group at KSC.

When the science rack began malfunctioning, the fault was traced to a set of fuses inside it. The rack wasn't designed to be serviced in orbit, but thanks to some excellent preparations on the ground by a team at KSC and project leader David Cox, they came up with a way to troubleshoot the problem, locate the fuses, and replace them with new ones. It was T. J. Creamer's responsibility to take those procedures for what was essentially depot-level maintenance and modify them for microgravity. According to Creamer, Cox's team conveyed just how much precision was needed for the task, to the point that he felt sort of like an explosives technician trying to defuse a bomb who has been told, "Don't cut the red wire!" Creamer got the rack up and running again in quick fashion, and the KSC team was given an award by NASA for their excellent work. The KSC scientists and engineers credit much of the success of the ABRS to Creamer and his orbital repair work.

A Room with a View

During T. J. Creamer's time on orbit, the ISS hosted three shuttle missions. STS-130 arrived first, with *Endeavour* carrying the Tranquility node with the cupola attached to one of its CBMs. The cupola is a large segmented window. It has one central window and six windows spread around it in a hexagonal pattern. It sort of resembles a turret station on a World War II bomber but without the gun. From the cupola, astronauts could look down on the world as well as at the ISS almost from end to end. When not in use, external covers are folded over to protect the windows from micrometeoroid debris. Rather than just being a place for astronauts to enjoy the view, the cupola also is equipped with a control station for the SSRMS and Dextre.

The Tranquility node was berthed on the port-side CBM of Unity directly opposite the Quest airlock. The cupola occupies Tranquility's nadir port. To help protect the outermost CBM of Tranquility from micrometeoroid debris, PMA-3 was docked to the end of it for storage. PMA-3 acts as a hardware backup in case PMA-2 on the front of Harmony is damaged, and the pair can be switched as needed. After berthing by the arm and three EVAs, the new node was opened for business, and it became a popular place for crewmembers to hang out in during their break times.

The view generated by this new seven-window array is very dramatic. On a couple of occasions, Creamer would bring newcomers into the node while the shutters were closed and have them close their eyes. Creamer said, "I would then open the shutters and have [the newcomers] open their eyes, and every single time I did that, I watched people cry. It is astoundingly, overwhelmingly, awesomely humbling to look out that window."

When STS-130 arrived, T. J. Creamer had been in space for about five weeks, and most veteran long-duration crewmembers say that five weeks is about the time when one finally has "arrived" in space. He had achieved a level of station system awareness and body control where he could get from place to place and find things without wasting energy or time. So Creamer was treated to something special when the next shuttle crew visited. "To watch the shuttle guys come up who have only been on orbit for two days and [who] have not been free floating in anything other than their tiny little cabin basically . . . it is really kind of funny to watch," as Creamer recalled.

During the time period with a docked shuttle, the priority is helping the shuttle crew to get their assigned tasks done as quickly as possible. It meant that the station crew fell behind as the newcomers needed assistance from ISS crewmembers to find things, conduct short-term experiments, and learn how to move efficiently from place to place in the cavernous interior of the ISS. Therefore, to compensate for this on the next two shuttle missions, the Expedition 23 crew got as much work done as they could ahead of time so that they didn't fall behind.

Williams and Surayev returned home in late March, turning over work to Expedition 23 with Oleg Kotov in command. The second half of Expedition 23 in the form of crewmates Aleksandr Skvortsov, Mikhail Korniyenko, and Tracy Caldwell-Dyson arrived on 4 April aboard *Soyuz TMA-18*. The crew didn't have long to wait for their next visitors, as STS-131 lifted off from KSC in the predawn darkness of Monday, 5 April 2010, on the final night launch of the shuttle program. *Discovery* docked two days later. Having three new ISS crewmembers still getting their space legs in addition to seven shuttle crewmembers on board made things a little more hectic, but the team made it work.

Discovery's cargo bay held the Leonardo MPLM for more supply off-loads. This was the last shuttle mission to have seven crewmembers on board, and it was the second-to-last mission for *Discovery* before retirement. The crew included Naoko Yamazaki, a JAXA astronaut who helped engineer the Kibo laboratory; Clay Anderson; and educator-astronaut Dottie Metcalf-Lindenburger. In addition to being the third educator-astronaut to fly from the 2004 class, Metcalf-Lindenburger was also the first alumna from Space Camp in Huntsville, Alabama, to become an astronaut. She also set another record; with nineteen letters in her last name, she has the longest name on a mission patch. When Metcalf-Lindenburger was selected to become an astronaut candidate, the people in charge of flight gear at Ellington Field near JSC, where astronauts fly T-38 jets, just labeled her storage locker and equipment "M-18" to represent the first letter of her last name and eighteen additional letters. So among some of her colleagues, she is known as Dottie M-18.

Upon reaching the ISS, Clay Anderson felt kind of like he was home again; although the ISS had grown a bit in the past three years, much of it was familiar. Having Oleg Kotov on orbit also helped. Anderson had also cultivated a friendship with cosmonaut Mikhail Korniyenko, as the two men had joined their respective countries' astronaut programs in 1998. Anderson nicknamed Korniyenko "Corn," in reference to his own Nebraska heritage; Korniyenko nicknamed Clay Anderson "Glue," since "clay" in Russian means "glue." So when the two men met each other on dock day, Clay called out "Corn!" while Mikhail called out "Glue!" before the two men embraced one another.

Anderson and lead spacewalker Rick Mastracchio conducted three space walks to replace an ammonia tank on the station, retrieve a seed experiment from the Kibo external rack, and replace a faulty gyro. Anderson's experience conducting staged EVAS from the Quest airlock came in handy at the conclusion of their EVAS. Both he and Mastracchio took time to stow equipment properly since Anderson knew where it all went, having done it before on Expedition 15. So when the airlock door was opened and T. J. Creamer entered to give them a hand, he was pleasantly surprised that most everything was already put away.

NanoRacks

Another cargo inside *Discovery* was the first elements of the NanoRacks platform. NanoRacks is a commercial firm with the intended goal of making space research available to everyone. Inside the MPLM were their first two payloads, FirstLab and Cubelab-2. Both systems were installed into one of the EXPRESS racks inside the Kibo laboratory. The Cubelab system allows customers to fly small scientific payloads to the ISS in sealed cubical compartments measuring about three inches per side. The cubes are plugged into the rack via a USB port and a cooling water hookup.

Data from the experiments is sent automatically to the ground. Once the modules are plugged in, there is no need for a crewmember to tend to them. And once each experiment is concluded, the cube is disposed of entirely, making room for the next one and keeping its contents from contaminating the station's environment. For further study, some NanoRacks payloads have been returned home aboard Soyuz or Dragon flights, but a round trip is more expensive than one way. As of 2013, several companies and government agencies have used NanoRacks' system on the ISS for research. While the experiments themselves are kept confidential, each proposal has to pass a strict review process for safety and to ensure proper research is being conducted, as opposed to just flying collectibles into space.

The Final Assembly Push

Discovery undocked on 17 April and returned home on the twentieth. On the ground, everyone was getting the sense that the shuttle program was nearing its end. When *Discovery* returned from STS-131, there were only three more missions scheduled on the manifest. *Atlantis* would fly next on STS-132. Another mission was added to the manifest for *Endeavour* to fly the final ISS module, and there was a good chance that yet another mission would be added as well. But after that, the shuttle would be retired permanently.

On 14 May 2010, *Atlantis* lifted off for the ISS. In its payload bay was the Russian Rassvet (a name meaning "Dawn") Mini-Research Module. Like Poisk and Pirs, it was a docking compartment with storage bays. Rassvet was a little larger than the other two modules. Its internal size meant that it could be used for additional storage while also providing a docking port for use by Soyuz and Progress craft. At the time Rassvet was launched, it would be the last Russian module flown to the ISS, although there are plans to replace the Pirs docking compartment with a science module called Nauka (a Russian name meaning "Science") in the near future. Since Pirs is about a decade old, the Russians want to decommission and release it before systems on board start deteriorating. If and when Nauka flies, it will become the primary science module for conducting Russian experiments. Like Zarya, Nauka is based on the TKS spacecraft and designed to dock autonomously with the station.

At the end of Expedition 23, it was time for Kotov, Noguchi, and Creamer to come home. In orbit, everyone's spine stretches a little as the lack of gravity causes expansion. In Creamer's case, he was at just the upper limit for sitting in a Soyuz crew couch, and NASA took steps to ensure that he would fit before beginning his ISS training in 2002. As the crew boarded *TMA-17* to come home, Creamer strapped in and noticed that he didn't quite fit in his suit liner anymore. He was no longer an egg in its carton but rather an egg teetering on the edge. But as the g-forces built up on reentry, Creamer suddenly felt a thump sensation as the gravitational forces caused his body to pop into the couch properly. He was now properly protected and pulled the straps in tight, feeling happy and confident that he could survive reentry and landing with no problems. *Soyuz TMA-17* touched down safely.

Permanent Logistics Module

It would be nine months before the next shuttle would visit the ISS. *Discovery* was scheduled to fly its last mission, STS-133, in November 2010, but large cracks in the foam of the shuttle's ET, found after a launch scrub, were traced back to cracked metal stringers on the structure of the tank itself. This forced the stack to get rolled back to the vehicle assembly building for repairs. During the wait, astronaut Tim Kopra, who was supposed to be the lead spacewalker for this flight, was injured in a bike crash and had to be scrubbed from the mission. Astronaut Stephen Bowen, who was a spacewalker on STS-132, took his place since he was already up to speed on the assigned mission tasks. *Discovery* lifted off into a clear Florida sky on 24 February 2011 for its final mission into space.

Discovery's primary payload was the Permanent Logistics Module (PLM) Leonardo. Leonardo had flown as an MPLM on many previous flights, but it was decided by NASA to outfit it as a permanent storage locker for the station since internal space would be at a premium in the coming years. The PLM would also allow equipment previously stored at various locations inside the ISS to be moved to one central area. As part of the PLM's modifications, coverings were removed from the never-flown Donatello MPLM. They were reinforced with Kevlar and installed on Leonardo. Additional micrometeoroid debris shields were also fitted, and the CBM seal was replaced. These alterations gave Leonardo the same level of protection as other ISS modules.

Discovery spent nine days docked with the ISS. Two EVAs were conducted by Stephen Bowen and Alvin Drew to place some new equipment outside the station and stow a failed ammonia coolant pump that had been replaced on a previous mission. The failed pump would be returned to Earth on STS-135, a mission that had only just been added to NASA's manifest a couple of weeks before *Discovery*'s final launch.

Robonaut 2

Another "crewmember" sent to the ISS aboard *Discovery*'s final flight was Robonaut 2. The Robonaut is a machine resembling a human upper torso with two arms and two hands, plus a head with cameras and sensors located inside it. But instead of a lower torso with legs, the Robonaut features a special giant "finger" that allows it to firmly anchor itself to parts of the station. It is hoped that the Robonaut will allow for maintenance to be performed on the station without the need for risky EVAs, as an ISS crewmember can operate it by telepresence from a special workstation. Future versions of the Robonaut could one day perform critical tasks on future space missions, perhaps to the moon, Mars, or an asteroid, allowing the operator to remain safely inside the spacecraft.

Twins in Orbit . . . Almost

Discovery's return from orbit marked the final flight of the space shuttle with the most time in the fleet. *Endeavour* would fly next on STS-134. When the mission assignment came up over a year earlier, it had been hoped that its commander, astronaut Mark Kelly, might mark a space first by visiting his twin brother, Expedition 26 commander Scott Kelly, on orbit. Both men were naval aviators. They graduated from the Naval Academy in the same class, and both got their pilots' wings the same year. Mark Kelly went on to fly A-6 Intruders, while Scott Kelly flew F-14 Tomcats. Both men joined NASA in 1996 as shuttle pilots. At NASA their careers took slightly different paths. Though Mark Kelly flew two missions as pilot before Scott did, Scott Kelly flew as a shuttle commander first. After commanding STS-118, Scott Kelly switched over to the ISS program in preparation to fly on Expedition 25 as a flight engineer, from which he assumed command of Expedition 26.

However, with the launch delays due to STS-133's tank problems, the window for a meeting of twins in orbit never came about. For the Kelly brothers, the meeting became the least of their concerns. On 8 January 2011 Mark Kelly's wife, U.S. congressional representative Gabrielle Giffords, was shot in the head during an assassination attempt by a lone gunman in Tucson, Arizona, at a public gathering. The attack left six people dead and Giffords critically injured. Scott Kelly received regular updates in orbit, while Mark Kelly took a leave of absence to be at his wife's side during her recovery. To help continue the training flow for the mission, NASA assigned astronaut Fred Sturckow as backup commander. Giffords was eventually transferred from Tucson to a rehabilitation facility in Houston, where she continued to make steady and almost miraculous progress. So Mark Kelly made the difficult decision to go back to work and fly the assigned mission with the rest of his crew. Scott Kelly returned from orbit in March and, after his rehabilitation period, helped out with Giffords's recovery while Mark Kelly continued his training.

Even though the Kelly twins never flew together, having had an identical twin in orbit did help in one aspect. After *Endeavour* docked with the ISS on 18 May 2011, Mark Kelly went searching for something. While Mark was helping with his wife's recovery, he wasn't on hand at NASA to be fitted for his in-flight clothing. Therefore, the measurements used for his pants were totally wrong, and the sets of pants provided were too large. Thankfully, *Endeavour*'s commander located a used set of pants sized for his brother, and he wore those for the remainder of the mission, as they fit perfectly.

Never Say "Never": The Alpha Magnetic Spectrometer

The final U.S. module flown aboard *Endeavour* was the Alpha Magnetic Spectrometer (AMS). It was the brain child of Nobel Prize–winning physicist Samuel Chao Chung Ting. Ting was born in Michigan to parents of Chinese ancestry who had both attended the University of Michigan. The family moved back to China not long after he was born and got caught up in the invasion by Japan during the 1930s. After the war years and the Communists' seizing of power in mainland China, Ting continued his schooling in Taiwan before being invited to attend the University of Michigan, where he excelled in mathematics and physics, earning bachelor's degrees in both. As a physicist, he won a Nobel Prize with linear accelerator researcher Burton Richter for discovery of a new meson particle.

In 1995, as a research physicist at MIT, Ting proposed that a cosmic-ray particle detector placed in orbit could be used to help in the discovery of unusual materials found in the universe, perhaps including dark matter. At the heart of the new detector is a large magnet, used to help detect these cosmic-ray particles and analyze their chemical makeup. The proposal was accepted, and an early version of what became known as the Alpha Magnetic Spectrometer was flown on STS-91 in June of 1998. Data from the early system was used to help flesh out the design of a larger and more capable AMS.

It was felt that the power and time requirements needed for the newer AMS were beyond the capabilities of the shuttle, so plans were made to fit the new system aboard the ISS. But when *Columbia* was destroyed in February 2003 and the decision was made to retire the shuttle program by 2010, the newly built AMS was not part of the flight manifest. Many scientists and engineers had worked on the AMS, and millions of dollars had already been spent. So Ting didn't take no for an answer. He successfully lobbied Congress, and these efforts got STS-134 added to NASA's mission manifest with the AMS as its payload.

The original plans were for the AMS to be equipped with a superconducting magnet cooled with cryogenic liquid helium. This would have given the instrument excellent sensitivity but a mission life of only about three years until the helium supply was used up. Problems were also encountered with the magnet design, so it was decided, instead, to fly the instrument with less powerful magnets. The instrument wouldn't be as sensitive, but it could stay in operation for a decade or more, allowing for more data collection.

On 19 May 2011 *Endeavour*'s RMS handed off the AMS to the SSRMS, which berthed the instrument at its new permanent location on top of the S3 truss segment. It has been collecting readings ever since. STS-134 also delivered a couple of more EXPRESS racks with equipment spares. Finally, *Endeavour* handed off its OBSS, which was then hooked up to the station's power grid during one of the EVAs. The OBSS would be available for use as an extension to the station's SSRMS, in case a future problem cropped up needing an arm extension to fix it. With that, ISS assembly was complete.

The Final Shuttle Visit

On 8 July 2011 the space shuttle *Atlantis* stood on launchpad 39A at KSC for the final time. The mission hadn't existed in NASA's manifest a year earlier, but lobbying work in Washington and at the NASA centers got it added to the roster. Part of the reason for NASA's decision to fly *Atlantis* came down to having hardware available. For all the missions after STS-107, NASA made sure to have a second shuttle undergoing launch preparation just in case critical damage to the orbiting shuttle's heat shield prevented a safe return. For STS-134, *Atlantis* served as the launch-on-need shuttle for *Endeavour*, but for this final flight, there would no shuttle to back up *Atlantis*.

STS-135 had flown thanks to a successful refurbishment of an ET damaged by Hurricane Katrina in 2005 at NASA's Michoud facility. The tank was flown on STS-134. That left one more tank and a set of SRBs available for use on STS-135, but a question remained regarding what should be done

53. Paolo Nespoli on *Soyuz TMA-20* photographed the completed ISS with the space shuttle *Endeavour* docked to it. Visible modules include the Tranquility node with cupola, the Leonardo PLM, and Rassvet on Zarya's nadir port. The Pirs and Poisk modules hosting Progress and Soyuz craft are also visible, as is the Johannes Kepler ATV.
Courtesy NASA.

if *Atlantis* were to become stranded at the ISS. The crew for STS-135 was reduced to four people, an all-veteran crew consisting of Chris Ferguson, Doug Hurley, Rex Walheim, and Sandy Magnus. The crew visited Russia to be fitted for Sokol pressure suits and Soyuz seat liners in case they had to come home on Soyuz capsules.

The mission of STS-135 would primarily be stockpiling. The Raffaello MPLM was packed with enough supplies and spares to keep the ISS going through 2012. A robot refueling experiment was also sent up. This experiment rack would allow the Dextre manipulator unit to test the feasibility of using a robot system to refuel satellites in orbit. *Atlantis* would also bring back the busted ammonia pump module that was replaced in 2010 so that engineers could study it and try to find out what prematurely failed on it.

The day before launch, the weather was very gloomy with steady rain and lightning at KSC for most of the morning and afternoon. All indications were that the flight would be postponed. But countdown proceeded as normal as conditions on 8 July improved. All along the Space Coast, the viewing sites, roads, and even some waterways were packed with massive crowds of spectators. It was estimated that nearly one million people turned out to watch this final launch of the shuttle program.

For many contract workers at KSC, STS-135 would be their final job. Some would remain to help prepare the orbiters for display in museums. But for many more, their careers at KSC would end once the shuttle landed for the final time. It would be many years before another manned spacecraft would rise into the sky from KSC's Launch Complex 39.

Liftoff occurred at 11:29:03 EDT, and the launch was a couple of minutes late due to an indication that one of the access arms on the launchpad had not retracted fully. The launch window to rendezvous with the ISS is only five minutes long, and liftoff is typically targeted for the middle of the window. Once controllers verified that the arm was out of the way, *Atlantis* roared off the pad with only a few seconds to spare. The shuttle stack did its characteristic roll to the proper heading and exceeded the speed of sound just before it punched through a low cloud deck and disappeared from sight. About ten minutes later *Atlantis* was in orbit. Everyone who saw the launch was left with fond memories of that day, colored with uncertainty about the future of America's space program.

Two days later, *Atlantis* performed its backflip rendezvous pitch maneuver and docked with the ISS one final time. No EVAs were conducted by the shuttle crew, although ISS crewmembers Mike Fossum and Ron Garan performed an EVA on the fifth flight day to secure the ammonia pump and attach the fueling experiment. Rex Walheim underwent special training for this mission. When STS-135 was manifested, he tested and refined the hands-on EVA procedures in the NBL himself after Fossum and Garan launched into orbit, in order to help walk them through the tasks. During stowage of the ammonia pump, Walheim said, "Take a look around, Ronny [Garan], you're the last EVA person in the payload bay of a shuttle." Garan acknowledged and continued his work. After conducting additional maintenance tasks on the Russian segment, the space walk was completed successfully.

Several more days were dedicated to equipment stowage, with the shuttle and ISS crews acting as one to facilitate these final transfers. The MPLM was loaded back into the shuttle's cargo bay with almost three tons of cargo for return to Earth. As one final ceremony before the crews parted company, the shuttle crew unveiled a small American flag aboard the ISS. The flag had flown on the first shuttle mission, STS-1 in 1981, and it would remain on the ISS until the next manned U.S.-built spacecraft visited the station to retrieve it and take it home.

The next day, *Atlantis* undocked from the ISS. Ron Garan rang the station's bell twice as part of the naval tradition and said, "*Atlantis* departing the International Space Station for the last time." The shuttle did a fly around inspection of the station before extending its distance until it was just a tiny dot visible from the ISS. After a couple of more days in orbit, *Atlantis* landed safely, concluding the shuttle program after three decades of service. From that point on, the ISS would be America's only manned presence in Earth orbit.

China's First Space Station

In September 2011 China launched a new module, Tiangong 1 (a Chinese name meaning "Heavenly Palace"), aboard a Long March CZ-2F rocket. It is reported to be a test module of a future Chinese space station design. The module is powered by solar arrays and apparently has an operational lifespan of two years. In November of the same year, *Shenzhou 8* launched unmanned into orbit and successfully docked with the Tiangong module, giving China confidence in their growing spaceflight capabilities.

Superficially, the Shenzhou spacecraft resembles the Soyuz, as it features orbital, descent, and propulsion modules similar in shape to their Russian counterparts. Although there are some differences, China based their design heavily on the Soyuz, as they had Russian help early on. The descent module is almost identical in shape yet larger. The orbital module is more cylindrical than spherical. Chinese pressure suits for launches and EVAs are based on Russian Sokol and Orlan suits respectively.

Tiangong 1 superficially resembles an early Salyut station, as it features a stepped cylinder design with a pair of solar arrays. But due to weight limitations with the CZ-2F rocket used to launch it, Tiangong 1 is smaller than Salyut. Inside, it only has enough room for two sleeping racks along with some science and exercise equipment.

After a successful unmanned *Shenzhou 8* docking, three taikonauts were launched to Tiangong 1 on 16 June 2012 aboard *Shenzhou 9*. The crew included veteran commander Jing Haipeng on his second spaceflight; rookie flight engineer Liu Wang; and the first female Chinese space traveler, Liu Yang. The crew of *Shenzhou 9* occupied Tiangong 1 for about two weeks before returning home to a touchdown site in Mongolia on 29 June 2012. Allegedly, there are plans to fly future modules based on the Tiangong module to develop a more capable Chinese space station complex, but until China comes up with a fully automated resupply capability, missions to Tiangong-based stations will likely be shorter in duration than missions to the ISS.

There have been some calls internationally for China to perhaps take part in the ISS program, but the invitation to join has not been extended by NASA. Indeed, many ISS participants have looked at China's program with caution since it seems to be controlled by the military rather than by a civilian agency. Another incident that hasn't helped Chinese space relations was a test by its antisatellite program (ASAT) that took place on 11 January 2007. A Chinese missile was launched at a deactivated weather satellite in a five-hundred-mile-high polar orbit, and it scored a direct hit. While the test was successful, it was not announced ahead of time and produced a very large debris cloud with over two thousand pieces of debris large enough to be tracked by the ground and potentially thousands more too small to be detected.

Over time, several pieces of debris have drifted lower into the path of the ISS; on numerous occasions, the station has had to perform orbital adjustment burns to avoid potential collisions with this debris. Debris avoidance has become a fact of life in Earth orbit, with the remains of dead satellites and spent rocket stages being common. Periodic increase of solar activity causing Earth's atmosphere to expand helps to sweep the sky of debris in lower orbits, but it can still take many years before debris from a high orbit drops low enough to burn up in Earth's atmosphere, producing a potential hazard to space navigation in the meantime.

Donald Pettit's Return to Orbit

On 21 December 2011 at the Baikonur Cosmodrome, an R-7 rocket stood on the same launchpad where, over fifty years earlier, Yuri Gagarin had taken mankind's first trip into space. The craft perched on top was *Soyuz TMA-03M*. The craft was a new variant of Soyuz that had begun flying the previous year. This new design weighed about seventy kilograms lighter

54. On 21 December 2011 Expedition 30 crewmembers Kononenko, Pettit, and Kuipers lift off from the Baikonur Cosmodrome's Site No. 1 aboard *Soyuz TMA-03M*. Courtesy NASA.

than the previous variants. This particular vehicle would be only the second Soyuz to fly in space since the shuttle's retirement. On board was a veteran crew. Commanding the craft was Oleg Kononenko on his second spaceflight. To his left sat Dutch ESA astronaut André Kuipers, taking his second trip into space. To their right sat probably the keenest mind of the group, NASA astronaut Don Pettit, on his third spaceflight. All three men were part of Expedition 30 and would join their colleagues Dan Burbank, Anton Shkaplerov, and Anatoli Ivanishin already in orbit.

Expedition 30 had some great plans in store. If all went well, Pettit would use the SSRMS to snag the first SpaceX Dragon spacecraft to visit the station in the coming months. There were also plans to take part in an experiment with amateur and professional astronomers in San Antonio, Texas, to see if spotlights and a laser could be used to signal the ISS from the ground. In orbit that day, Dan Burbank was shooting photographs of Comet Lovejoy, which passed within 140,000 kilometers of the Sun's surface without burning up. The AMS was continuing to produce data, as were the racks in three scientific modules and additional experiments fitted outside the station. The next six months were shaping up to be very busy ones aboard Earth's manned outpost.

It was very cold at the time of launch, with a temperature of minus eighteen degrees Celsius (zero degrees Fahrenheit). But at least it was a beautiful night to launch, with no clouds in the sky, unlike the day on which Burbank's crew launched, which lifted off into orbit under gray, overcast skies during a snow shower. In the early evening darkness at 13:16 GMT, the Soyuz lifted off successfully. Two days later, the Soyuz docked with the ISS, and the three men got right to work, continuing the job started by countless crews who had come before them. Mankind had learned a lot, but it still had much more to learn. And it was time to get down to the business of unlocking but a few of the secrets of Earth, humanity, and the universe.

Epilogue

At the close of 2011, over fifty years had passed since the first manned flights of Yuri Gagarin and Alan Shepard. Two countries that had once started out as adversaries in a contest for global influence had become well-established partners in spaceflight, even if they hadn't become firm friends.

Periodically, the relationship has been an uneasy one. It was tested in early 2014 when Russia annexed Crimea from Ukraine after unrest in that country led to the ouster of its Russian-backed president. To this day, ongoing tensions in the region continue to cast a shadow over political and economic relationships between the United States, Europe, and Russia. There have been calls to have NASA assume full control of the ISS, but that type of intervention is simply not possible. Put simply, the Russians know their modules best and have the expertise to keep them operational, while NASA does not. Those modules form the backbone of the ISS, and they can't be replaced or bypassed easily.

The Russians do not have the capability to operate or maintain NASA's hardware, either, for any more than a brief period of time. A split of the ISS partnership means Roscosmos would no longer have a destination for their manned space program or enough income to build the additional Soyuz and Progress spacecraft needed to support it. The Soyuz currently is the only method for sending people up to the station, but it is believed that commercial manned spacecraft for NASA's use will be ready for flight testing in 2017 and operational missions by 2018.

But how long can the ISS remain operational? The Russians recently stated that they plan to support the ISS through 2024. By that time, the RSM will be nearly twenty-five years old, twice the age of *Mir* when it started having problems. The United States wants to operate the ISS well into the 2020s also. But the oldest NASA modules will be pushing two decades in age, and critical systems will likely need refurbishment or perhaps replacement in that time.

In preparation for that, the ISS already has an electronics soldering station aboard the ISS, and in September 2014 a SpaceX Dragon resupply capsule launched a prototype 3D printer to the station. It is hoped that this printer will prove useful in manufacturing spare parts on an as-needed basis. But even with these capabilities, there likely will still be problems that crop up that will require more-creative solutions to repair them.

There are additional concerns about other equipment, such as the EMU and Orlan space suits, as they get used more often. Astronauts Luca Parmitano and Chris Cassidy had to cut short an EVA in July 2013 when water from Parmitano's cooling system backed up into his helmet, threatening to drown the young Italian astronaut.

Even with the uncertainties about the station's long-term future, the International Space Station can easily be tracked in the sky both by the naked eye at night and by software applications for computers and smartphones. As of 20 November 2013, the fifteenth anniversary of the launch of the Zarya module, the station had travelled 85,916 orbits during 5,479 days, with it being occupied for 4,766 of those days. It has conducted over 1,500 scientific experiments thanks to partnerships with over sixty-eight countries.

In popular culture, the ISS has replaced the space shuttle as NASA's flagship manned space program. It has been featured in educational programs, music videos, fictional television shows, and documentaries. In the late summer of 2013, the ISS also briefly shared movie screens with actors Sandra Bullock and George Clooney in the space disaster motion picture *Gravity*, which became a worldwide hit with the general public. Several ISS astronauts have become well known in their home countries. A prime example is Chris Hadfield, who became the first Canadian commander of the ISS during Expedition 35 in early 2013.

But even with this spotlight, most of the day-to-day activities in orbit aren't covered all that much by the major news organizations, leaving it instead to various websites and online blogs to provide coverage unless anything "newsworthy" happens. NASA television provides daily coverage of the station, allowing viewers to periodically peek in to see what is happening on board, thanks to cameras mounted both inside and out.

Both before and during its construction, the ISS was criticized as being a waste of money and the only project NASA could take part in because it lacked the focus and budget to do the "exciting" missions to other worlds.

While there is some validity in those points, they tend to gloss over the fact that a big key to the success of future missions beyond Earth orbit will likely lie not just with the exciting bits of the launch, landing, and return but also with a journey itself, potentially lasting weeks, months, or even years.

There are still a lot of questions that need to be answered as to how well a human body will tolerate such journeys both physically and mentally. Even though both the Russians and the Americans have answered many of those questions by breaking endurance records on previous missions, it always helps to collect more data as science gathering tools improve. To that end, in March 2015 both NASA and Roscosmos launched cosmonaut Mikhail Korniyenko and NASA astronaut Scott Kelly into orbit on a year-long mission to the ISS. Scott's twin brother, Mark Kelly, was monitored medically to see what physiological changes took place during that time. The data collected from their mission is intended for use to help understand the space environment better and potentially reduce the risks associated with long-duration journeys in space.

Regardless of whether one holds the ISS program in high regard or considers it an expensive diversion from other types of space exploration, the International Space Station has done one important thing: it has given many people the opportunity not just to visit space for a few days but rather to live and work there for long periods.

The famed explorer Roald Amundsen may have led the first expedition to successfully reach the South Pole in Antarctica and return, but it took individuals like Richard Byrd to spend months and years on the continent, trying to figure out how to live there while conducting sustained science and data gathering. These early efforts led to the permanent establishment of research bases on Antarctica that remain in use to this day. The ISS and future space stations have the potential to offer similar benefits for space research.

Financial support and public interest will help dictate where mankind goes from here in future space endeavors. It could be a few years yet, or perhaps even decades, before humans set foot on the moon again or other worlds for the first time. But contributions from the stations and laboratories that have flown before will be no less important to space exploration's future than missions that commanded the headlines of decades past.

Sources

Books

Abramov, Isaak P., and A. Ingemar Skoog. *Russian Spacesuits*. Chichester, UK: Praxis, 2003.

Burrough, Bryan. *Dragonfly: An Epic Adventure of Survival in Outer Space*. New York: HarperCollins, 1998.

Chertok, Boris. *Rockets and People*. Vol. 4, *The Moon Race*. Edited by Asif Siddiqi. NASA History Series, NASA SP-2011-4110. Washington DC: National Aeronautics and Space Administration, 2011.

Compton, W. David, and Charles D. Benson. *Living and Working in Space: A History of Skylab*. NASA History Series, NASA SP-4208. Washington DC: National Aeronautics and Space Administration, 1983.

Ezell, Edward Clinton, and Linda Neuman Ezell. *The Partnership: A History of the Apollo-Soyuz Test Project*. NASA History Series, NASA SP-4209. Washington DC: National Aeronautics and Space Administration, 1978.

Hagerty, Jack, and Jon C. Rogers. *Spaceship Handbook: Rockets and Spacecraft Designs of the 20th Century; Fictional, Factual and Fantasy*. Livermore CA: ARA, 2001.

Hale, Wayne, ed. *Wings in Orbit: Scientific and Engineering Legacies of the Space Shuttle*. NASA History Series, NASA SP-2010-3409. Washington DC: National Aeronautics and Space Administration, 2010.

Hall, Rex D., and David J. Shayler. *Soyuz: A Universal Spacecraft*. Chichester, UK: Praxis, 2003.

Harland, David M. *The Story of Space Station Mir*. Chichester, UK: Praxis, 2005.

Hitt, David, Owen Garriott, and Joe Kerwin. *Homesteading Space: The Skylab Story*. Lincoln: University of Nebraska Press, 2008.

Ivonovich, Grujica S. *Salyut—the First Space Station: Triumph and Tragedy*. Chichester, UK: Praxis, 2008.

Jones, Chris. *Too Far from Home: A Story of Life and Death in Space*. New York: Doubleday, 2007.

Jones, Tom. *Sky Walking: An Astronaut's Memoir*. New York: Smithsonian Books, 2006.

Khrushchev, Sergei N. *Nikita Khrushchev and the Creation of a Superpower*. University Park: Pennsylvania State University Press, 2000.

Kitmacher, Gary, ed. *Reference Guide to the International Space Station*. NASA SP-2006-557. Washington DC: National Aeronautics and Space Administration, 2006.

Launius, Roger D. *Space Stations: Base Camps to the Stars*. Old Saybrook CT: Konecky and Konecky, 2003.

Linenger, Jerry M. *Letters from Mir*. New York: McGraw-Hill, 2003.

———. *Off the Planet*. New York: McGraw-Hill, 2000.

Lord, Douglas R. *Spacelab: An International Success Story*. NASA SP-487. Washington DC: National Aeronautics and Space Administration, 1987.

Mullane, Mike. *Riding Rockets: The Outrageous Tales of a Space Shuttle Astronaut*. New York: Scribner, 2006.

Peebles, Curtis. *Guardians: Strategic Reconnaissance Satellites*. Presidio, 1987.

Scott, David, and Alexei Leonov. *Two Sides of the Moon*. With Christine Toomey. New York: Thomas Dunne, 2004.

Stafford, Thomas P. *We Have Capture: Tom Stafford and the Space Race*. With Michael Cassutt. Washington DC: Smithsonian Books, 2002.

Interviews and Personal Communications

Anderson, Clayton C. Interview by Jay Chladek. 16 July 2008. Johnson Space Center, Houston TX.

———. Interview by Jay Chladek. 6 April 2011. Johnson Space Center, Houston TX.

Anderson, Susan H. Interview by Jay Chladek. 16 July 2008. Johnson Space Center, Houston TX.

Ashby, Jeffery S. Interview by Jay Chladek. 8 October 2008. Omaha NE.

Cassidy, Christopher J. Interview by Jay Chladek. 22 October 2009. Johnson Space Center, Houston TX.

Creamer, Timothy J. Interview by Jay Chladek. 16 July 2008. Johnson Space Center, Houston TX.

———. Interview by Jay Chladek. 6 April 2011. Johnson Space Center, Houston TX.

Dake, Jason R. Interview by Jay Chladek. 21 October 2009. Johnson Space Center, Houston TX.

Kloeris, Vickie L. Interview by Jay Chladek. 21 October 2009. Johnson Space Center, Houston TX.

Melroy, Pamela A. Interview by Jay Chladek. 15 July 2008. Johnson Space Center, Houston TX.

Parazynski, Scott E. Interview by Jay Chladek. 15 July 2008. Johnson Space Center, Houston TX.

Petit, Donald R. Interview by Jay Chladek. 6 April 2011. Johnson Space Center, Houston TX.

Tani, Daniel M. Interview by Jay Chladek. 16 July 2008. Johnson Space Center, Houston TX.

Thomas, Andy S. W. Interview by Jay Chladek. 22 October 2009. Johnson Space Center, Houston TX.

Walheim, Rex J. Interview by Jay Chladek. 22 October 2009. Johnson Space Center, Houston TX.

Whitson, Peggy A. Interview by Jay Chladek. 16 July 2008. Johnson Space Center, Houston TX.

Other Sources

Bamford, James. "Astrospies." *Nova*, directed by Andreas Dirr and Scott Willis, aired 12 February 2008. Boston: WGBH Boston for PBS, 2008. DVD, 56 min.

Columbia Accident Investigation Board Report. Vol. 1. Washington DC: National Aeronautics and Space Administration, 2003.

Columbia *Crew Survival Investigation Report.* NASA SP-2008-565. Houston TX: Lyndon B. Johnson Space Center, 2008.

Congressional Staff Briefing on the Soyuz Launch Abort of April 5 1975. 29 May 1975. Box 1231, no. 132469. ASTP Program Series. Johnson Space Center History Collection, University of Houston–Clear Lake.

Exchange of Remarks between the President and ASTP Astronauts. 18 July 1975. Box 1232, no. 133075. ASTP Program Series. Johnson Space Center History Collection, University of Houston–Clear Lake.

Fullerton-Smith, Jill. "Terror in Space." *Nova*, aired 27 October 1998. Boston: WGBH Boston for PBS, 2000. Videocassette (VHS), 60 min.

Kalina, Jon, and Gary Lang. "The Human Factor." *Mars Rising*, directed by Michael Jorgensen, aired 21 October 2007. Montréal: Galafilm Productions, in association with Discovery Channel Canada, 2010. DVD.

Oral History Interview with Thomas P. Stafford about ASTP/Skylab. 12 April 1976. Box CD-S, no. 480974. ASTP Program Series. Johnson Space Center History Collection, University of Houston–Clear Lake.

Skylab Experiments. Vol. 3, *Materials Science.* Washington DC: National Aeronautics and Space Administration, 1973.

Skylab Experiments. Vol. 4, *Life Sciences*. Washington DC: National Aeronautics and Space Administration, 1973.

Skylab Experiments. Vol. 5, *Astronomy and Space Physics*. Washington DC: National Aeronautics and Space Administration, 1973.

Skylab Experiments. Vol. 6, *Mechanics*. Washington DC: National Aeronautics and Space Administration, 1973.

Skylab Experiments. Vol. 7, *Living and Working in Space*. Washington DC: National Aeronautics and Space Administration, 1973.

SL 1/2 Final EVA Checklist and Change "A." Box 626. Skylab Series. Johnson Space Center History Collection, University of Houston–Clear Lake.

Spacelab Data Handbook. ESA BR-14. Paris: ESA Scientific and Technical Publications Branch, 1983.

Technical Letter Report, Quick Look Evaluation Space Shuttle Orbiter Ninth Orbital Flight, Descent Phase, AFFTC. January 1984. Box 8. Shuttle Series, STS-9 Documents. Johnson Space Center History Collection, University of Houston–Clear Lake.

von Braun, Wernher. *Space Superiority as a Means for Achieving World Peace*. Washington DC: Business Advisory Council, 1952. Published form of presentation given in Washington DC, 17 September 1952.

Index

Page numbers in italics refer to illustrations.

Bold They Rise
David Hitt and Heather R. Smith
Foreword by Bob Crippen

Go, Flight! The Unsung Heroes of Mission Control, 1965–1992
Rick Houston and Milt Heflin
Foreword by John Aaron

Infinity Beckoned: Adventuring Through the Inner Solar System, 1969–1989
Jay Gallentine
Foreword by Bobak Ferdowsi

Fallen Astronauts: Heroes Who Died Reaching for the Moon, Revised Edition
Colin Burgess and Kate Doolan with Bert Vis
Foreword by Eugene A. Cernan

Apollo Pilot: The Memoir of Astronaut Donn Eisele
Donn Eisele
Edited and with a foreword by Francis French
Afterword by Susie Eisele Black

Outposts on the Frontier: A Fifty-Year History of Space Stations
Jay Chladek
Foreword by Clayton C. Anderson

To order or obtain more information on these or other University of Nebraska Press titles, visit nebraskapress.unl.edu.